'Engaging farmers and their families in all stages of the global effort to make agricultural systems more sustainable and productive is a vital prerequisite to success. This timely and remarkable book focuses on collaborations between farmers and plant breeders across all types of agroecosystems, and offers real hope for widespread redesign and transformation'.

Jules Pretty, *University of Essex, UK*

'Addressing the Grand Challenges faced by agriculture under a global changing climate requires a highly client-oriented, participatory research-for-development approach. This book provides insights on, and lessons learnt from 'citizen science' for both decentralized plant breeding and on-farm variety testing. It further shows how genetic diversity may remain in agroecosystems by involving the farming community in crop betterment and deployment. This publication also highlights the hurdles to overcome when pursuing participatory plant breeding. Its analytical and clearly written text along with the examples given by various authors in each chapter demonstrates the advantages of this citizen science approach for variety replacement elsewhere'.

Rodomiro Ortiz, *Swedish University of Agricultural Sciences*

'This book will be of interest not only to researchers working directly on crop improvement, but also to policy makers and other professionals involved in food and nutritional security. The authors amalgamate knowledge from decades of farmers' participation in plant breeding, as a major component of improving various crops around the world. For generations, farmers have been advancing their varieties through on-farm mass selections while also conserving seeds. Therefore, it is highly prudent to enhance the partnership between farmers and plant breeders. The publication contributes to this goal by providing insight into approaches, concerns, new perspectives and legal contexts for effective farmer-breeder collaborations'.

Segenet Kelemu, *International Centre of Insect Physiology and Ecology, Kenya*

Farmers and Plant Breeding

This book presents the history of, and current approaches to, farmer–breeder collaboration in plant breeding, situating this work in the context of sustainable food systems, as well as national and international policy and law regimes.

Plant breeding is essential to food production, climate-change adaptation and sustainable development. This book brings together experienced practitioners and researchers involved in collaborative breeding programmes across a diversity of crops and agro-ecologies around the world. Case studies include collaborative sorghum and pearl millet breeding for water-stressed environments in West Africa, participatory rice breeding for intensive rice farming in the Mekong Delta, and evolutionary participatory quinoa breeding for organic agriculture in North America. While outlining the challenges, the volume also highlights the positive impacts, such as yield increases, farmers' empowerment in the innovation and development processes, contributions to maintenance of crop genetic diversity and adaptation to climate change. This collection offers a range of perspectives on enabling conditions for farmer–breeder collaboration in plant breeding in relation to biodiversity agreements such as the Plant Treaty, trade agreements and related intellectual property rights (IPR) regimes, and national seed policies and laws.

Relevant to a wide audience, including practitioners with experience in plant breeding and management of crop genetic resources and those with a broader interest in agriculture and development, as well as students of international cooperation and development, this volume is a timely addition to the literature.

Ola Tveitereid Westengen is an Associate Professor at the Department of International Environment and Development Studies (Noragric), Faculty of Landscape and Society, Norwegian University of Life Sciences, Norway.

Tone Winge is a Senior Research Fellow at the Fridtjof Nansen Institute, Lysaker, Norway.

Issues in Agricultural Biodiversity

Series editors: Michael Halewood and Danny Hunter

This series of books is published by Earthscan in association with Bioversity International. The aim of the series is to review the current state of knowledge in topical issues associated with agricultural biodiversity, to identify gaps in our knowledge base, to synthesize lessons learned and to propose future research and development actions. The overall objective is to increase the sustainable use of biodiversity in improving people's well-being and food and nutrition security. The series' scope is all aspects of agricultural biodiversity, ranging from conservation biology of genetic resources through social sciences to policy and legal aspects. It also covers the fields of research, education, communication and coordination, information management and knowledge sharing.

Published titles:

Diversifying Food and Diets
Using Agricultural Biodiversity to Improve Nutrition and Health
Edited by Jessica Fanzo, Danny Hunter, Teresa Borelli and Federico Mattei

Community Seed Banks
Origins, Evolution and Prospects
Edited by Ronnie Vernooy, Pitambar Shrestha and Bhuwon Sthapit

Farmers' Crop Varieties and Farmers' Rights
Challenges in Taxonomy and Law
Michael Halewood

Tropical Fruit Tree Diversity
Good Practices For *In Situ* and On-Farm Conservation
Edited by Bhuwon Sthapit, Hugo A.H. Lamers, V. Ramanatha Rao and Arwen Bailey

Farmers and Plant Breeding
Current Approaches and Perspectives
Edited by Ola Tveitereid Westengen and Tone Winge

Farmers and Plant Breeding

Current Approaches and Perspectives

**Edited by Ola Tveitereid Westengen
and Tone Winge**

LONDON AND NEW YORK

from Routledge

First published 2020
by Routledge
2 Park Square, Milton Park, Abingdon, Oxon OX14 4RN

and by Routledge
52 Vanderbilt Avenue, New York, NY 10017

Routledge is an imprint of the Taylor & Francis Group, an informa business

First issued in paperback 2021

British Library Cataloguing-in-Publication Data
A catalogue record for this book is available from the British Library

Library of Congress Cataloging-in-Publication Data
Names: Westengen, Ola Tveitereid, editor. | Winge, Tone, editor.
Title: Farmers and plant breeding : current approaches and perspectives /
edited by Ola Tveitereid Westengen and Tone Winge.
Description: Abingdon, Oxon ; New York, NY : Routledge, 2020. | Series:
Issues in agricultural biodiversity | Includes bibliographical references and
index.
Identifiers: LCCN 2019028191 (print) | LCCN 2019028192 (ebook) |
ISBN 9781138580428 (hbk) | ISBN 9780429507335 (ebk)
Subjects: LCSH: Plant breeding.
Classification: LCC SB123 .F34 2020 (print) | LCC SB123 (ebook) | DDC
631.5/2–dc23
LC record available at https://lccn.loc.gov/2019028191
LC ebook record available at https://lccn.loc.gov/2019028192

ISBN: 978-1-138-58042-8 (hbk)
ISBN: 978-1-03-208887-7 (pbk)
ISBN: 978-0-429-50733-5 (ebk)

Typeset in Bembo
by Wearset Ltd, Boldon, Tyne and Wear

Contents

Figures

Tables

Contributors

Editors

Ola Tveitereid Westengen, Associate Professor, Department of International Environment and Development Studies (Noragric), Norwegian University of Life Sciences, Norway.

Tone Winge, Senior Research Fellow, Fridtjof Nansen Institute (FNI), Norway.

Contributing authors

Flavio Aragón-Cuevas, Investigador en Recursos Genéticos, Instituto Nacional de Investigaciones Forestales, Agrícolas y Pecuarias, Oaxaca, Mexico.

Alejandro Argumedo, Director, Asociación para la Naturaleza y el Desarrollo Sostenible (ANDES), Peru.

Carlos Avila, Agronomist, La Fundación para la Investigación Participativa con Agricultores de Honduras (FIPAH), Honduras.

Merida Barahona, Agronomist, La Fundación para la Investigación Participativa con Agricultores de Honduras (FIPAH), Honduras.

Trygve Berg, Associate Professor Emeritus, Department of International Environment and Development Studies (Noragric), Norwegian University of Life Sciences, Norway.

Åsmund Bjørnstad, Professor, Faculty of Biosciences, Norwegian University of Life Sciences, Norway.

Merideth Bonierbale, former Head of Crop Improvement, International Potato Center, Peru.

Kirsten vom Brocke, Breeder, Centre de coopération internationale en recherche agronomique pour le développement (CIRAD), Madagascar.

Fernando Castillo-Gonzalez, Profesor Investigador Titular PREGEP-Genética, Colegio de Postgraduados, Mexico.

Raul Ccanto, Leader of Agronomic Research, Grupo Yanapai, Peru.

Salvatore Ceccarelli, Independent Consultant, Italy, formerly of The International Center for Agriculture Research in the Dry Areas (ICARDA).

Anja Christinck, Consultant, Seed4Change (Research & Communication), Germany

Bram de Jonge, Seed Policy Advisor SD = HS, Oxfam, The Netherlands; Researcher, Wageningen University, The Netherlands.

Paola De Santis, Dipartimento di Biologia Ambientale, Sapienza Università di Roma/Bioversity International.

Abdoulaye Diallo, Sorghum Breeder/Head of Sorghum Programme, Institut d'Economie Rurale (IER), Mali.

Bocar Diallo, Sorghum Breeder, Institut d'Economie Rurale (IER), Mali.

Anita Dohar, Oxfam Novib, The Hague, the Netherlands.

Devendra Gauchan, Nepal Country Office, Bioversity International, Nepal.

Krishna Hari Ghimire, National Agriculture Genetic Resources Centre, Nepal Agricultural Research Council (NARC), Nepal.

Marvin Gomez, Agronomist, La Fundación para la Investigación Participativa con Agricultores de Honduras (FIPAH); Regional Facilitator, USC Canada Latin America.

Stefania Grando, Independent Consultant, Italy, formerly of The International Center for Agriculture Research in the Dry Areas (ICARDA).

Bettina Haussmann, apl. Professor, University of Hohenheim, Germany.

F. Humberto-Castro Garcia, Profesor Investigador, Universidad Autónoma de Chapingo Centro Regional Universitario Sur, Oaxaca, Mexico.

Sally Humphries, Professor, Department of Sociology and Anthropology, University of Guelph, Ontario, Canada.

Normita G. Ignacio, Director, Southeast Asia Regional Initiatives for Community Empowerment (SEARICE), Vietnam.

Devra I. Jarvis, Bioversity International/Department of Crop and Soil Sciences, Washington State University, USA.

Jose Jimenez, Director, La Fundación para la Investigación Participativa con Agricultores de Honduras (FIPAH), Honduras.

Bal Krishna Joshi, National Agriculture Genetic Resources Centre, Nepal Agricultural Research Council (NARC), Nepal.

Krishna D. Joshi, Country Representative (Nepal), International Rice Research Institute.

Patrick Kasasa, Community Technology Development Trust, Harare, Zimbabwe.

Julianne Kellogg, Research Assistant, Sustainable Seed Systems Lab, Washington State University, USA.

Theodore Kessy, Rice Breeder, Department of Research & Development (DRD), Ministry of Agriculture, Tanzania.

Abdourasmane Konate, Rice Breeder, Institut de l'Environnement et Recherches Agricoles (INERA), Burkina Faso.

Selim Louafi, Senior Research Fellow, CIRAD, UMR AGAP, Montpellier, France, and AGAP, University Montpellier, CIRAD, INRA, Montpellier SupAgro, Montpellier, France.

Gigi Manicad, Programme Leader SD = HS, Oxfam, The Netherlands.

Baboucarr Manneh, Irrigated Rice Breeder; Coordinator of STRASA and GSR Projects in Africa, AfricaRice, Senegal Regional Centre, Senegal.

Daniele Manzella, Technical Officer, Secretariat of the International Treaty on Plant Genetic Resources for Food and Agriculture, Food and Agriculture Organization of the United Nations, Rome.

Hilton Mbozi, Community Technology Development Trust, Harare, Zimbabwe.

Kevin Murphy, Associate Professor, Sustainable Seed Systems Lab, Washington State University, USA.

Andrew Mushita, Director, Community Technology Development Trust (CTDT), Zimbabwe.

Baloua Nebie, Sorghum Breeder, ICRISAT (International Crops Research Institute for the Semi-Arid Tropics) – Mali.

Paola Orellana, Agronomist, La Fundación para la Investigación participativa con Agricultores de Honduras (FIPAH), Honduras.

Lanqiu Qin, Breeder/Genetic Researcher, Guangxi Maize Research Institute, Nanning, China.

Fred Rattunde, Honorary Associate, Agronomy Department, University of Wisconsin-Madison; formerly ICRISAT (International Crops Research Institute for the Semi-Arid Tropics) – Mali.

Juan Carlos Rosas, Research Director, La Escuela Agricola Panamericana, Zamorano, Honduras; Leader, Bean Research Programme, Honduras.

Rene Salazar, Oxfam Novib, The Hague, the Netherlands.

Maria Scurrah, Director, Grupo Yanapai, Peru.

Mamourou Sidibe, Scientific Officer, ICRISAT (International Crops Research Institute for the Semi-Arid Tropics) – Mali.

Moussa Sie, Principal Scientist/Rice Breeder, Institut de l'Environnement et Recherches Agricoles (INERA), Burkina Faso.

Fredy Sierra, Professor, El Centro Universitario Regional del Litoral Atlántico (CURLA), La Ceiba, Honduras; Administrator, La Fundación para la Investigación Participativa con Agricultores de Honduras (FIPAH), Honduras.

Xin Song, Local Coordinator, Farmer Seeds Network (China), Guangxi, China.

Yiching Song, Programme Leader/Senior Researcher, Centre for Chinese Agriculture Policy, Chinese Academy of Sciences, China.

Bhuwon Sthapit, Nepal Country Office, Bioversity International (passed away in August 2017).

Sajal Sthapit, Local Initiatives for Biodiversity, Research and Development (LI-BIRD), Nepal.

Milin Tian, Local Coordinator, Farmer Seeds Network (China), Guangxi, China.

Álvaro Toledo, Treaty Technical Officer, Secretariat of the International Treaty on Plant Genetic Resources for Food and Agriculture, Food and Agriculture Organization of the United Nations, Rome.

Aboubacar Toure, Sorghum Breeder, ICRISAT (International Crops Research Institute for the Semi-Arid Tropics) – Mali.

Seydou Alexis Traore, Rice Breeder, Institut de l'Environnement et Recherches Agricoles (INERA), Burkina Faso.

Nicholas Tyack, Ph.D. candidate in Development Economics, Graduate Institute of International and Development Studies, Geneva, Switzerland.

Ronnie Vernooy, Genetic Resources Policy Specialist, Bioversity International, Rome.

D.S. Virk, Senior Research Fellow and Professor, School of Environment, Natural Resources and Geography (SENRGY), Bangor University, Wales.

Bert Visser, Oxfam Novib, The Hague, the Netherlands.

Eva Weltzien, formerly of ICRISAT (International Crops Research Institute for the Semi-Arid Tropics) – Mali; Honorary Associate, University of Wisconsin-Madison, USA.

Martha C. Willcox, Maize Landrace Coordinator, CIMMYT (International Maize and Wheat Improvement Center), Mexico.

John R. Witcombe, Professor, School of Environment, Natural Resources and Geography (SENRGY), Bangor University, Wales.

Hexia Xie, Breeder/Genetic Researcher, Guangxi Maize Research Institute, Nanning, China.

Foreword

Should plant breeders collaborate with farmers? On one level, the answer seems obvious: why not – what do they have to lose? After all, plant breeders are there to help farmers, to make their lives easier. Surely it makes sense for them to ask farmers what their problems are, and to work together to solve them, if only on the principle that two heads are better than one. And who could know crop diversity better than the farmers who rely on it for their livelihoods, if not their very lives?

On another level, however, the response might well be less positive. For many breeders, working closely with farmers introduces into their beloved breeder's equation an extra, unwelcome, unpredictable factor. Which farmers? Old or young? Female or male? Where? When? In what language? What happens if farmers disagree? What happens if I disagree with them? It all sounds suspiciously like social science. Plant breeding is difficult enough already.

But we do these things not because they are easy, but because they are hard – isn't that so? In fact, there is an irrefutable case to be made for some form of farmer involvement in the plant breeding process. Name an industry which does not at the very least consult its customers at some point. The real question – as Ola Westengen and Tone Winge point out in their introductory chapter to this landmark tenth volume in the Routledge series Issues in Agricultural Biodiversity – is not so much *if*, but *how*.

There are indeed many ways for farmers to work together with breeders to their mutual advantage, as shown in the contributions to this book. It is particularly poignant to note that one of the most successful models is one with which the late Dr Bhuwon Sthapit was closely involved for many years. Bhuwon sadly passed away before seeing this volume, including his own contribution, in print. He would have been proud that the pioneering efforts in Nepal are so prominently included, not as a curiosity or outlier, but as one of many examples from all around the world.

From the very beginning, Bhuwon recognized the importance of having the right policies and legislation in place to complement innovative technical solutions and partnerships. Therefore, he would also have been gratified to read of the progress that has been made, at national and international levels, in developing an enabling, albeit not always directly supportive, environment for collaboration between farmers and breeders.

For our part, we are encouraged to note how genebanks are recognized as key facilitators of that collaboration. We are tempted to say that they have a foot in both camps. But it is perhaps more accurate to say that both farmers and breeders have a foot in genebanks – or they should have. However, that is perhaps another story.

As is almost inevitable with books of this type, the take-home message is that, despite the considerable progress described here, much remains to be done. While increased productivity is still an important aim of plant breeding, it has been joined on the development altar, and quite rightly so, by the imperatives of sustainability and, crucially, of empowerment – and gender-sensitive empowerment at that. It is hard to see how these can be achieved unless farmers are at the centre of the plant breeding process.

Indeed, there are many challenges ahead. Much may go wrong. But in this book we have an important guide, for the converted and the sceptics alike. And the alternative to the vision set out in this book is that everybody loses.

Luigi Guarino, Director of Science, Crop Trust
Hannes Dempewolf, Head of Global Initiatives, Senior Scientist, Crop Trust

Acknowledgements

We wish to thank all those who have contributed to the realization of this volume. What made this book possible in the first place was a grant from the Global Crop Diversity Trust (the Crop Trust) to conduct the study 'From base broadening to enhancing crop adaptation to climate change'. The study was part of the project 'Adapting agriculture to climate change: collecting, protecting and preparing crop wild relatives', managed by the Crop Trust in collaboration with the Millennium Seed Bank (MSB) at Royal Botanical Gardens, Kew, and funded by the Government of Norway. We are grateful to the Crop Trust for the grant to carry out the study and to the Norwegian Agency for Development Cooperation (NORAD) for a grant to support the preparation and editing of this book. Publication in the Routledge series 'Issues in Agricultural Biodiversity' was made possible by a joint grant from Bioversity International and the Crop Trust.

Further, we would like to thank the series editors Danny Hunter and Michael Halewood for helpful feedback along the way. At Routledge, Tim Hardwick, Hannah Ferguson and Amy Johnston have provided valuable guidance and advice. We are also very grateful to Susan Høivik for her professional language-editing, and for internal institutional support from our respective institutions, the Department of International Environment Studies, NORAGRIC, at the Norwegian University of Life Sciences and the Fridtjof Nansen Institute.

Finally, we wish to thank all the contributing authors. Many of them participated in a seminar organized at Staur Farm, near Lake Mjøsa in Norway, in September 2017. Other contributors have joined the book project since. We would like to extend a special thanks to Trygve Berg for being a continuous source of knowledge and encouragement. As editors, we feel humbled and honoured to have had the opportunity to team up with experts who can draw on such a wealth of knowledge and experience from working with farmers pivotal to the ongoing efforts to achieve sustainable food systems. We are proud to present this collaborative effort!

Part I

Introduction

1 New perspectives on farmer–breeder collaboration in plant breeding

Ola Tveitereid Westengen and Tone Winge

Adapting plant breeding: productivity, sustainability and empowerment

Crop improvement is essential for adapting agriculture to changing environmental conditions and human needs. Modern plant breeding has succeeded in developing higher-yielding varieties of staple crops, contributing substantially to the productivity increase of global agriculture – first in the modernization of agriculture in the Global North in the early twentieth century and later during the 'Green Revolution' underway in the Global South since the 1960s (Evenson and Gollin, 2003). Today, however, plant breeding faces a different and arguably more complex agricultural research and development agenda than during earlier waves of agricultural modernization.

In the last decade, sustainability science has produced compelling evidence of the need for agri-food systems research and development to address a broader suite of goals than traditionally. In addition to providing the genetic basis for continued productivity increase, crop research must also contribute to agri-food systems that improve the nutritional value and environmental and societal sustainability of food systems (Foley *et al.*, 2011; Foresight, 2011; Godfray *et al.*, 2010; Pingali, 2012; Pretty *et al.*, 2018; Springmann *et al.*, 2018; Willett *et al.*, 2019). A more mundane but no less consequential framing of the objectives for agri-food system development is reflected in the UN Sustainable Development Goal (SDG) 2: 'End hunger, achieve food security and improved nutrition and promote sustainable agriculture' (UN, 2018). The SDGs provide new perspectives, as well as new metrics for gauging development. Here it is not only countries with low GDPs that face development and sustainability challenges, but also those with high GDPs and large ecological footprints.

In virtually all high-level calls for making food systems more sustainable, the development of new crop varieties features prominently as an important part of the strategy – but, with the overarching focus on system outcomes, the sustainability science literature has remained largely silent on the modalities of how these varieties are to be developed and disseminated. However, ideas and evidence on approaches to agricultural innovation are available from other academic and practice fields. Research on technology development and dissemination has shown that agricultural innovation consists of complex, non–linear processes embedded in social, institutional and economic processes, better conceptualized as 'innovation systems' (Hall *et al.*, 2001; Klerkx *et al.*, 2012; Thompson and Scoones, 2009).

Together, the food security, sustainability and innovation–system perspectives indicate three dimensions that plant breeding must relate to as part of the new agenda for sustainable agro-food system development: productivity, sustainability and empowerment. Thus, the starting point for this book is not *whether* plant breeding should play a key role in the necessary transformation of the food system, but *how* plant breeding can be organized to contribute to the development agenda of our times. We seek to showcase the diversity and achievements of types of plant breeding that involve farmers in the breeding process.

Ever since they were first described in the literature in the early 1990s (see Berg and Westengen, Chapter 2 in this volume), programmes referred to as 'Participatory Plant Breeding' (PPB), have had the objectives of releasing high-yielding varieties, along with enhancing sustainability and farmer empowerment. The contributions of PPB have since been widely recognized, including by institutions and actors associated with the conventional Green Revolution approach to plant breeding. For example, the 2007 World Development Report favourably described PPB as 'a complementary institutional development' to conventional crop improvement programmes; further noting: 'decentralised and participatory approaches allow farmers to select and adapt technologies to local soil and rainfall patterns and to social and economic conditions, using indigenous knowledge as well' (World Bank, 2007: 160).

Yet, a decade after the 2007–2008 food-price spikes which brought food security and agriculture back onto the international policy agenda, the Green Revolution model and the accompanying 'feed the world' narrative have continued to be the dominant agricultural development pathway promoted (Andersson and Sumberg, 2017; Clapp, 2017; Frison, 2016; Westengen and Banik, 2016). Do dominant actors today see farmer involvement as a luxury we cannot afford, now that population growth and climate change are requiring swift returns on investments from plant breeding? The cases presented in this book, diverse in their socio–economic and agro–ecological settings, firmly reject that premise. On the contrary, the chapters show that much is to be gained by these approaches across the agri-food system dimensions that we need to address in the coming decades.

This book is divided into four main parts. This first part presents the history of farmer involvement in plant breeding; the second part showcases the diversity of current plant-breeding programmes that promote farmer involvement. The third part focuses on overarching concerns related to programme sustainability, while the fourth and final part situates participatory plant-breeding approaches within national and international legal contexts.

We present these parts and their chapters in further detail, after a brief overview of different approaches to farmer involvement and existing frameworks for distinguishing among them.

Farmer–breeder collaboration in plant breeding: definitions and differences

Various interrelated frameworks for categorizing and assessing plant-breeding approaches have been developed (Almekinders and Hardon, 2006; Ashby, 2009; Jones *et al.*, 2014; Lilja and Ashby, 1999; Morris and Bellon, 2004; Sperling *et al.*, 2001; Weltzien *et al.*, 2003). Ashby (2009) has distinguished five categories, based on the division of decision-making between farmers and scientists, and to what extent co-production of new knowledge is promoted. Conventional plant breeding is classified as plant breeding without farmer participation: the decisions are taken solely by scientists. At the other end of the spectrum is what Ashby (2009) calls farmer experimentation, also known as farmer breeding (Morris and Bellon 2004): farmers make the decisions, without scientist participation. Almekinders and Hardon (2006) remind us that farmers have historically been the major plant breeders, and that it is actually a question of 'bringing farmers back into breeding'. It is between these two opposing ends of the spectrum – conventional breeding and farmer breeding – that various types of farmer–breeder interaction are classified. Sperling *et al.* (2001) and Ashby (2009) operate with three major categories: consultative, collegial, and collaborative. In consultative plant breeding, scientists are the decision-makers, but there is organized communication with farmers about their opinions, varietal preferences and priorities. In collegial plant breeding, it is the other way around: farmers make the decisions, but are informed about the priorities and research hypotheses of scientists through organized communication. It is the last category, collaborative plant breeding, which best promotes co-production of knowledge, as 'decision-making authority is shared between farmers and scientists based on organized communication between the two groups' (Ashby, 2009: 655).

Farmer involvement can take place at various stages of the plant-breeding process. If varietal release/distribution is included among these (Ashby, 2009), three main stages emerge: design, testing and diffusion. In Sperling *et al.* (2001) the stages are more technically defined as follows: setting breeding targets; generating or accessing variation; selecting in segregating populations; variety testing and characterization; interaction with seed systems. One of the most common types of farmer involvement in plant breeding, Participatory Varietal Selection (PVS), focuses mainly on the variety testing stage.

In addition to the five categories and the stages of farmer involvement, Sperling *et al.* (2001) and Weltzien *et al.* (2003) distinguish among breeding programmes, in terms of the institutional context, the biosocial environment, the 'degree' or 'nature' of participation achieved, and the goals in question. It makes a difference whether a breeding programme is organized by a non-governmental organization (NGO) or an International Agricultural Research Centre (IARC); whether it is designed for poor farmers in low-potential areas or for better-off farmers in high-potential areas; and whether the main goal is productivity increase, maintenance of genetic diversity, or farmer empowerment. In addition, crop type and its reproduction biology and agroecology shape the options for farmer involvement.

It is also important to recognize that the term 'participatory' often is misused in development discourse. Issues of legitimacy, justice, power relationships and gender and other group differences are sometimes concealed when actors claim to use 'participatory methods' (Kapoor, 2002). If participatory plant breeding becomes merely an approach for testing varieties developed by professional breeders in order to gain legitimacy with regard to donor policies etc., it might more correctly be called 'manipulative', not 'consultative' (Jones *et al.*, 2014).

Nevertheless, PPB remains the most commonly used term when referring to plant breeding programmes that involve some type of farmer–breeder collaboration – but its precise meaning must be outlined on a case-by-case basis. Comparing programmes that differ along all these axes can too easily become a matter of comparing apples and pears. Thus, in this volume we have deliberately not opted for one common terminology: instead, we attempt to showcase the diversity of the field by presenting perspectives and approaches on their own terms.

Current approaches to farmer–breeder collaboration

The case chapters in Part II describe programme contexts and the approaches taken, including objectives, types and stages of farmer participation, impacts and outcomes. The cases presented differ in their objectives for involving farmers in plant breeding – from production-oriented objectives such as higher adoption rates of high-yielding varieties, to enhanced *in-situ* conservation of crop diversity and farmer empowerment. The role of farmers also varies along that spectrum, from participation in varietal evaluation to active participation in all stages of the breeding process.

Together, the eight chapters in Part II cover a wide range of crops in various agro-ecological zones around the world. Although these cases represent merely a selection of all the plant breeding with farmer involvement being undertaken across the globe, they illustrate both the variety of approaches taken and the impressive results that are possible. Programmes and projects also vary in geographical reach: some operate in one country, or a part of one country, whereas others are regional in scope. Also with regard to their timelines, the cases presented vary greatly: some have been implemented for around two decades, whereas others have been initiated much more recently.

The regional initiatives are presented first. In Chapter 3, Weltzien *et al.* describe a long-term programme for farmer–breeder collaboration on sorghum and pearl-millet breeding in West Africa. The programme was initiated in 1998, and the chapter focuses on the efforts and results in Burkina Faso, Mali and Niger. The authors argue that the long-term nature of variety development necessitates long-term collaborative efforts. They also explain that the long-term perspective of the programme in question required institutional collaboration with farmer cooperatives and other farmer organizations, which helped to ensure programme sustainability. Programme results include formally released varieties in all three countries, large-scale dissemination of seed, and enhanced varietal diversity in farmers' fields.

In Chapter 4, Witcombe, Virk and Joshi report from programmes on maize and rice breeding in India and Nepal. Their chapter presents the pragmatic approach to farmer involvement which they call 'highly client-oriented plant breeding', which they see as related to the aim of widespread adoption. The focus is on the scientific argument for using few, but carefully chosen, crosses. Such an approach is central to participatory methods, according to these authors, and is linked to greater likelihood of success. The breeding programmes they describe have led to well-documented, high adoption rates of the resulting varieties, some of which have been officially released and are now widely cultivated.

Another programme with both regional outreach and long-standing experience is AfricaRice, with its PVS efforts. In Chapter 5, Tyack *et al.* explain how the poor results achieved by the Green Revolution in tropical Africa led AfricaRice to pursue a participatory approach to rice breeding from 1996. Together with its national partners, AfricaRice has conducted widespread PVS trials in member countries for the purpose of rapid multi-location evaluation and dissemination of new varieties. As a result of these regional PVS efforts, more than 100 varieties have been released during the last decade alone. Other results reported include increased yields, higher farmer incomes and reduced poverty.

Maize is an essential crop in Mexico, and in Chapter 6, Willcox *et al.* present various participatory approaches to maize breeding there. They focus on what they refer to as Community-Based Landrace Improvement and 'the three-step method', but other approaches to involving farmers – for example, in mass selection, limited backcrossing, and breeder-family formation and selection – are discussed as well. Their chapter also provides thought-provoking examples of how connecting with culinary markets in general, and chefs at high-end restaurants in particular, can provide useful pathways for improving the livelihoods of the small-scale custodian maize-diversity farmers.

In Honduras, Local Agricultural Research Committees (CIALs), a concept first launched in the country in 1993, have been central in the participatory breeding of bean varieties. As of 2019, 165 CIALs are operative in Honduras. Through the joint efforts of these farmer groups, NGOs and plant breeders from the state Pan American Agricultural School (EAP-Zamorano), several bean varieties have been developed and released since 2000. The key aspects of this approach, including the teaching of randomized replication trials to farmer groups, are outlined by Gomez *et al.* in Chapter 7. They also note successes and challenges related to seed production, and bureaucratic bottle-necks encountered along the way.

Chapter 8 by Scurrah, Ccanto and Bonierbale tells a recent story of farmer involvement in the evaluation and selection of potatoes with traits from crop wild relatives. This participatory varietal selection was conducted in two communities in the Central Andes of Peru, to see whether material developed in pre-breeding efforts with wild relatives could be useful and acceptable to farmers. The findings were promising, indicating both the potential usefulness of crop wild relatives and how farmers can play a role in such efforts.

Another long-term programme on farmer involvement is presented in Chapter 9, by Song *et al.* They present the participatory maize breeding undertaken in southwest China in terms of the changing societal and environmental conditions in China, focusing specifically on the role of women in Chinese agriculture and the need for gender-sensitive participatory plant breeding. Results of this programme include the release of 12 farmer-preferred varieties in research villages, the adaptation of 5 CIMMYT varieties to local conditions, the joint improvement of 30 landraces and 15–20 per cent yield increases compared with average yields of local varieties. This chapter also shows how successful programmes can evolve and grow further: in 2013, the programme in question became a multi-actor platform which now encompasses 36 villages in 10 provinces, as well as 10 breeding/policy institutes/universities at national and provincial levels.

Also of recent origin is the Evolutionary Participatory Breeding (EPB) of quinoa conducted in the US Pacific Northwest, presented by Kellogg and Murphy in Chapter 10. They explain the rationale, the methods used and the results achieved in breeding quinoa populations and varieties adapted to local agro-ecological conditions and organic cultivation. Further, the authors show that EPB resulted in yields higher than evolutionary breeding alone; they also note the usefulness of a farmer-driven selection index for determining the best-performing breeding lines.

Overarching concerns and new perspectives on implementing collaborative practices

The implementation of farmer involvement in plant breeding can be viewed and analysed from various standpoints. Economic viability, long-term sustainability, models for expansion, diffusion of products and knowledge, empowerment and social dynamics are all aspects that are relevant when reviewing historical experiences with farmer–breeder collaboration in plant breeding, and asking how such practices can best be employed in the future. Each of the five chapters in Part III deals with these issues in its own way.

The human development and social reform perspectives of collaborative plant breeding are investigated by Salazar *et al.* in Chapter 11. They hold that such breeding efforts cannot merely be about higher yields or increased adoption rates, but should address social inequities, improve livelihoods and aim for greater empowerment as well. Gender sensitivity and the co-development of knowledge are central. PPB as envisioned here should also address base-broadening of crop genetic diversity, while seeking to maintain farmer-based innovation and seed systems rather than the traditional cultivars per se. Chapter 11 draws on the experiences of PPB work conducted in the Mekong Delta in Vietnam – an important example of successful PPB in a high-potential environment.

Drawing on experiences from West Africa, in Chapter 12 Christinck, Rattunde and Weltzien examine how collaborative advantages can be created through long-term collaboration between plant breeders from national and international institutions as well as farmers' organizations and farmer cooperatives. The long-term nature of the programme is here seen as central to the advantages realized; institutional collaboration, rather than relying solely on individuals, proved essential to long-term sustainability.

In Chapter 13, Sthapit *et al.* present and analyse methods for sourcing and deploying new crop varieties, drawing on experiences from mountainous production environments, mainly in Nepal. Distinguishing between two main categories of approaches – conventional and participatory – these authors conclude that participatory tools such as PVS, diversity kits and informal research and development kits can accelerate the adoption of new varieties – but that, because of their cost-effectiveness and simplicity, diversity kits and informal research and development kits tend to be preferred as sourcing and deployment tools in the risk-prone and complex mountain systems of Nepal.

Another approach to achieving long-term sustainability is explored by Visser *et al.* in Chapter 14, where the Farmer Field School (FFS) approach to the management of genetic diversity in general, and collaborative plant breeding in particular, is evaluated in terms of costs, benefits and ability to be self-sustaining. As presented by these authors, the FFS approach is about education, capacity-building and empowerment; further, although costs are high initially, they fall later. They argue that one way to lower the costs, while also promoting community ownership, is to train and involve farmer-trainers.

In the final Chapter in Part III, Ceccarelli and Grando present scientific arguments for collaborative plant breeding, analyse the institutional barriers to expansion, and argue that EPB is the most promising way forward. The authors view this approach, which aims to bring genetic diversity and knowledge back into the hands and fields of farmers through the use of evolutionary populations, as more self-sustaining because it is much less reliant on formal institutional support.

Collaborative plant breeding within international and national legal contexts

Relevant legislation and policy may create opportunities but also barriers to farmer–breeder collaboration and the status of the varieties developed. The final part of this book places farmer-breeder collaboration in the context of current national and international policy and law regimes. A host of international agreements and national policies and laws affect the operation of farmer–breeder collaborative plant breeding programmes, including – but not limited to – biodiversity agreements (most notably the Convention on Biological Diversity (CBD) and the International Treaty on Plant Genetic Resources for Food and Agriculture (the Plant Treaty)); trade agreements and related intellectual property rights (IPR) regimes, phytosanitary regulations, biotechnology and biosafety laws, national seed policies and laws.

Much in the Plant Treaty is relevant to participatory approaches to plant breeding. Particularly significant are its Articles 6 and 9, on sustainable use and Farmers' Rights, respectively. In Chapter 16, Winge analyses the relationship between these Articles and farmer involvement in plant breeding. She shows that farmer–breeder collaboration in plant breeding can indeed serve as a tool for implementing Farmers' Rights and sustainable use, as partly reflected in the recognition accorded to PPB in the Plant Treaty system.

The Plant Treaty is also central to the focus of Chapter 17. Here, Toledo presents and analyses how PPB relates to the international agricultural development agenda, the Funding Strategy of the Plant Treaty and its Benefit-sharing Fund (BSF) in particular. He contextualizes PPB with the intertwined challenges confronting agriculture and the new SDG agenda, and shows that there is considerable donor interest and support available for such breeding programmes.

In Chapter 18, De Jonge *et al.* analyse the effects of seed legislation on PPB, with examples from legislations in selected countries in Africa and Asia. The authors argue that national seed laws and regional harmonization processes aim primarily at regulating and stimulating the formal, commercial seed sector: they are therefore often at odds with PPB programmes, which typically require more flexible registration and certification criteria.

Also intellectual property rights can affect PPB programmes. In Chapter 19, Manzella and Louafi analyse the relationship between plant variety protection, or plant breeders' rights, and participatory approaches to plant breeding. Taking as their point of departure the *sui generis* option offered by the Agreement on Trade-related Aspects of Intellectual Property Rights (TRIPS) of the World Trade Organization (WTO), they analyse the options for collaborative breeding projects with regard to accessing protected material, recognizing collective innovation and ensuring the continued availability of resulting material.

Unlike the multilateral system of access and benefit sharing of the Plant Treaty, the access and benefit-sharing system under the CBD and the Nagoya Protocol from 2010 is based on case-by-case agreements between users and source countries, according to the principles of prior informed consent and mutually agreed terms. In Chapter 20, Bjørnstad and Westengen analyse how access and benefit-sharing regimes and IPR regimes affect access to genetic diversity in plant breeding programmes. Arguing that both the CBD/Nagoya

Protocol and the patent/contract systems represent a tightening of the 'strait-jacket' of collaborative plant breeding programmes, the authors discuss alternative instruments and approaches to ensure that breeding programmes have access to diversity while also maintaining the objective of fair benefit sharing.

The future of farmer–breeder collaboration

As outlined above, the new plant breeding agenda must contribute to the traditional objective of increased productivity, and also to sustainability and farmer empowerment. The chapters of this volume offer new insights on how approaches to and perspectives from farmer–breeder collaboration can contribute to this agenda as well as to dealing with the challenges that remain.

Productivity

Food security assessments point to the need for 'closing the yield gap', which implies that yields must increase in developing countries in particular (Godfray *et al.*, 2010; Tester and Langridge, 2010). The old but constantly evolving *yield reducers* in crop cultivation – crop diseases, weeds and pests – continue to be major targets for resistance breeding, but breeding must also increasingly deal with the *yield-limiting* impact of factors like higher temperatures and shorter growing seasons caused by climate change (Challinor *et al.*, 2014; Challinor *et al.*, 2016). The chapters in this volume point out that continued productivity increase is indeed an objective for much famer–breeder collaboration. Several authors show that farmers rank yield among the most important breeding objectives or selection criteria (e.g. Weltzien *et al.* in Chapter 3, Song *et al.* in Chapter 9, Kellogg and Murphy in Chapter 10, all this volume); some also show that collaboratively developed varieties exhibit superior yields compared to the varieties commonly grown in the areas in question. Weltzien *et al.* report that the new sorghum hybrids developed in Mali yield on average 30 per cent more than popular landraces, across diverse low- and high-productivity environments. Song *et al.* report that maize 'PPB varieties' have achieved yield increases of 15 per cent to 20 per cent over average yields of local varieties. Gomez *et al.* (Chapter 7) find that PPB bean varieties outcompete nationally released improved varieties; and Willcox *et al.* (Chapter 6) report that the three-step model for community-based maize landrace improvement has resulted in a 30 per cent+ increase in grain yield in only three to five years. Also the chapters describing the development of heterogeneous populations through EPB report outperforming the common improved varieties (Kellogg and Murphy (Chapter 10), Ceccarelli and Grando (Chapter 15), this volume).

Cases presented in this book also show that, although the biosocial environments most commonly addressed by participatory projects are high-stress environments with subsistence orientation, there is also an increasing number of projects that address low-stress environments and market-driven economic contexts (Kellogg and Murphy (Chapter 10), Visser *et al.* (Chapter 14), this volume).

Another factor influencing the productivity outcomes of plant breeding is the extent to which the new varieties are adopted. The argument that PPB programmes are more efficient because adoption starts already during the selection process (Ceccarelli, 2015) finds support in several chapters in this book (Gomez *et al.* (Chapter 7), Song *et al.* (Chapter 9), Visser *et al.* (Chapter 14), Weltzien *et al.* (Chapter 3)). Moreover, farmer involvement in itself helps ensure popular demand for the varieties developed. Farmer participation in the breeding process also has the potential to embed varietal development in local food culture and culinary preferences – an aspect that, as many of the chapters show, is central to achieving acceptance and adoption of new varieties.

Sustainability

The 'Borlaug hypothesis' – the argument that plant breeding contributes to keeping the ecological footprint of food production down through intensification of existing cropland instead of extensification into wild habitats – still enjoys considerable traction in sustainability science (Springmann *et al.*, 2018). However, there is also widespread concern over the Green Revolution's 'package approach' to intensification – with negative environmental consequences in the form of fertilizer run-off, overuse of freshwater resources and loss of biodiversity. The challenge is therefore to continue contributing to intensification while at the same time developing resilient varieties that are less dependent on external inputs. Several of the chapters in this volume take climate change and other environmental problems as their point of departure (e.g. Scurrah *et al.*, Song *et al.*). Weltzien *et al.* explicitly target the need for varieties adapted to soils with low phosphorus availability. Management of agrobiodiversity is reported as a central objective in several programmes discussed in this book (Song *et al.*, Sthapit *et al.*, Weltzien *et al.*, Willcox *et al.*). The fact that many of these programmes manage to combine the objective of increasing food security in the face of environmental change with the need to maintain agrobiodiversity can probably explain why as much as 44 per cent of the funding disbursed from the BSF of the Plant Treaty thus far has been allocated to projects that incorporate PPB in some way (Toledo, Chapter 17).

Empowerment

The idea that plant breeding should contribute to empowerment or to more socially fair agri-food systems might sound strange to those who wish to maintain a clear distinction between science and the use of science in society. However, the uneven distribution of the benefits and costs of the Green Revolution shows clearly that there is no such thing as an a-political agronomy (Sumberg and Thompson, 2012).

As outlined in Chapter 2, the participatory agenda in agronomic research is intertwined with the objective of empowering farmers in the technology development process. This objective is formulated variously, but explicitly, in many chapters in this volume (see Gomez *et al.*, Salazar *et al.*, Song *et al.*, Tyack *et al.*, Weltzien *et al.*). The root causes of the situation that disempowered the farmers in the first place are framed in various ways: as a consequence of Structural Adjustment Programmes (Gomez *et al.*), of increasing market orientation and centralization (Song *et al.*), of the dominance of agrochemical corporations (Kellogg and Murphy), of globalization, with the weakening of institutions and marginalization of the world's poor (Salazar *et al.*) and as a general trend whereby farmers' stewardship over crop evolution has been removed, and delegated to specialized scientists (Ceccarelli and Grando). Some of the programmes described in this book have the empowerment of farmers as a central ideological superstructure, as with the programmes in Vietnam and Zimbabwe currently organized under the Sowing Diversity = Harvesting Security (SD = HS) programme (Salazar *et al.*, Visser *et al.*, both this volume) and EPB programmes. Most programmes frame the empowerment aspect in terms of the roles accorded to farmers in setting breeding targets and actively participating in the stages of the breeding process (Salazar *et al.*, Song *et al.*, Weltzien *et al.*). EPB programmes shift the focus away from the process of crossing and selection in early generations to farmer-led on-farm selection in heterogeneous populations (Ceccarelli and Grando, this volume).

The importance of gender sensitivity in participatory breeding is another recurrent theme in this book (Salazar *et al.*, Song *et al.*, Weltzien *et al.*). Understanding local gender dynamics is essential to achieving broad participation of women. When farmer–breeder collaboration succeeds in incorporating gender perspectives in the breeding process, that demonstrates the potential for influencing power dynamics and promoting more equitable power relations. However, a range of institutional factors will also have to be addressed if collaborative programmes are to fulfil their transformative potential.

Institutionalization

IPR legislation and seed laws are often at odds with the logic and needs of collaborative plant breeding programmes like those described in this book. The varieties and populations developed in these programmes are often disseminated mainly through farmers' seed systems – sometimes by design, sometimes informally. A common pattern is described in Song *et al.*: that in the first phase of participatory breeding efforts, there was no focus on policy and legal issues, but as the programme grew several concerns emerged, as did practical and legal obstacles. As De Jonge *et al.* explain, many seed regulatory frameworks do not allow farmer groups to register as seed producers, or register their new varieties – or even permit them to exchange farm-saved seed legally. In practice, such seed legislation is often not enforced with regard to farmers who save and exchange seeds – but the reality is that many farmer–breeder collaborative breeding programmes have little or no official policy or legal *support* for their production and distribution of varieties.

The chapters in this book describe the various ways in which breeding programmes have overcome legal and practical limitations. A fundamental challenge for all plant breeding programmes concerns institutional sustainability in the face of time-limited funding. The importance of connecting with national agricultural research organizations and state extension services is underlined by Visser *et al.*, who report on some of the longest-lasting NGO-organized programmes. In a considerable number of the programmes reported here, international agricultural research organizations – the CGIAR centres in particular – have played a long-term central role. Further, several chapters showcase the value of linking the CGIAR system to national institutions, farmers and local communities. Collaborative breeding projects can thus function as important bridge-builders. In addition, linking up with existing farmer organizations and cooperatives or contributing to the establishment of such groups is another ingredient for success, particularly concerning long-term viability as well as the dissemination stage. The strength of the connection between a farmer–breeder collaborative programme and effective seed-production and seed-supply systems can have substantial effects on programme outreach.

The contributions in this book show that different crops and different areas require different solutions. Some programmes have chosen to take varieties through the official registration process, with distribution through formal public and private seed suppliers; others have opted for variants of FAO's Quality Declared Seeds approach – a decentralized and moderate quality certification that allows for seed sales within defined geographical areas. Yet others have chosen entirely non-commercial, informal distribution of seeds. Despite the differences between the programmes and approaches described here, they all recognize that involving farmers and other end-users in crop variety development strengthens the links with the rest of the food chain. In that sense, the contributions to this volume highlight how embedding plant breeding in local, societal and environmental contexts can promote achievement of the ambitious, but compellingly necessary, objectives of developing productive, sustainable and societally just agri-food systems for our world.

References

Almekinders C, Hardon J. 2006. *Bringing Farmers Back into Breeding: Experiences with Participatory Plant Breeding and Challenges for Institutionalisation.* AgroSpecial 5. Agromisa, Wageningen, The Netherlands.

Andersson JA, Sumberg J. 2017. Knowledge politics in development-oriented agronomy. In Sumberg J, (ed) *Agronomy for Development: The Politics of Knowledge in Agricultural Research*, 1–13. Abingdon: Routledge.

Ashby JA, 2009. The impact of participatory plant breeding. In Ceccarelli S, Guimarães EP, Weltizien E, (eds) *Plant Breeding and Farmer Participation*, 649–671. Rome: FAO.

Ceccarelli S, 2015. Efficiency of plant breeding. *Crop Science* 55, 87–97.

Challinor A, Watson J, Lobell D, Howden S, Smith D, Chhetri N, 2014. A meta-analysis of crop yield under climate change and adaptation. *Nature Climate Change* 4, 287–291.

Challinor AJ, Koehler A-K, Ramirez-Villegas J, Whitfield S, Das B, 2016. Current warming will reduce yields unless maize breeding and seed systems adapt immediately. *Nature Climate Change* 6, 954–958.

Clapp J, 2017. The trade-ification of the food sustainability agenda. *Journal of Peasant Studies* 44, 335–353.

Evenson RE, Gollin D, 2003. Assessing the impact of the Green Revolution, 1960 to 2000. *Science* 300, 758–762.

Foley JA, Ramankutty N, Brauman KA, Cassidy ES, Gerber JS, Johnston M, Mueller ND, O'Connell C, Ray DK, West PC, Balzer C, Bennett EM, Carpenter SR, Hill J, Monfreda C, Polasky S, Rockström J, Sheehan J, Siebert S, Tilman D, Zaks DPM, 2011. Solutions for a cultivated planet. *Nature* 478, 337–342.

Foresight, 2011. *The Future of Food and Farming*. Final Project Report, London: Government Office for Science.

Frison EA, 2016. *From Uniformity to Diversity: A Paradigm Shift from Industrial Agriculture to Diversified Agroecological Systems*. Louvain-la-Neuve (Belgium): IPES.

Godfray HCJ, Beddington JR, Crute IR, Haddad L, Lawrence D, Muir JF, Pretty J, Robinson S, Thomas SM, Toulmin C, 2010. Food security: The challenge of feeding 9 billion people. *Science* 327, 812–818.

Hall A, Bockett G, Taylor S, Sivamohan M, Clark N, 2001. Why research partnerships really matter: Innovation theory, institutional arrangements and implications for developing new technology for the poor. *World Development* 29, 783–797.

Jones K, Glenna LL, Weltzien E, 2014. Assessing participatory processes and outcomes in agricultural research for development from participants' perspectives. *Journal of Rural Studies* 35, 91–100.

Kapoor I, 2002. The devil's in the theory: A critical assessment of Robert Chambers' work on participatory development. *Third World Quarterly* 23, 101–117.

Klerkx L, Van Mierlo B, Leeuwis C, 2012. Evolution of systems approaches to agricultural innovation: Concepts, analysis and interventions. In Darnhofer I, Gibbon D, Dedieu B, (eds) *Farming Systems Research into the 21st Century: The New Dynamic*, 457–483. Dordrecht: Springer.

Lilja N, Ashby JA, 1999. Types of participatory research based on locus of decision making. Consultative Group on International Agricultural Research (CGIAR), Participatory Research and Gender Analysis (PRGA), Cali, Colombia. (PRGA Working Document no. 6).

Morris ML, Bellon MR, 2004. Participatory plant breeding research: Opportunities and challenges for the international crop improvement system. *Euphytica* 136, 21–35.

Pingali PL, 2012. Green Revolution: Impacts, limits, and the path ahead. *Proceedings of the National Academy of Sciences* 109, 12302–12308.

Pretty J, Benton TG, Bharucha ZP, Dicks LV, Flora CB, Godfray HCJ, Goulson D, Hartley S, Lampkin N, Morris C, Pierzynski G, Vara Prasad PV, Reganold J, Rockström J, Smith P, Thorne P, Wratten S, 2018. Global assessment of agricultural system redesign for sustainable intensification. *Nature Sustainability* 1, 441–446.

Sperling L, Ashby JA, Smith ME, Weltzien E, McGuire S. 2001. A framework for analyzing participatory plant breeding approaches and results. *Euphytica* 122, 439–450.

Springmann M, Clark M, Mason-D'Croz D, Wiebe K, Bodirsky BL, Lassaletta L, de Vries W, Vermeulen SJ, Herrero M, Carlson KM, Jonell M, Troell M, Declerck F, Gordon LJ, Zurayk R, Scarborough P, Rayner M, Loken B, Fanzo J, Godray CJ, Tilman D, Rockström J, Willett W, 2018. Options for keeping the food system within environmental limits. *Nature* 562, 519–525.

Sumberg J, Thompson J. 2012. *Contested Agronomy: Agricultural Research in a Changing World*. Abingdon: Routledge.

Tester M, Langridge P. 2010. Breeding technologies to increase crop production in a changing world. *Science* 327, 818–822.

Thompson J, Scoones I. 2009. Addressing the dynamics of agri-food systems: An emerging agenda for social science research. *Environmental Science & Policy* 12: 386–397.

UN, 2018. *Sustainable Development Goals*. www.un.org/sustainabledevelopment/

Weltzien R, Smith ME, Meitzner LS, Sperling L, 2003. Technical and institutional issues in participatory plant breeding – from the perspective of formal plant breeding: A global analysis of issues, results, and current experience. CGIAR Systemwide Program on Participatory Research and Gender Analysis for Technology Development and Institutional Innovation; Centro Internacional de Agricultura Tropical (CIAT), Cali, Colombia. (PPB Monograph no. 1).

Westengen OT, Banik D, 2016. The state of food security: From availability, access and rights to food systems approaches. *Forum for Development Studies* 43, 112–134.

Willett W, Rockström J, Loken B, Springmann M, Lang T, Vermeulen S, 2019. Food in the Anthropocene: the EAT–Lancet Commission on healthy diets from sustainable food systems. *The Lancet* 393, 447–492.

World Bank, 2007. *World Development Report 2008: Agriculture for Development*. Washington.

2 Origins and evolution of participatory approaches in plant breeding

Trygve Berg and Ola Tveitereid Westengen

The origin and its context

Plant breeders may listen to opinions and feedback from farmers and involve them in consultative ways. Truly functional involvement, however, requires organized participation in one or more of the stages of a variety-formation process. Such participation, now known as Participatory Plant Breeding (PPB), was first reported in the 1980s, followed by considerable experimentation from the early 1990s. By the first decade of this century, accumulated evidence had established the approach as workable (Walker, 2006), and had clarified its organizational requirements, policy implications and potential impacts (Weltzien *et al.*, 2003).

The 1980s had seen the spectacular success of the 'Green Revolution' which, since its breakthrough in 1968, had impressed the world with steady progress in the productivity of the primary target crops, wheat and irrigated rice. The Consultative Group on International Agricultural Research (CGIAR) was established in 1971 to follow up and extend the Green Revolution to important crops and areas that had not yet been covered. The story of the origin and evolution of international agricultural research before and under the CGIAR is told in a book issued by the World Bank in 1986: *Partners against Hunger* (Baum and Pimentel, 1987). In spite of the title, there is no mention of farmers' involvement. Farmers were seen as the end-users of new technology, not partners in its development. The founding fathers were obsessed with the idea of a world without hunger, and needed the best that science could offer. They mobilized eminent scientists; for technical advisors, they sought 'men of towering stature'. The story of the Green Revolution is commonly told with Norman Borlaug and M.S. Swaminathan as the main characters (e.g. Hesser, 2006). The successes were indeed impressive, but they were limited to the chosen target crops and high-potential areas. Moving on to other crops and areas, researchers encountered complex low-resource farming systems and found it difficult to develop improved technology suited to local conditions (Pingali, 2012).

As noted in the chapter by Ceccarelli and Grando in this volume, one of the earliest traces of the thinking behind PPB was the seminal *Farmer-back-to-Farmer:*

A Model for Generating Acceptable Agricultural Technology by Rhoades and Booth (1982). They argued that agricultural research should start and end with ana-lysing farmers' perspectives; further, that all technical problems should be socially situated (Crane, 2014). This was an early sign of a new trend of 'parti-cipation' and 'farming system research', as opposed to the 'transfer of techno-logy' model that had characterized the Green Revolution approach in the 1960s and 70s (Chambers and Ghildyal, 1985; Chambers, 1983). Rhoades and Booth were working at the International Potato Research Centre in Peru; at around the same time, researchers at several CGIAR centres had realized the need to include social scientists in research teams and locate more of the research in physical conditions resembling those experienced by resource-poor farmers (Chambers and Ghildyal, 1985; Ashby, 1986). Also beyond the CGIAR several researchers proposed a 'reversal of learning and location'. Working in West Africa, the anthropologist Paul Richards, in his *Indigenous Agricultural Revolution* (1985) provided a strong, empirically based argument for recognizing the merits of traditional knowledge and innovation skills (Richards, 1985). The new Zeitgeist proclaimed: start with farmers' priorities and perceptions, learn from them and build the research on that! This way of thinking and working became known as the 'Farmers' First' approach.

The first operationalization of this approach was in agronomic technolo-gies such as fertilizer application (Ashby, 1986) and post-harvest technologies (Rhoades and Booth, 1982), but soon influenced crop improvement as well. The first publication on systematic involvement of farmers in plant breeding programmes was Maurya *et al.* (1988). Working on rice for rainfed areas in India, they involved farmers in line selection and ended up with several new varieties that were quickly adopted in the target areas. Also in 1988, a partici-patory bean selection programme was initiated in Rwanda involving the Institut des Sciences Agronomique du Rwanda (ISAR) and the CGIAR-institute CIAT (Centre for Tropical Agriculture). Researchers involved farmers in on-station screening and later devolved the programme to community-based screening. The research included the identification of expert farmer collaborators, which resulted in the introduction of gender as an issue in plant breeding (Sperling *et al.*, 1993). This project became a pio-neering case in the experimental stage that followed.

Experimental stage

The new ideas and pioneering projects of the 1980s showed an alternative way, but most plant breeders within and outside the CGIAR still used cen-tralized approaches while working on programmes of making higher-yielding varieties for crops and areas not yet served by formal breeding and seed supply. Noting that their new varieties tended to be rejected under local farm conditions, some of them started experimenting with decentralized selection.

Working on barley breeding at the International Center for Agricultural Research in the Dry Areas (ICARDA) in Syria and targeting dryland stress environments, Ceccarelli and Grando compared selection in high- and low-yield environments. They found that low-yield environments are better served by repeated selection under low-yield conditions when locally adapted materials are included (Ceccarelli and Grando, 1991a, Ceccarelli and Grando, 1991b). Initially this research concerned decentralized vs. centralized selection. However, since decentralized selection would be easier with farmer participation, the work was expanded to include participatory methods. In later publications, these authors describe a breeding approach characterized by testing and selection on farm rather than on station, decisions made jointly by farmers and breeders, and using a large number of locations (Ceccarelli and Grando, 1997). With inclusion of village-based seed production, this became a complete model of a breeding and seed supply system.

A group led by John Witcombe, of the Centre for Arid Zone Studies in Bangor (Wales), working primarily with rice in India and Nepal, involved farmers in selection among available released varieties (Participatory Variety Selection) as well as in breeding of new varieties through PPB. Their early research showed that farmer participation increased the likelihood of identifying the locally best choice, resulting in higher adoption and use of the varieties developed (Joshi and Witcombe, 1996; Sthapit *et al.*, 1996; Witcombe *et al.*, 1996). This group of researchers has since argued for a pragmatic combination of the most effective elements of 'conventional breeding' with participatory approaches in what they call 'highly client-oriented' plant breeding (Witcombe and Yadavendra, 2014).

Based in ICRISAT (the International Crops Research Institute for the Semi-Arid Tropics), India, Weltzien *et al.* (1998) involved farmers in work on pearl millet for the desert state of Rajasthan (Weltzien *et al.*, 1998). This research generated a wealth of new information on farmers' knowledge and preferences that appeared to differ by social variables (gender and wealth), also included varied and mutually incompatible needs that could be met only by cultivating mixtures.

These early PPB efforts primarily involved research on varieties for crops and target areas where conventional centralized breeding had failed. Researchers consistently reported that decentralization and involvement of farmers increased the likelihood of identifying materials that met local agroecological and socioeconomic requirements. They also found that participatory methods could be cost-effective, shortening the time needed for variety-formation (Virk, *et al.*, 2003; Ashby, 2009; Ceccarelli, 2015) and increasing their knowledge about farmers' preferences in diverse target areas (Mangione *et al.*, 2006). When scientists showed that decentralized selection with farmer participation could work, another, and perhaps surprising, group of actors came to embrace PPB.

The genetic resources movement

Loss of crop diversity became an issue on the environmental agenda already in the 1970s when concerns about *genetic erosion* were included in the 1972

UN Declaration on the Human Environment (Pistorius, 1997). The main cause of genetic erosion was identified as displacement of local varieties by modern varieties (Frankel and Bennett, 1970; Harlan, 1975); activists also noted the links between genetic erosion and the concentration of market power and the increasing use of Intellectual Property Rights (IPRs) in the private plant breeding industry (Fowler and Mooney, 1990; Mooney, 2011). The rapid spread of modern varieties and breeders' rights were seen as central elements in the threat to a natural resource of crucial importance for global food security. However, there were few in the genetic resources movement who questioned the importance of continued development of well-adapted crop varieties. PPB emerged as a technology alternative that could avoid both genetic erosion and IPR, while producing and distributing useful new varieties. Assuming that PPB could also meet productivity needs, the promotion of PPB became a practical means of pursuing the agenda of the genetic resources movement.

By connecting to farming communities where traditional plant breeding persisted, PPB also became a way of making farmers' breeding more effective. This 'farmer-led PPB' could be supported by external experts, by means of access to a broader genetic resource base, more effective breeding and selection skills, links, and organizational matters. McGuire and Manicad reviewed this sector with 11 case studies from Asia, Africa, and North and South America, involving a range of crops (McGuire and Manicad, 2003). Despite the limited local impact, closer study revealed a considerable potential for improvement through external support to local community-based breeding; moreover, functional local breeding tends to make communities more resilient to disasters when some farmers lose their seeds.

However, it would be an over-simplification to give the impression that the adoption of PPB was purely a pragmatic approach for reconciling genetic resources conservation and plant breeding. There has been a significant ideological and political rationale behind the Farmers First and farmer-led PPB efforts implemented by NGOs around the world. The perspectives of the educator and philosopher Paulo Freire in his seminal book *The Pedagogy of the Oppressed* (1968, many re-issues) have played a key role. Freire's liberation pedagogy focuses on avoiding asymmetric knowledge-transfer models (from rich to poor, from North to South) and replacing them with models that empower the poor to develop their own knowledge (Freire, 2018). Groups working on an agenda for empowering farmers begin to use participatory approaches for a range of rural development issues. The greatest success here was achieved through participatory methods of pest management in southeast Asia. Farmers' Field Schools (FFS) evolved through years of trials and failures, eventually proving to be a well-functioning way of educating farmers in effective pest management as well as for empowering farming communities (Röling and van de Fliert, 1994). This inspired others, and the FFS pedagogy was applied to a range of issues in many countries. Some projects developed curricula for FFSs in plant breeding, boosting farmers' motivation and capacity for involvement in PPB (Salazar *et al.*, this volume; Visser *et al.*, this volume).

The community empowerment aspect of PPB has been discovered not only by community organizers and civil society activists in the Global South, but also increasingly among actors in the food movement in the Global North, concerned about the 'supermarketization' of diets and engaged in creating food systems rooted in local ecology and food culture (GAfFF, 2016). In Europe and North America it is especially within the organic farming sector that PPB in various forms has been conducted (Chable *et al.*, 2014). This is partly because organic and other low external input farming is more concerned about adapting crops to the environment than actors involved with input-intensive farming, but it is also because organic farming, like other low-input farming, views itself as a counter-movement to 'industrial agriculture'. Involving farmers and other end-users in plant breeding is seen as a way to regain control over a part of the food system that has undergone increasing corporate power concentration in recent years (Brouwer *et al.*, 2016; Luby *et al.*, 2015).

PPB and seed system development

As described above, the Farmers' First and participatory research agendas were from the outset about developing more appropriate technology with higher uptake potential among farmers. The genetic resources movement framed participatory plant breeding in terms of conservation of diversity, with promotion of farmers' rights and community empowerment. Some projects and actors involved with PPB arguably came to emphasize other objectives above the objective of achieving a high uptake of the varieties developed. Many NGO-led projects developed varieties that never reached beyond the group of farmers involved in these projects, partly because of weak plans for connecting the variety development with seed production and dissemination. The more recent literature has shown a renewed focus on the importance of connecting variety development with seed production and dissemination.

The attention paid to farmers' seed systems in the writings of Paul Richards (Richards *et al.*, 2008; Richards, 1985) inspired a strong school of research and education at Wageningen UR in the Netherlands, focused on understanding seed use from the farm-level perspective (Almekinders *et al.*, 1994; McGuire, 2008). This school has elaborated development-oriented frameworks for 'integrated seed system development' (de Boef *et al.*, 2013; Louwaars and de Boef, 2012; de Boef *et al.*, 2010; Thijssen *et al.*, 2008), and, more recently, the use of 'citizen science' in seed-system development (Beza *et al.*, 2017; Steinke and van Etten, 2017). Awareness of the importance of involving farmers in crop improvement programmes is present, but the focus is placed 'downstream' – on farmers' access to suitable seeds – rather than on their role in the breeding of new varieties or conservation of traditional varieties. Thus, the seed-system framing of participatory plant breeding has brought the distribution and uptake aspect to the forefront.

With their limited attention to seed distribution, early PPB projects tended to have impact only within the participating communities. The present volume and other recent work on PPB offer many examples of how PPB can be integrated into larger seed-system development efforts. The hybrid sorghum varieties developed in the programmes described by Weltzien and colleagues (this volume) are bulked by farmer seed cooperatives that originated from prior participatory variety-testing activities. The example of the Seed Clubs in Vietnam demonstrates that such seed production is not necessarily small-scale only. The participatory breeding undertaken by the Mekong Delta Institute has succeeded in producing competitive farmer-bred varieties and supplying commercial-scale amounts of high-quality seeds (Tin *et al.*, 2011; Visser *et al.*, this volume). These examples show that there is no conflict between community empowerment and professional seed production and dissemination – quite the contrary. 'PPB initiatives must be linked to a secure diffusion strategy within and beyond the participating communities if the technical and social benefits of this approach are to be fully realized' (Bishaw and Turner, 2008: 31). Constraints to the development of PPB-linked seed distribution systems seem to be a matter of government regulatory requirements (de Jonge *et al.*, this volume) more than of the enterprising capabilities of the farmers themselves.

Conclusions

Plant breeding has made indisputable contributions to greater agricultural production and food security. The central role of plant breeding in the Green Revolution in Latin America and Asia has shown the immense potential for enhancing yields through genetic means. However, the Green Revolution also revealed several shortcomings of conventional on-station research, particularly with regard to reaching farmers in low-resource stress environments. The fundamental rationale for integrating farmers and other end-users in breeding has always been to increase the efficiency of breeding: to ensure that the varieties developed are considered sufficiently useful to be taken up by farmers. In the course of its historical development, participatory plant breeding has been embraced by various actors and agendas, most notably the genetic resources movement agendas of conservation and farmers' rights. With differing objectives come differing emphases on when and how farmers should participate. A wide range of approaches has emerged, taking the concept of PPB beyond the scope of a common definition today. Such diversity is in itself desirable for all those who are concerned with maintaining pluralistic seed systems in the face of power concentration and uniformity in the seed sector.

References

Almekinders CJM, Louwaars NP, de Bruijn GH, 1994. Local seed systems and their importance for an improved seed supply in developing countries. *Euphytica* 7–8, 207–216.

Ashby JA, 1986. Methodology for the participation of small farmers in the design of on-farm trials. *Agricultural Administration* 22, 1–19.

Ashby JA, 2009. The impact of participatory plant breeding. In *idem.*, *Plant Breeding and Farmer Participation*, 649–671. Rome: FAO.

Baum WC, Pimentel D, 1987. Partners against hunger. *Environment: Science and Policy for Sustainable Development* 29, 25–27.

Beza E, Steinke J, van Etten J, Reidsma P, Fadda C, Mittra S, Mathur P, Kooistra L, 2017. What are the prospects for citizen science in agriculture? Evidence from three continents on motivation and mobile telephone use of resource-poor farmers. *PLoS One* 12, e0175700.

Bishaw Z, Turner M, 2008. Linking participatory plant breeding to the seed supply system. *Euphytica* 163, 31–44.

Brouwer BO, Murphy KM, Jones SS, 2016. Plant breeding for local food systems: A contextual review of end-use selection for small grains and dry beans in Western Washington. *Renewable Agriculture and Food Systems* 31, 172–784.

Ceccarelli S, 2015. Efficiency of plant breeding. *Crop Science* 55, 87–97.

Ceccarelli S, Grando S, 1991a. Environment of selection and type of germplasm in barley breeding for low-yielding conditions. *Euphytica* 57, 207–219.

Ceccarelli S, Grando S, 1991b. Selection environment and environmental sensitivity in barley. *Euphytica* 57, 157–167.

Ceccarelli S, Grando S, 1997. Increasing the efficiency of breeding through farmer participation. In *Ethics and Equity in Conservation and use of Genetic Resources for Sustainable Food Security*. Proceedings of a workshop to develop guidelines for the CGIAR, 21–25 April 1997, 116–121. Foz de Iguacu, Brazil, IPGRI, Rome.

Chable V, Dawson J, Bocci R, Goldringer I, 2014. Seeds for organic agriculture: Development of participatory plant breeding and farmers' networks in France. In Bellon S, Penvern S, (eds) *Organic Farming, Prototype for Sustainable Agricultures*, 383–400. Dordrecht: Springer.

Chambers R, 1983. *Rural Development: Putting the Last First*. London, Lagos, New York: Longman.

Chambers R, Ghildyal B, 1985. Agricultural research for resource-poor farmers: The farmer-first-and-last model. *Agricultural Administration* 20, 1–30.

Crane TA, 2014. Bringing science and technology studies into agricultural anthropology: Technology development as cultural encounter between farmers and researchers. *Culture, Agriculture, Food and Environment* 36, 45–55.

de Boef WS, Dempewolf H, Byakweli JM, Engels JMM, 2010. Integrating genetic resource conservation and sustainable development into strategies to increase the robustness of seed systems. *Journal of Sustainable Agriculture* 34, 504–531.

de Boef WS, Subedi A, Peroni N, Thijssen M, O'Keeffe E, 2013. *Community Biodiversity Management: Promoting Resilience and the Conservation of Plant Genetic Resources*. London: Earthscan/Routledge.

Fowler C, Mooney PR, 1990. *Shattering: Food, Politics, and the Loss of Genetic Diversity*. Tucson: University of Arizona Press.

Frankel O, Bennett E, eds, 1970. *Genetic Resources in Plants: Their Exploration and Conservation.* Oxford: Wiley-Blackwell.

Freire P, 2018. *Pedagogy of the Oppressed* (50 anniversary re-issue, paperback). New York: Bloomsbury Academic.

Global Alliance for the Future of Food (GAfFF), 2016. *Seeds of Resilience: A Compendium of Perspectives on Agricultural Biodiversity from around the World.* https://futureof food.org/report/the-future-of-food-seeds-of-resilience/.

Harlan JR, 1975. Our vanishing genetic resources. *Science* 188, 617–621.

Hesser LF, 2006. *The Man Who Fed the World: Nobel Peace Prize Laureate Norman Borlaug and his Battle to End World Hunger: An Authorized Biography.* Dallas, TX: Durban House.

Joshi A, Witcombe J, 1996. Farmer participatory crop improvement. II. Participatory varietal selection, a case study in India. *Experimental Agriculture* 32, 461–477.

Louwaars NP, de Boef WS, 2012. Integrated seed sector development in Africa: A conceptual framework for creating coherence between practices, programs, and policies. *Journal of Crop Improvement* 26, 39–59.

Luby CH, Kloppenburg J, Michaels TE, Goldman IL, 2015. Enhancing freedom to operate for plant breeders and farmers through open source plant breeding. *Crop Science* 55, 2481–2488.

Mangione D, Senni S, Puccioni M, Grando S, Ceccarelli S, 2006. The cost of participatory barley breeding. *Euphytica* 150, 289–306.

Maurya D, Bottrall A, Farrington J, 1988. Improved livelihoods, genetic diversity and farmer participation: A strategy for rice breeding in rainfed areas of India. *Experimental Agriculture* 24, 311–320.

McGuire SJ, 2008. Securing access to seed: Social relations and sorghum seed exchange in eastern Ethiopia. *Human Ecology* 36, 217–229.

McGuire S, Manicad G, 2003. Technical and institutional issues in participatory plant breeding-done from a perspective of farmer plant breeding: A global analysis of issues and of current experience. CGIAR Systemwide Program on Participatory Research and Gender Analysis, Cali, Columbia, 109 p. PPB monograph No 2.

Mooney P, 2011. International non-governmental organizations. The hundred year (or so) seed war: Seeds, sovereignty and civil society – a historical perspective on the evolution of the law of the seed. In Frison C, López F, Esquinas-Alcázar JT, (eds) *Plant Genetic Resources and Food Security: Stakeholder Perspectives on the International Treaty on Plant Genetic Resources for Food and Agriculture*, 135–148. London: Earthscan.

Pingali PL, 2012. Green Revolution: Impacts, limits, and the path ahead. *Proceedings of the National Academy of Sciences* 109, 12302–12308.

Pistorius R, 1997. *Scientists, Plants and Politics – A History of the Plant Genetic Resources Movement.* Rome: IPGRI.

Rhoades RE, Booth RH, 1982. *Farmer-Back-to-Farmer: A Model for Generating Acceptable Agricultural Technology.* Lima: International Potato Center.

Richards P, 1985. *Indigenous Agricultural Revolution: Ecology and Food Production in West Africa.* London: Hutchinson.

Richards P, De Bruin-Hoekzema M, Hughes SG, Kudadlie-Freeman C, Offei SW, Struik PC, 2008. Seed systems for African food security: Linking molecular genetic analysis and cultivator knowledge in West Africa. *International Journal of Technology Management* 45, 196–214.

Röling N, van de Fliert E, 1994. Transforming extension for sustainable agriculture: The case of integrated pest management in rice in Indonesia. *Agriculture and Human Values* 11, 96–108.

Sperling L, Loevinsohn ME, Ntabomvura B, 1993. Rethinking the farmer's role in plant breeding: Local bean experts and on-station selection in Rwanda. *Experimental Agriculture* 29, 509–519.

Steinke J, van Etten J, 2017. Gamification of farmer-participatory priority setting in plant breeding: Design and validation of 'AgroDuos'. *Journal of Crop Improvement* 31, 356–378.

Sthapit B, Joshi K, Witcombe J, 1996. Farmer participatory crop improvement. III. Participatory plant breeding, a case study for rice in Nepal. *Experimental Agriculture* 32, 479–496.

Thijssen MH, Bishaw Z, Beshir A, de Boef WS, 2008. *Farmers, Seeds and Varieties: Supporting Informal Seed Supply in Ethiopia*. Wageningen: Wageningen International, Centre for Development Innovation (ISBN 9789085852155-347).

Tin HQ, Cuc NH, Be TT, Ignacio N, Berg T, 2011. Impacts of seed clubs in ensuring local seed systems in the Mekong Delta, Vietnam. *Journal of Sustainable Agriculture* 35, 840–854.

Virk D, Singh D, Prasad S, Gangwar J, Witcombe J, 2003. Collaborative and consultative participatory plant breeding of rice for the rainfed uplands of eastern India. *Euphytica* 132, 95–108.

Walker TS, 2006. *Participatory Varietal Selection, Participatory Plant Breeding, and Varietal Change*. Washington, DC: World Bank.

Weltzien R, Smith ME, Meitzner LS, Sperling L, 2003. Technical and institutional issues in participatory plant breeding – from the perspective of formal plant breeding: A global analysis of issues, results, and current experience. Systemwide Program on Participatory Research and Gender Analysis, Cali, Columbia, 208 p. PPB monograph No 1.

Weltzien E, Whitaker M, Rattunde H, Dhamotharan M, Anders M, 1998. Participatory approaches in pearl millet breeding. pp. 143–170. In Witcombe JR, Virk DS, Farrington J, (ed) *Seeds of Choice: Making the Most of New Varieties for Small Farmers*. Oxford and IBH Publishing Co. Pvt Ltd, Delhi, India.

Witcombe J, Yadavendra J, 2014. How much evidence is needed before client-oriented breeding (COB) is institutionalised? Evidence from rice and maize in India. *Field Crops Research* 167, 143–152.

Witcombe JR, Joshi A, Joshi K, Sthapit B, 1996. Farmer participatory crop improvement. I: Varietal selection and breeding methods and their impact on biodiversity. *Experimental Agriculture* 32, 445–460.

Part II

Current approaches to farmer–breeder collaboration

3 Long-term collaboration between farmers' organizations and plant breeding programmes

Sorghum and pearl millet in West Africa

Eva Weltzien, Fred Rattunde, Mamourou Sidibe,
Kirsten vom Brocke, Abdoulaye Diallo,
Bettina Haussmann, Bocar Diallo, Baloua Nebie,
Aboubacar Toure and Anja Christinck

Introduction

Sorghum and pearl millet are cultivated across West Africa in a great many contexts, due both to the differing agro-ecological conditions and the specific, contrasting production objectives of individual farmers, women or men, with more or less access to resources. These contrasting contexts lead to highly differentiated varietal preferences and needs.

The ancestors of todays' farmers in West and Central Africa contributed to domesticating sorghum and pearl millet. Also today, these farmers use and manage varietal diversity of these crops to optimize household productivity and minimize risks (Haussmann *et al.*, 2012). These crops, vital for survival and livelihoods, are closely intertwined with the cultures of the various peoples of West Africa (Weltzien *et al.*, 2006a).

This great diversity of varietal preferences and needs of West African farmers, as well as farmers' deep knowledge of these crops, calls for approaches that can support both the co-creation of contextual knowledge and collaborative farmer–researcher implementation of variety development activities. A farmer–researcher joint network approach, with decentralized collaborative activities, offers many possibilities for significant and tangible genetic gains that can contribute to improving the livelihoods of farm families.

Documentation of experiences of participatory breeding efforts has generally been limited to individual projects or specific research designs. Longer-term experiences have rarely been described. However, the long-term nature of variety development, along with challenges for seed production and dissemination, calls for collaboration between farmers and plant breeders over longer periods.

The joint activities conducted by sorghum and pearl millet breeders together with farmers in West Africa from 1998 onwards represent one case of such long-term collaboration. Activities have included the joint development of experimental varieties, the testing and validation of those varieties in a decentralized manner, and building the seed production and marketing capacities of farmers' organizations.

A wide range of farmers' organizations have been engaged in these collaborative activities – from village-based cooperatives like the Cooperative des Producteurs Semenciers du Mande (COPROSEM) in Siby, Mali, with less than 100 members, to unions of farmer cooperatives, like the Union Locale de Producteurs de Cereales (ULPC) in Mali, and to large farmer federations like MOORIBEN and Fuma Gaskiya in Niger, with several thousand members coming from large geographical regions. Also individual large cooperatives like Association Minim Song Panga (AMSP) and Union des Groupements pour la Commercialisation des Produits Agricoles (UGCPA) in Burkina Faso have been involved in this work.

Participating sorghum and pearl millet breeders have come from the International Crops Research Institute for the Semi-Arid Tropics (ICRISAT), the French Agricultural Research Centre for International Development (CIRAD), and the national agricultural research organizations in Mali (Institut d'Economie Rurale, IER), Burkina Faso (Institut de l'Environnement et Recherches Agricole, INERA) and Niger (Institut National de Recherche Agricole du Niger, INRAN). Teams of national and international breeders have interacted with the various farmers' organizations in planning and implementing activities.

This chapter documents and discusses the methodologies used, the experiences gained, and some of the achievements of this long-term farmer–breeder collaboration. These collaborative activities have covered the full range of breeding stages (Weltzien and Christinck, 2017) and are here discussed as follows: (a) setting and revising breeding objectives, (b) generating diversity, and continuing through (c) selecting in segregating materials and (d) evaluating experimental varieties, followed by (e) variety registration, seed production and dissemination. Farmer–researcher collaboration for developing hybrid varieties is presented separately, as an overarching engagement through all stages of the breeding cycle. Many of the methods and tools used for facilitating farmer–researcher collaboration are described by Christinck *et al.* (2005).

Collaboration and main achievements

Setting breeding objectives

Translating overarching goals into breeding objectives

The goals of the sorghum and pearl millet breeding programmes in West Africa have been broad, targeting the improvement of food security, income generation, farmer capacity-building and the conservation of agrobiodiversity.

As the research team came to realize that nutritional improvement could be integrated with this collaborative breeding work (Isaacs *et al.*, 2018; Rattunde *et al.*, 2018), that became a deliberate goal, based on the growing understanding of farmers' targeted management and use of varietal diversity and of the nutritional role of these crops, especially for children and women (Bauchspies *et al.*, 2017; Christinck and Weltzien, 2013).

Translating these rather broadly defined goals into breeding objectives has called for continual reflection and revision of priorities. These efforts have been spurred by the limited adoption of varieties from prior breeding work, as has been documented for sorghum (Yapi *et al.*, 2000), and by the researchers' desire to create a range of varietal options in line with farmers' diverse needs and opportunities for improving their overall farming systems.

Priority setting, identifying key traits

Setting priorities in breeding has been pursued through specific studies of farmers' production systems and goals, seed systems and trait preferences, also taking gender roles into account, as well as through continual interaction and discussions between farmers and breeders as part of routine participatory breeding activities conducted both on-farm and on-station. Work that contributed to the formulation of more specific breeding objectives has included studies of farmers' seed system for sorghum in Mali (Siart, 2008), women's roles in sorghum production (Rattunde *et al.*, 2018), the contributions of specific traits to the adaptation of plants to farmers' production contexts, e.g. photo-period sensitivity (Clerget *et al.*, 2008) and adaptation to low soil-phosphorous conditions (Leiser *et al.*, 2012).

The priorities for sorghum breeding in Mali, as determined by this process, were to develop new open-pollinated varieties with higher grain yield and adaptation to the predominant growing conditions. Additionally, the grain quality needs to be appropriate for storage, processing, and food uses, as farmers want to increase the amount of food produced, not just the weight of grain harvested per unit area (Weltzien *et al.*, 2018).

The formulation of general priorities and the breeders' evolving understanding of farmers' objectives and needs resulted in a growing list of specific traits to be targeted in variety development for sorghum. These traits include early maturity, with photo-period sensitivity helping to ensure timely maturity and stability despite variable sowing dates (Haussmann *et al.*, 2012); adaptation to soils with low phosphorous availability, typical of sorghum cultivation, especially by women (Leiser *et al.*, 2018); drooping panicles to reduce damage from insect, bird and grain-mould attacks (Weltzien *et al.*, 2018); open glumes for clean threshing, also under drought conditions (Diallo *et al.*, 2018); ease for women to decorticate the grain, and hard grain to minimize losses during storage and decortication (Isaacs *et al.*, 2018). Further: high food yield (Weltzien *et al.*, 2018); and intermediate plant heights for ease of harvesting – reduced from 4 m but not shorter than 2.5 m, where risks of transhumant cattle grazing become serious.

The traits targeted for pearl millet in the Sahelian zone of West Africa include higher grain yield; yield stability in the unpredictable and variable climate; earlier maturity than prevalent local varieties, to provide grain during the hunger period; plasticity of tillering, to respond to moisture availability; compact panicles that can resist insect attack; resistance to downy mildew (*Sclerospora graminicola* (Sacc.) Schroet.) and the parasitic weed *Striga hermonthica;* and adaptation to poor soil fertility.

Generating diversity

The choice of parental material for producing specific crosses or new populations was carefully considered by the sorghum and millet breeding teams, to assure variability for the farmers' priority traits. The sorghum team emphasized use of local germplasm to provide a strong foundation for the diverse adaptation traits wanted by farmers (Weltzien *et al.*, 2018). Some farmers contributed to the crossing programme specific varieties with which they were familiar (see also below). Exotic germplasm was introgressed approx. 12–25 per cent, to increase variability for traits related to productivity.

Diverse breeding approaches have been used to create new sources of useful diversity for breeders and farmers, to derive new lines and experimental varieties. Approaches for which long-term farmer–breeder collaboration has been of particular importance include the use of introgression with a single backcross generation; population improvement; and dynamic genepool management. These approaches are described briefly in the following sections.

Introgression with a single backcross generation

Efforts to create new varieties by crossing preferred local varieties with exotic donor parents gave unsatisfactory results, as the derived lines approached one or the other parent, with frustratingly few novel, desirable types emerging. A useful alternative was to create backcross-derived sub-populations by using a farmer-preferred variety chosen for improvement as recurrent parent, and conducting a single backcross generation. Multiple exotic donors were used, to generate diversity within the context of the farmer-preferred variety type, based on experiences in Australia (Jordan *et al.*, 2011). The diverse set of lines or sub-populations derived from these backcrosses, with approximately 75 per cent preferred variety background, has provided new variability with acceptable levels for many required adaptation traits. Farmers have shown great interest in selecting within these sub-populations, because of the diversity and acceptable levels of key farmer-preferred panicle, grain and glume traits. This method is seen as effectively creating useful diversity, from the perspective of farmers and breeders alike.

Population improvement through recurrent selection

The improvement of broad-based populations, created by crossing multiple parents to make a population bulk, is a cyclical process that results in new gene combinations from mixing the genetic backgrounds of all the parents. Each cycle involves selecting promising progenies from a population bulk and then inter-mating them to produce the next, improved population bulk. This selection and random mating in rapid cycles is expected to concentrate favourable genes and thus increase the frequency of progenies with superior performance for the targeted traits, while retaining genetic variation for future gains from selection.

This approach was used to create a sorghum population for West Africa based on tall Guinea-race landrace varieties (Rattunde *et al.*, 1997). The retention of genetic diversity for plant height in this tall (plant heights of 3–4 m) population enabled deriving a new shorter Guinea-race population, the 'Population Guinea Naine Diversifié' (Weltzien *et al.*, 2018). This shorter population offers a unique source material for responding to farmers' interest in new dual-purpose grain and fodder varieties that combine good quality Guinea-race grains for food and better stem-digestibility for feeding their livestock during the dry-season. The population has been improved for grain yield as well as panicle form and glume opening, to develop highly produc-tive and short plant types with key farmer-preferred traits. One of the vari-eties derived from this population, 'NafalenP6', has been top-ranked for grain yield and is superior in both phosphorous uptake and phosphorous use effi-ciency (Leiser *et al.*, 2014).

Farmers' collaboration with breeders in the various stages of recurrent selection aimed at improving this population has evolved over the years, based on joint learning and inspiration from engagement in diverse activities (Table 3.1). Farmers' engagement in deriving new progenies from the popu-lation bulk has been a major contribution here. A few (5–15) farmers with particular expertise and interest opted to grow plots of the population bulk from the most recent random-mating, and conducted single plant selections, which the breeders could then use to derive further progenies (Table 3.1). These farmers, with their eye for the diverse panicle, grain and glume charac-teristics, frequently detecting differences that breeders fail to notice, helped to retain genetic materials with important traits and trait combinations before the range of variability was narrowed. Their involvement has helped to increase genetic gain: firstly by increasing the scale of operations and the population size; secondly, by helping to focus the limited breeding resources on more acceptable plant types during the later stages of selection and testing. Furthermore, farmers' selections contributed to widening the diversity of progenies, as each person selected using his or her own criteria, and the farmers' selections complemented those of the breeders. The farmers shared half of the seed of each selected panicle with researchers and kept the remain-der for their own use.

Table 3.1 Breeding scheme for recurrent selection, random-mating 'Population Guinea Naine Diversifié', Mali

Year	Material sown/main step	Farmer activity	Breeder activity
1	Random-mated S0 population bulk (3–10,000 plants/ field) to derive new progenies	Sow bulk in isolated fields, thin to single plants, label male-fertile plants at flowering, select desirable male-fertile panicles to derive S1 lines.	
2	S1 progenies (500–750) for evaluation	On-station scoring of grain desirability, panicle appreciation, and threshability in S1 Trial; Contribute to selection of panicles to derive S2 lines.\n\nExperienced women farmers score grain quality.	Manage S1 progeny trials, evaluate for maturity, disease resistances, grain yield and overall appreciation, chose progenies for further testing based on index of farmer- and breeder-observations; Conduct S1 Nursery to derive S2 lines.
3	S2 progenies (approx. 125) for stage 2 testing and selection	Score grain and panicle desirability, threshability, label desired progenies in S2 trial; Contribute to selection of panicles to derive S3 lines for on-farm testing.	Manage S2 progeny trials, evaluate for maturity, disease resistances, grain yield and overall appreciation, create selection index of to choose best progenies to recombine; Conduct S2 Nursery to derive S3 lines for line/variety development.\n\nPlant remnant S1-seed of the selected progenies in isolation for an initial random mating and increasing the frequency of the male sterility gene. Only seed from sterile plants will be harvested.
4	S1 progenies derived in S0 (approx. 30) for random-mating	Contribute to selection of desirable panicles based on form, grain and glume characteristics.	Sow remnant S1-seed of selected progenies in isolation, I alternating rows with seed harvested from sterile plants from the first random mating, label male-sterile and male-fertile plants, select panicles of desirable sterile-plants and bulk to create next cycle.

Farmers also contributed their skills during on-station progeny evaluation activities (Table 3.1). For example, one or two farmers with special expertise in observing panicle form and free-threshing characteristics scored hundreds of progeny plots prior to harvest. Women farmers recognized for their skill at judging grain quality contributed by visually scoring grain desirability and grain hardness after harvest. These farmers were paid for contributing their special expertise. Farmers participating in annual field days also volunteered to evaluate population progeny trials, attaching labels to more desirable progenies and afterwards discussing the strengths and weaknesses of the new materials they observed. These forms of farmer involvement have helped breeders to gain both quantitative data and qualitative understanding for informing various breeding decisions.

Farmers' collaboration with breeders in population improvement has offered advantages for addressing the large number of plant, panicle and grain traits required for farmer acceptability and adaptation to local conditions. It has also facilitated sharing of the improved and genetically diverse populations with other breeders or farmers interested in selecting for adaption and quality traits for their own conditions and objectives (Weltzien *et al.*, 2018). For example, one farmer-breeder in a zone of Mali more humid than that of the experiment stations conducted multiple years of his own on-farm selection, starting from his population bulk grow-out. He later shared the lines he had developed with the breeders, for further variety development work targeting the more humid regions in Mali and other West African countries.

Dynamic genepool management

Another option for recurrent selection that may rely more heavily on farmers' selection activities in broad-based populations is 'Dynamic Genepool Management'. It aims at simultaneous *in situ* conservation and facilitating adaptation of plant genetic resources to climatic changes, specific site characteristics and evolving farmers' needs. As it is performed in close cooperation with farmers, it increases their access to and use of crop genetic resources, thereby increasing intra-species or intra-varietal genetic diversity, which in turn can contribute to yield stability.

In this approach, a diverse base population is built, through crossing and recombining genetically highly diverse genetic materials (see above for sorghum). After phenotypic (and possibly genetic) characterization of the base population, representative seed lots are distributed to farmers and grown at contrasting sites in a diverse target region. Selection by farmers and breeders, as well as natural selection, lead to the development of locally adapted, farmer-preferred subpopulations with new trait combinations (via recombination) not previously available. The sum of all subpopulations grown in the contrasting sites can be considered as a 'mass reservoir' of genetic adaptability. This concept has been applied successfully in several ways. A population of pearl millet adapted to the Sahelian agro-ecologies has been significantly improved for *Striga* resistance in Niger and Mali. Several types of *Striga*-resistant experimental varieties have been extracted from it (for example, one with shorter and one with longer panicles) in line with the preferences of various farmers (Kountche *et al.*, 2013). Farmers' seed-cooperatives and breeders in Niger have developed new varieties that combine improved yield with local adaptation. These varieties have been registered on the national variety list of Niger; the seed is disseminated by the farmers' seed cooperatives which selected them.

Selection in segregating materials

Taking selection decisions for complex requirements and needs

Conducting selection within and among segregating progenies or sub-populations is the critical bridge between the newly generated genetic diversity on the one hand, and, on the other, the identification of experimental varieties that are sufficiently uniform to effectively discriminate and test yet retain a degree of intra-varietal variability necessary for yield stability. Selection over several generations in this phase operates like a funnel in narrowing down total genetic diversity, so as to focus on a more limited number of progenies/sub-populations of greatest interest for achieving the breeding objectives. Decisions made during this phase are pivotal. In conducting this selection, the breeding programmes have worked hard to apply their evolving understanding of farmers' needs and preferences and defined breeding objectives. The definition of a variety- or product-profile (Ragot *et al.*, 2018) to guide the selection provides a framework for planning and documenting this approach.

The breeding programmes have experimented with diverse options for involving farmers in this selection process. Farmers' direct evaluation of progenies using various methods, as described for the population improvement activities, has become an integral part of the breeding programmes, contributing to the final selection of progenies to use in developing experimental varieties. Farmers have also sowed nurseries of 30 to 50 early-generation progenies, to select within these segregating materials using their own criteria and growing conditions. However, the breeding programmes have not managed to integrate this approach in their regular activities due to various factors, including farmers' propensity to select within nearly all progenies and their reluctance to eliminate any progenies, and the difficultly of effectively selecting among hundreds of progenies with only subsets, and under highly contrasting conditions with individual farmers.

Although farmers' contributions to selection of progenies for key, farmer-preferred, morphological and phenological traits have been of obvious and high value, it seems less clear whether or how they could contribute to early-generation selection for improving grain yield – the top priority for the breeding programme.

Early generation progeny selection for yield

The selection among early generation progenies for grain yield solely on the basis of on-station testing has been of questionable utility for achieving yield gains in farmers' fields under different, lower-input, conditions (Bänzinger and Cooper, 2001). However, the ability to achieve genetic gains for a complex and environmentally sensitive trait like yield through on-farm testing of early-generation progenies has been uncertain due to the obscuring effects of uncontrolled within-field, site-to-site, and year-to-year heterogeneity (Atlin *et al.*, 2001).

Sorghum breeders and 34 volunteer farmers set out to test the feasibility of large-scale early generation yield testing on-farm. The team tested a set of 150 early-generation progenies (S2/S3) for grain productivity using a trial design with sub-sets of 50 progenies and two common repeated check varieties tested per farmer. The testing of just 50 progenies per farmer and use of single-row plots was intended to make the trial manageable for participating farmers and to help locate the trials in somewhat more homogeneous portions of the farmers' field. Progenies were selected using combined on-farm or on-station results, and their on-farm yield performance was tested in subsequent years in a series of replicated on-farm trials.

Although farmers' management practices and field conditions have differed greatly among the various on-farm selection trials, early generation on-farm yield testing has been found effective (Rattunde *et al.*, 2016). Significant genetic variation and acceptable heritability for grain yield have been obtained in combined analyses of all on-farm yield trials. Moreover, although the genetic gains for yield from on-farm evaluations had been predicted to be greater than those expected from only on-station testing, combining yield results from on-farm and on-station, particularly under low soil-phosphorus conditions, emerged as more effective for selecting for on-farm yield performance than using solely on-farm or on-station results.

Evaluating experimental varieties

Protocols and procedures for variety evaluation

Many farmers have been keen to evaluate new varieties, which led the West African sorghum and pearl millet breeding programmes to develop protocols for variety evaluations by and with farmers. These protocols address two main objectives: achieving a common understanding of the advantages and disadvantages of specific varieties (and traits); and assessing productivity and its stability across a wide range of growing conditions within the priority target-zone for the new varieties (Rattunde *et al.*, 2013; Weltzien *et al.*, 2006a, 2008).

Initial understandings of sorghum farmers' variety and trait preferences were obtained by using standard tools like open-ended evaluations, matrix ranking and occasionally pairwise ranking (Ashby, 1990; Quirós *et al.*, 1991). The team subsequently developed a system where farmers (sometimes assisted by a village facilitator) could score a standard set of priority traits (crop duration, adaptation to the local conditions, panicle appreciation, overall appreciation) (Weltzien *et al.*, 2006a). Researchers, with assistance from farmers, measured grain yields and yield components. All observations were recorded in the field book kept by the farmer who conducted the trial, with researchers entering the data for analysis and keeping a photocopy of the field book.

Gradually, methods developed for ascertaining preferences for traits and specific experimental varieties from a wider base of men and women farmers. Field days were arranged, for visiting and evaluating the trials at harvest in at least ten participating villages (Weltzien *et al.*, 2006a). The procedures for the farmers' evaluations evolved, from scoring a set of fixed criteria across all villages, to having farmers chose the three most important traits for which they observe varietal differences pertinent to their context. Men and women farmers first discussed separately their choice of traits, followed by plenary presentations by both men and women, and joint negotiations to agree which three traits would be evaluated by all participating farmers (vom Brocke *et al.*, 2010). The participants, divided into small groups of women or men farmers, scored all plots for the three chosen traits and for overall appreciation. This approach has facilitated inclusion of gender-specific trait preferences. The technician recording the scores has also noted the reasons mentioned for scoring certain varieties especially high or low. A scoring system of 5 ('much better' than the local check) to 1 ('very poor') was used; farmers could readily relate to this, as it corresponded with the school grading system.

Developing and adapting trial designs

The initial design for the sorghum variety trials in Mali was a single replicate of 32 entries (Weltzien *et al.*, 2006a). Joint discussions of breeders with participating farmers led to a modified design, where 2 separate trials were conducted, each with 16 entries: 1 for taller height entries and the other for shorter entries. A two-replicate alpha-lattice design with sub-blocks of four plots was used to enable repeatability estimates and control error, with trial size maintained constant at 32 plots. The experimental varieties to be tested were contributed by both the Malian IER and the West and Central African ICRISAT sorghum breeding programmes, on the basis of on-station and on-farm progeny trials (as described in Table 3.1), as well as contributions coming directly from farmers. Both trials included a common landrace check variety (Tieblé, initially registered as CSM 335). Both trials were conducted in 10 to 12 villages, with each trial conducted by two farmers per village, as well as in two to three research stations in a single year. Each set of experimental varieties was tested for two years, based on the farmers' and researchers' desire to see varietal performance over years. All experimental varieties were given short vernacular names that could easily be remembered, but without any suggestive connotations, to facilitate farmers' discussions and feedback. The analysis of individual farmers' trials and combined over-environment analyses were conducted by the breeders, and the results were presented to the farmers for discussion and selection of entries for post-harvest grain quality tests and second-stage, fully farmer-managed, on-farm testing. This made it possible to collect robust data on grain yield performance and farmer appreciation of the experimental varieties over a large number of diverse environments (Kante *et al.*, 2017; Rattunde *et al.*, 2013).

Developing post-harvest quality tests

Post-harvest culinary tests of experimental varieties were generally conducted in each of the participating villages. These tests were necessary since grain quality, grain storage, and food-processing attributes are critical for variety adoption, and the breeders lacked capacity to assess these traits on their own. Farmers chose four of the experimental varieties and a local check to include in the tests based on their preferences from field observations and the yield results of both the tall and short variety trials presented in the feedback sessions. Procedures for the culinary tests evolved such that teams of women could provide quantitative and qualitative measurements of varietal differences for grain-quality attributes like the ease and amount of time involved in various processes, decortication losses, flour-to-grit ratios, and swelling potential (the capacity to absorb water) of the stiff porridge *(tô)*, a local staple dish (Isaacs *et al.*, 2018). A village-based panel of men and women taste-testers evaluated the colour, taste and consistency of the prepared *tô*.

Large-scale testing of varieties for adaptation

The second stage on-farm adaptation trials were conducted to give a much larger number of farmers the opportunity to evaluate, under their own field and management conditions, varieties chosen from the first-stage tests (Welt-zien *et al.*, 2006a). The testing procedure evolved to include the option of splitting the plots with one-half fertilized, so that farmers could assess the performance of each variety with and without fertilizer. Adaptation trials were conducted in villages where four or more farmers were interested. Farmers had to agree on the specific objective for their trials – like finding varieties with good performance even with *Striga* infestation, late or early sowing, more or less weeding, or in a specific intercropping situation. Demand for these trials was very high, so, for several years, trials were allocated to villages only if also a group of at least four women conducted a variety trial in their own fields. The trials were designed with three, four, five or sometimes six varieties, always including a common local check widely grown in the village. Separate protocols with single-row test plots, intercropped with groundnuts, were developed to facilitate testing by women (Rattunde *et al.*, 2018). Farmers recorded their observations directly or with help of village facilitators, and each group of farmers jointly discussed and documented its choice of varieties. Researchers visited some of the trials, and assisted some groups, particularly women's groups, with harvesting and weighing of plot yields. Yield data from these trials, extracted from the field books, enabled broad assessments of the relative performance, profitability, and risks of not recouping investments in seed of improved open-pollinated varieties and hybrid varieties and fertilizer relative to the farmers' local varieties, over diverse environments for men and women farmers (Weltzien *et al.*, 2018).

Preparing varieties for release and seed production

Some experimental varieties tested during the first stage of evaluation showed superior levels of productivity, but were not sufficiently homogeneous for traits readily noticed by farmers such as panicle form, glume opening, or grain colour. In such cases, one or two farmers known for their expertise grew plots of 100 to 200 single plants per variety and then visually selected desirable panicles that were uniform and conformed to the farmers' standard for those traits. The selected panicles were grown head to row by the breeders; uniform comparable rows were bulked together for initiating seed multiplication.

Variety registration, seed production and dissemination

Variety registration as a step towards official seed marketing

Following the introduction of harmonized seed legislation in all countries of the Economic Community of West African States (ECOWAS) region in 2008, variety registration has become a prerequisite for legal seed marketing in West Africa (for an analysis of the opportunities and barriers created by seed legislation for PPB, see de Jonge *et al.*, this volume). The two-stage procedure for yield testing (explained earlier) provided high-quality data, which, together with farmers' decisions as to which of the selected varieties to disseminate, facilitated the identification of varieties for official release and the preparation of release proposals. In addition, some farmers' varieties included in the trials have been formally released in Mali, such as 'Boboje' and 'Sakoykaba', just as direct selections from germplasm accessions, such as IS 15401 ('Soumalemba') originating from Cameroon or 'Gnossiconi' from Burkina Faso (vom Brocke *et al.*, 2014). Six sorghum varieties derived through population improvement with farmer–researcher collaboration have been released in Mali to date, in addition to releases of varieties selected mostly on-station. One of these, 'Lata3 (Bala Berthe)', is now cultivated by Malian farmers as a variety per se and used as the male-parent for hybrid breeding (Kante *et al.*, 2017; Rattunde *et al.*, 2013). In Burkina Faso and Niger, varieties from these collaborations have also been released, seeds produced and disseminated by farmers' organizations who were involved in variety development earlier on.

Farmer-managed seed production and marketing

Farmers' access to seed of the newly bred varieties was not assured initially, since West Africa's large, state-run seed production and distribution agencies had basically collapsed at the turn of the century (Coulibaly *et al.*, 2008). Sorghum and pearl millet seed of non-traditional varieties was distributed in Mali through one or two remaining sub-centres in low quantities, mostly of older varieties, with even nearby farmers rarely acquiring seed. This is why training in production of certified seed began at this time (starting in 2002), supporting participating farmers' organizations in following through, from variety testing to seed production of newly identified and released varieties. The farmers' organizations proved very capable of producing seed; their members were highly motivated to produce seed, which for them was a new 'cash crop' involving less investment and risk as compared to cotton, for example.

Emerging for-profit seed companies (Dalohoun *et al.*, 2010), with minimal seed production capacities of their own, bought the seed produced by the farmers' organizations. Although the expectation was that these seed companies would sell seed to farmers, years passed, with ever-increasing quantities of seed produced by farmer cooperatives but without detectable seed sales or distribution networks for rural sorghum producers. It became apparent that closer examination of the sorghum seed system was needed to identify entry points for large-scale dissemination surpassing traditional exchanges between neighbours and relatives.

A detailed study of sorghum seed systems in the Sudan Savannah of Mali revealed strong cultural norms regarding sorghum seed (Siart, 2008). 'Good farmers' are expected to produce their own seed; moreover, it is a social obligation to give seed to anyone who asks, in exchange for an equal volume of any cereal grain after harvest – but never for money. Collaborating farmers and farmers' organizations explained that although it is unacceptable for individuals to sell seed and profit from someone else's need, a farmer *group* could sell seed, as the benefit then goes to the group. This understanding motivated efforts aimed at strengthening the capacity of farmer seed-cooperatives to market their seed directly to local farmers.

Seed dissemination in accordance with farmers' values and needs

The marketing of sorghum seed by farmer cooperatives was hampered because the farmers were not familiar with purchasing seed. A breakthrough came when farmer demand for seed-sets from adaptation trials soared, exceeding the breeders' capacity. When single variety seed 'mini-packs' of just 100g were offered for sale by the extension office to interested farmers, this resulted in initial sales of 2,500 seed packets, with minimal advertising. This approach was then pursued with a continually increasing number of farmer cooperatives, until farmer demand for larger seed packages began to develop. A decentralized model of seed dissemination by farmer cooperatives is now emerging in West Africa (Access to Seeds Foundation, 2018; vom Brocke *et al.*, 2014; Christinck *et al.*, 2014): this model is in line with local socio-cultural values, the farmers' need for trust and knowledge of the seed seller, and the provision of diverse varieties to men and women farmers in contrasting agro-ecologies. Opportunities for strengthening the capacities of farmer seed-cooperatives to market seed include linking demonstration plots with local radio coverage, threshing and yield comparisons of new and old varieties, testimonials by locally respected farmers, use of photos and 'farmer friendly' varietal descriptions, and ensuring that seed sellers have experience cultivating the varieties they sell.

Sorghum hybrid varieties – learning and actions across breeding stages

Initially, West African breeders focused on open-pollinated varieties, conscientiously deciding not to breed hybrids due to farmers' desire to have varieties that they could reproduce by means of traditional, culturally rooted practices. Sorghum breeders reconsidered this decision after learning from a 1999 farmer survey that higher grain yield was farmers' top priority for new varieties (Weltzien and Christinck, 2017), and the fact that yield gains with open-pollinated varieties were limited (15 per cent or less) and too small for farmers to notice. A proof-of-concept effort was launched to determine whether hybrid varieties based on Guinea-race germplasm with the required adaptation and grain characteristics could deliver markedly higher yields under farmers' diverse and typically low-input conditions. An initial step was to breed the first Guinea-race cytoplasmic male-sterile seed parents (near-isogenic pairs of a male-sterile seed parent line used as the female for producing hybrid seed and its male-fertile 'maintainer' for multiplying female parent seed). The new hybrid parents, based on Malian farmers' Guinea-race landrace varieties and breeders' Guinea-Caudatum inter-racial materials, were used to produce the first Guinea-race experimental hybrids.

Farmer participatory trials were conducted in three zones of contrasting agricultural intensification over several years to evaluate these experimental hybrids. The yield superiorities of these new hybrids over the popular Malian landrace variety 'Tieblé', averaging over 30 per cent across diverse low- and high-productivity environments (Rattunde *et al.*, 2013), and the farmers' appreciation of the grain qualities of these hybrid varieties (Kante *et al.*, 2017) were very positive results. The hybrid 'Pablo', although not a top choice of the breeders due to its tall height and its yield superiorities under productive conditions being less than those of shorter hybrids, is currently the most popular hybrid in the intensified cotton-production region of Mali, as it closely resembles the local varieties but has significantly higher yields. Thus, the results obtained through a decade-long proof-of-concept effort provided solid justification for pursuing hybrid development as a complement to open-pollinated variety breeding for sorghum in West Africa.

There were, however, substantial concerns about how hybrid seed would be produced and reach the farmers, as there was little commercial sale of seed and no production of hybrid seed of any crop in Mali at that time. Therefore, in 2009, experimentation and training in hybrid seed production of several farmers' organizations already engaged in seed production were initiated, parallel to hybrid variety testing and the release of the first hybrids. Formal training courses were conducted jointly by IER and ICRISAT prior to providing seed and protocols for interested farmers to start up small (e.g. 600 m²) seed-production plots to gain experience. Training booklets in the local language were developed (Rattunde *et al.*, 2014) and trainings were repeated for at least two years, for each cooperative to absorb new concepts of sowing male and female 'varieties' (hybrid parents) in alternating blocks, managing them so as to obtain simultaneous flowering, and ensuring precision harvest and threshing of each parent separately.

Three years after the initial trainings, farmers began larger-scale hybrid seed production by using plots of one ha or larger, and modifying the planting design (e.g. by increasing the ratio of female- to male-parent rows). Farmers became increasingly interested in producing hybrid seed when they learned that they could 'add a cash crop to their food production plots', as explained by a village facilitator, referring to the female parent being harvested for sale of hybrid seed while the grain of the male parent could be kept for food. Some farmers reported producing as much grain for home consumption with the male-parent in their hybrid seed-production plot as they would have previously achieved on the entire plot thanks to the fertilizer obtained on credit from their cooperative for producing hybrid seed. The amount of hybrid seed produced doubled annually (Kante *et al.*, 2017) as more and more farmers and farmers' organizations started producing hybrid seed.

In fact, disseminating the seed of hybrid varieties proved more challenging than producing hybrid seed. Farmers had no prior contact with hybrid crop varieties; the term *hybride* in French was widely feared as it was associated with genetically modified organisms. Discussions led farmers, already familiar with hybrid chickens, to coin the term '*si woloso*' ('si' means 'seed, and 'woloso' is a term used to refer to chickens or goats produced by crossing a local breed with an exotic breed) to identify hybrid seed in the local language Bambara. Farmer field days, trainings, radio and video messages were used to communicate the expected advantages, disadvantages and differences of sorghum hybrids relative to open-pollinated varieties. One major topic for farmers was the issue of saving and re-sowing seed harvested from these hybrids: Would it actually germinate and grow? What would it look like? How much yield would it give? Farmers who grew hybrids for several years confirmed the on-station results, indicating that some hybrids could be re-grown for one or two years without noticeable yield disadvantages (both parents of Guinea race origin), whereas others could be re-grown for only one season, and others not at all.

Village-level culinary tests were conducted specifically with hybrids, confirming not only that hybrid grain was edible, but that food quality of several

hybrids equalled or surpassed the popular landrace check varieties. Farmers then related their own experiences with these new hybrids to their neighbours. For example, one farmer reported that five measures of grain of the hybrid Pablo (with high test-weight, hard grain and low decortication losses) were sufficient to feed his family, whereas six measures of his own variety would be needed to prepare sufficient food for one day.

That farmers' seed-cooperatives have been doubling the production of hybrid seed annually (Kante *et al.*, 2017) and that farmers are now starting to purchase and cultivate hybrid seed (Smale *et al.*, 2014, 2018) is ultimate proof of farmer acceptance. Such change has been possible only through sustained collaboration and joint learning between farmers and breeders.

Conclusions

The opportunity for breeders and farmers to work together over such a long period, and their shared interest in learning from each other, have been important factors enabling the extent and depth of knowledge increase. The capacities and skills of farmers and breeders alike have further contributed to genetic gains for the objectives targeted. Every joint interaction and project with farmer–researcher collaboration has advanced joint learning and effective farmer-oriented breeding. Closer examination of the long-term farmer–researcher collaboration in sorghum and pearl millet breeding in West Africa reveals crucial advantages for achieving transformational change, including:

- The possibility of follow-through, from understanding of farmer needs and conditions to changing the design and activities in the breeding programme: the long-term collaboration has helped the breeding programmes to evolve, linking lessons learned directly with actions at every stage of the breeding process in order to create new varieties that respond to identified priority needs and emerging opportunities.
- Establishing synergetic roles and sharing responsibilities for breeding: long-term engagement has made it possible to build on the strengths and expertise of farmers as well as scientists in clearly defined procedures, in turn helping to achieve the strong genetic correlation of test results, even with activities conducted at larger scales.
- Increased performance of new varieties for diverse users: the decentralized design of the breeding programmes, combining on-station selection and selection in target environments in an effective manner, has helped to achieve sizeable genetic gains despite limited resources and complex environmental and socio-economic contexts.
- Realizing impact and contributing to high-level development goals: the sustained collaboration of farmers and breeders has resulted in clearly detectable contributions to overarching goals such as:
    - Enhanced food security, with understanding of and breeding for traits required for reduced risk to climate variability (Haussmann *et*

al., 2012; Weltzien *et al.*, 2006b) and yield improvements also under poor soil-fertility and farmers' low-input management systems (Kante *et al.*, 2017; Leiser *et al.*, 2012, 2015; Rattunde *et al.*, 2013), helping to reduce the amounts of sorghum grain to be purchased for food while increasing the portion of harvest that is sold (Smale *et al.*, 2018).

- Improved nutrition, particularly in view of the micronutrient deficiencies widespread among women and children in West Africa (Bauchspies *et al.*, 2017; Christinck and Weltzien, 2013): understanding the women's practice of preparing separate meals for children with the grain they produce, along with selecting for vitreous grain and reducing micronutrient losses during decortication, has provided a pathway to enable the benefits of bio-fortified varieties to reach vulnerable groups, especially children.
- Empowerment of farmers: women and men initiated their own variety and cropping-system experimentation as well as methods of marketing seed and facilitating communication about seed among farmers and farmers' cooperatives, based on their learning and access to new types of varieties (Weltzien *et al.*, 2018), thereby also becoming co-owners of the process of variety development and dissemination.
- Conservation and sustainable use of agro-biodiversity: the extensive use of local germplasm in breeding programmes has resulted in an expanded range of variety types made available to farmers through a decentralized system for variety development, seed production and delivery that involves and responds to various types of farmers within and across differing agro-ecological zones (Weltzien *et al.*, 2018).

Acknowledgements

We gratefully acknowledge the commitment, interest and hospitality, with many frank and open discussions, on the part of members of the various farmers' organizations, village-level farmer facilitators, as well as technical staff associated in the districts. We also wish to thank the donors who supported this research: the McKnight Foundation Collaborative Crop Research Program, BMZ/GIZ, IFAD, EU, the Rockefeller Foundation, the Bill and Melinda Gates Foundation and USAID.

References

Access to Seeds Foundation, 2018. *The Rise of the Seed-producing Cooperative in Western and Central Africa*. Amsterdam: Access to Seeds Foundation.

Ashby JA, 1990. *Evaluating Technology with Farmers: A Handbook*. Cali: CIAT.

Atlin GN, Cooper M, Bjornstad A, 2001. A comparison of formal and participatory breeding approaches using selection theory. *Euphytica* 122, 463–475.

Bänzinger M., Cooper M, 2001. Breeding for low input conditions and consequences for participatory plant breeding: Examples from tropical maize and wheat. *Euphytica* 122, 503–519.

Bauchspies WK, Diarra F, Rattunde F, Weltzien E, 2017. 'An Be Jigi': Collective cooking, whole grains, and technology transfer in Mali. *FACETS* 2, 955–968. https://doi.org/10.1139/facets-2017-0033.

Christinck A, Weltzien E, 2013. Plant breeding for nutrition-sensitive agriculture: An appraisal of developments in plant breeding. *Food Security*. https://doi.org/10.1007/s12571-013-0288-2.

Christinck A, Diarra M, Horneber G, 2014. Innovations in seed systems: Lessons from the CCRP-funded project 'Sustaining Farmer-managed Seed Initiatives in Mali, Niger and Burkina Faso'. www.ccrp.org/sites/default/files/christinck_diarra_and_horneber_2014_1.pdf.

Christinck A, Weltzien E, Hoffmann V, 2005. *Setting Breeding Objectives and Developing Seed Systems with Farmers. A Handbook for Practical Use in Participatory Plant Breeding Projects*. Weikersheim, Germany/Wageningen, The Netherlands: Margraf Publishers/CTA.

Clerget B, Dingkuhn M, Goze E, Rattunde HFW, Ney, B, 2008. Variability of phyllochron, plastochron and rate of increase in height in photoperiod-sensitive sorghum varieties. *Annals of Botany* 101, 579–594. https://doi.org/10.1093/aob/mcm327.

Coulibaly H, Didier B, Sidibé A, Abrami G, 2008. Les systèmes d'approvisionnement en semences de mils et sorghos au Mali: Production, diffusion et conservation des variétés en milieu paysan. *Cahiers Agricultures* 17, 199–209.

Dalohoun DN, Van Mele P, Weltzien E, Diallo D, Guinda H, vom Brocke K, 2011. Mali: When government gives entrepreneurs room to grow. In Van Mele P, Bentley JW, Guéi RG, (eds) *African Seed Enterprises. Sowing the Seeds of Food Security*. FAO and AfricaRice.

Diallo C, Isaacs K, Gracen V, Touré A, Weltzien Rattunde E, Danquah EY, Sidibé M, Dzidzienyo DK, Rattunde F, Nébié B, Sylla A, 2018. Learning from farmers to improve sorghum breeding objectives and adoption in Mali. *Journal of Crop Improvement*, 32 (6), 829–846.

Haussmann BIG, Fred Rattunde H, Weltzien-Rattunde E, Traoré PSC, vom Brocke K, Parzies HK, 2012. Breeding strategies for adaptation of pearl millet and sorghum to climate variability and change in West Africa. *Journal of Agronomy and Crop Science* 198, 327–339. https://doi.org/10.1111/j.1439-037X.2012.00526.x.

Isaacs K, Weltzien E, Diallo C, Sidibé M, Diallo B, Rattunde F, 2018. Farmer engagement in culinary testing and grain-quality evaluations provides crucial information for sorghum breeding strategies in Mali. In Tufan HA, Grando S, Meola C, (eds) *State of the Knowledge for Gender in Breeding: Case Studies for Practitioners*. Lima, pp. 74–85. CGIAR Gender and Breeding Initiative. Working Paper. No. 3. http://hdl.handle.net/10568/92819.

Jordan DR, Mace ES, Cruickshank AW, Hunt CH, Henzell RG, 2011. Exploring and exploiting genetic variation from unadapted sorghum germplasm in a breeding program. *Crop Science* 51, 1444. https://doi.org/10.2135/cropsci2010.06.0326.

Kante M, Rattunde HFW, Leiser WL, Nebié B, Diallo B, Diallo A, Touré AO, Weltzien E, Haussmann BIG, 2017. Can tall guinea-race sorghum hybrids deliver yield advantage to smallholder farmers in West and Central Africa? *Crop Science* 57, 833. https://doi.org/10.2135/cropsci2016.09.0765.

Kountche BA, Hash CT, Dodo H, Laoualy O, Sanogo MD, Timbeli A, Vigourou Y, This D, Nijkamp R, Haussmann BIG, 2013. Development of a pearl millet Striga-resistant genepool: Response to five cycles of recurrent selection under Striga-infested field conditions in West Africa. *Field Crops Research* 154, 82–90. http:// dx.doi.org/10.1016/j.fcr.2013.07.008.

Leiser WL, Rattunde HFW, Weltzien E, Haussmann BIG, 2014. Phosphorus uptake and use efficiency of diverse West and Central African sorghum genotypes under field conditions in Mali. *Plant and Soil* 377, 383–394. https://doi.org/10.1007/ s11104-013-1978-4.

Leiser WL, Rattunde HFW, Weltzien E, Haussmann BIG, 2018. Sorghum tolerance to low-phosphorus soil conditions. In *Achieving Sustainable Cultivation of Sorghum: Genetics, Breeding and Production Techniques*, Burleigh Dodds Series in Agricultural Science. Burleigh Dodds Science Publishing, 247–272. https://doi.org/10.19103/ AS.2017.0015.30.

Leiser W, Rattunde F, Piepho H-P, Weltzien E, Diallo A, Touré A, Haussmann B, 2015. Phosphorous efficiency and tolerance traits for selection of sorghum for performance in phosphorous-limited environments. *Crop Science* 55, 1–11. https://doi. org/10.2135/cropsci2014.05.0392.

Leiser WL, Rattunde HFW, Piepho H-P, Weltzien E, Diallo A, Melchinger AE, Parzies HK, Haussmann BIG, 2012. Selection strategy for sorghum targeting phosphorus-limited environments in West Africa: Analysis of multi-environment experiments. *Crop Science* 52, 2517–2527. https://doi.org/10.2135/ cropsci2012.02.0139.

Quirós CA, Gracia T, Ashby JA, 1991. *Farmer Evaluations of Technology: Methodology for Open-ended Evaluation*. IPRA : CIAT, Cali, Colombia.

Ragot M, Bonierbale M, Weltzien E, 2018. From market demand to breeding decisions: A framework. CGIAR Gender and Breeding Initiative Working Paper 2. Lima (Peru): CGIAR Gender and Breeding Initiative. http://hdl.handle. net/10568/91275.

Rattunde HFW, Sidibé A, vom Brocke K, Diallo A, Weltzien E, Nebié B, 2014. Keninke Siwolosow Keninke siwolosow ani u silabugu feerew. 22p. ICRISAT Bamako, Mali.

Rattunde HFW, Weltzien E, Bramel-Cox PJ, Kofoid K, Hash CT, Schipprack W, Stenhouse JW, Presterl T, 1997. Population improvement of pearl millet and sorghum: Current research, impact and issues for implementation. In *Proceedings of the International Conference on Genetic Improvement of Sorghum and Pearl Millet*, 188–212. Presented at the International Conference on Genetic Improvement of Sorghum and Pearl Millet, Lubbock, Texas USA.

Rattunde HFW, Michel S, Leiser WL, Piepho HP, Diallo C, vom Brocke K, Haussmann BIG, Weltzien E, 2016. Farmer participatory early-generation yield testing of sorghum in West Africa: Possibilities to optimize genetic gains for yield in farmers' fields. *Crop Science* 56, 1–13. https://doi.org/10.2135/cropsci2015.12.0758.

Rattunde F, Sidibé M, Diallo B, van den Broek E, Somé H, vom Brocke K, Diallo A, Nebie B, Touré A, Isaacs K, Weltzien E, 2018. Involving women farmers in variety evaluations of a 'Men's Crop': Consequences for the sorghum breeding strategy and farmer empowerment in Mali. In Tufan HA, Grando S, Meola C, (eds) *State of the Knowledge for Gender in Breeding: Case Studies for Practitioners*, 95–107. Lima: CGIAR Gender and Breeding Initiative. Working Paper. No. 3. http://hdl.handle.net/10568/92819.

Rattunde HFW, Weltzien E, Diallo B, Diallo AG, Sidibe M, Touré AO, Rathore A, Das RR, Leiser WL, Touré A, 2013. Yield of photoperiod-sensitive sorghum hybrids based on guinea-race germplasm under farmers' field conditions in Mali. *Crop Science* 53, 2454. https://doi.org/10.2135/cropsci2013.03.0182.

Siart S, 2008. *Strengthening Local Seed Systems: Options for Enhancing Diffusion of Varietal Diversity of Sorghum in Southern Mali*. Kommunikation und Beratung, Band 85. Weikersheim, Germany, Margraf Publishers.

Smale M, Kernga A, Assima A, Weltzien E, Rattunde F, 2014. *An Overview and Economic Assessment of Sorghum Improvement in Mali*. Michigan State University International Development Working Paper. Working Paper 137, 40.

Smale M, Assima A, Kergna A, Thériault V, Weltzien E, 2018. Farm family effects of adopting improved and hybrid sorghum seed in the Sudan Savanna of West Africa. *Food Policy* 74, 162–171. https://doi.org/10.1016/j.foodpol.2018.01.001.

vom Brocke K, Trouche G, Weltzien E, Barro-Kondombo CP, Gozé E, Chantereau J, 2010. Participatory variety development for sorghum in Burkina Faso: Farmers' selection and farmers' criteria. *Field Crops Research* 119, 183–194. https://doi.org/10.1016/j.fcr.2010.07.005.

vom Brocke K, Trouche G, Weltzien E, Kondombo-Barro CP, Sidibé A, Zougmoré R, Gozé E, 2014. Helping farmers adapt to climate and cropping system change through increased access to sorghum genetic resources adapted to prevalent sorghum cropping systems in Burkina Faso. *Experimental Agriculture* 50, 284–305. https://doi.org/10.1017/S0014479713000616.

Weltzien E, Christinck A, 2017. Participatory plant breeding: Developing improved and relevant crop varieties with farmers. In Snapp S, Pound B, (eds) *Agricultural Systems: Agroecology and Rural Innovation for Development* (259–301) Academic Press, Burlington, MA, USA and London.

Weltzien E, Rattunde HFW, Van Mourik TA, Ajeigbe HA, 2018. Sorghum cultivation and improvement in West and Central Africa. In *Achieving Sustainable Cultivation of Sorghum: Sorghum Utilization around the World* (217–240). Burleigh Dodds Science Publishing, Cambridge, UK.

Weltzien E, Christinck A, Touré A, Rattunde F, Diarra M, Sangare A, Coulibaly M, 2006a. Enhancing farmers' access to sorghum varieties through scaling-up participatory plant breeding in Mali, West Africa. In Almekinders C, Hardon J, (eds) *Bringing Farmers Back into Breeding. Experiences with Participatory Plant Breeding and Challenges for Institutionalisation*, 58–69. Agromisa, Agromisa Foundation, Wageningen, The Netherlands.

Weltzien E, Rattunde HFW, Clerget B, Siart S, Touré A, Sagnard F, 2006b. Sorghum diversity and adaptation to drought in West Africa. In Jarvis D, Mar I, Sears L, (eds) *Enhancing the Use of Crop Genetic Diversity to Manage Abiotic Stress in Agricultural Production Systems*, 31–38. Rome/Budapest: IPGRI, Rome, Italy, Budapest, Hungary.

Weltzien E, Kanouté M, Toure A, Rattunde F, Diallo B, Sissoko I, Sangaré A, Siart S, 2008. Sélection participative des variétés de sorgho à l'aide d'essais multilocaux dans deux zones cibles. *Cahiers Agricultures* 17, 134–139.

Yapi AM, Kergna AO, Debrah SK, Sidibe A, Sanogo O, 2000. *Analysis of the Economic Impact of Sorghum and Millet Research in Mali*. Impact Series. International Crops Research Institute for the Semi-arid Tropics, Patancheru, Andhra Pradesh, India.

4   Rice PPB in India and Nepal

Client-oriented plant breeding using
few, carefully chosen crosses

*John R. Witcombe, D.S. Virk and
Krishna D. Joshi*

Introduction

Our participatory research in rice began in India, with participatory varietal
selection (PVS) trials on upland rice (Joshi and Witcombe, 1996). These trials
showed that farmers had not had access to seed of appropriate upland rice
varieties – variety Kalinga III was the most preferred, but had never been
recommended for that area. However, PVS could exploit only existing variation among finished cultivars, because it was not linked to a breeding programme. To achieve varieties better than Kalinga III we would have to go
beyond PVS, to the breeding of new varieties. We began programmes on
participatory plant breeding (PPB) in upland rice in India (Joshi and Witcombe, 1996; Witcombe *et al.*, 1996) and in high-altitude rice breeding in
Nepal (Sthapit *et al.*, 1996).

This rice breeding eventually covered upland and lowland areas in India,
as well as lowland and high-altitude areas in Nepal – and hence provided a
test on the wider applications of participatory methods.

All our breeding involved making only a few crosses. The method
described here applies to all breeding programmes, not just participatory ones.
Soundly based on theory, it is the only practicable method of involving
farmers without intensive inputs from scientists. We provide here an extensive explanation of this theory because the adoption of a few-cross approach
is essential to adopting participatory methods. We discuss the social construct
of many-cross breeding and the institutional constraints to the adoption of a
few-cross method. For several case studies, we also review the constraints to
popularizing varieties.

Breeding methods: using few, smart crosses

Why use fewer crosses?

The traditional approach is to make many crosses and involve scientists in a
non-participatory crossing programme. Once farmers participate, they have
to evaluate a huge number of lines that the crosses have produced. For this,

the research station must be effectively transferred to the farm – a decentralized rather than a participatory approach. To get farmers really involved without scientists planting material for them, the breeding programme must involve fewer crosses than has been common (Witcombe and Virk, 2001). The use of few crosses is strongly supported by theory (Witcombe *et al.*, 2013). It also forces the breeder to be more client-oriented, as it requires learning what farmers need in terms of new varieties and then choosing appropriate parents to meet this need; further, it enables farmer participation in evaluating advanced lines, with limited inputs from breeders.

Theory behind cross number

The model of Yonezawa and Yamagata, 1978 has a constant P1 but should it vary?

Theoretical models, exemplified by Yonezawa and Yamagata (1978), that examined optimum cross numbers assumed that all crosses would have the same probability of success (*P1*). Hence, the first cross that a breeder makes is assumed to be no better than the 1,000th or even 10,000th one. However, this assumption means either that the breeder knows nothing about potential parents, or has this knowledge but does not use it. Possible reasons why using parental information may be considered as not improving the likelihood of earlier choices being better than later ones include:

- there is no correlation between the performance per se of a parental variety and its value as a parent. However, for this to be true, a disease-resistant or drought-tolerant parent should be no better at producing progeny that have these traits than parents that lack them.
- Parental information is irrelevant because if one makes enough crosses, the best one is bound to be included by chance. However, with constant resources, as more crosses are made, the size of the population from each cross diminishes. Hence, even though the 'right' cross might be made, the probability of its succeeding is very low, because the rare desirable transgressive segregants are much less likely to be found in a small population.

It is more reasonable to assume that breeders should use knowledge about potential parents to help in choosing crosses more likely to succeed than randomly selected ones. Hence, *P1* will start at an above-average level, declining as more crosses are made. The first choices utilize plentiful knowledge on the parents, allowing the breeder to make clever or smart crosses – for example, between a parent reliably known to be highly adapted to the target environments and the other with a complementary trait such as disease resistance. As more crosses are made, the breeder has to use parents where there is less useful information to guide the choice (e.g. lack of reliable data on disease resistance or yield potential). Consequently, the first crosses that breeders choose are more likely to succeed, and the probability of success of each cross declines as more are made. With fewer crosses, then, it is possible to grow larger populations from each cross.

Even using the model of Yonezawa and Yamagata, 1978 with a constant
P1 more crosses are not always better

Using the (incorrect) assumption of Yonezawa and Yamagata (1978) of a constant *P1* (the probability of a cross succeeding) as well as a constant *P2* (the probability of a plant within a cross succeeding) it is mathematically impossible for fewer crosses to have a higher overall probability success than the case with many crosses. However, even with a constant *P1* and *P2*, many crosses are not always better because, it is possible, without using more resources, to increase the overall population size *K* of the breeding pro-gramme. More crosses involve more labour for crossing, and more plots, labels, seed packets, parental checks, and data collection and management; when fewer crosses are made, *K* can be increased without an increase in overall costs. Hence, Yonezawa and Yamagata (1978) argued that the relative costs of crossing versus growing plants determines the optimum number of crosses, particularly when *P1* is 0.1 or more and *K* is also large. This is because, at higher values of *P1*, increasing the number of crosses becomes less effective in increasing the overall probability of success (see Figure 4.1). After *K* is more than 25,000 then increasing the number of crosses has little effect. For example when *K* = 30,000 and when *P1* = 0.005, 200 crosses are 2.8 times more likely to succeed (a 180 per cent increase) than 10, but when *P1* = 0.2, this falls to a 13 per cent increase. Under reasonable assumptions on costs, making an extra 190 crosses would not be worth a 13 per cent increase when the resources saved could be used to grow more plants from each cross.

It still may not be better with a *P1* = 0.005 to make 190 more crosses even to get a 180 per cent increase, but to increase *K* instead. A few crosses with large populations (giving a large *K*) are more likely to succeed than many more crosses with a smaller *K*. For example, 10 crosses with *K* = 25,000 are more likely to succeed than 200 crosses where *K* = 5000 (Figure 4.2). Which one is cheaper and simpler will depend on the balance of costs between making crosses and growing plants.

Using the model of Yonezawa and Yamagata (1978) but allowing P1 to
vary demonstrates that fewer crosses are more likely to succeed

So far, *P1* has been assumed to be a constant. However, *P1* declines as more crosses are made, if we include the ability of breeders to choose superior crosses using parental data. Witcombe *et al.* (2013) evaluated the impact of *P1* declining as more crosses were made. Three declines were modelled – linear, S-shaped and exponential. Under these scenarios, fewer crosses were always

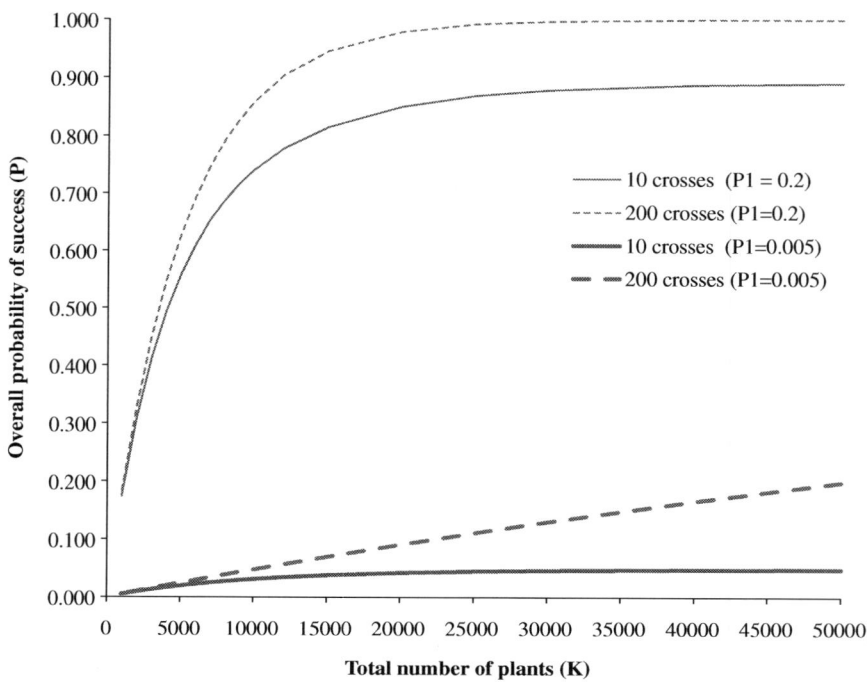

Figure 4.1 Probability of a cross succeeding with varying K at two levels of P1.

Source: Witcombe *et al.*, 2013.

Note: Using the assumption of Yonezawa and Yamagata (1978), comparison of 10 and 200 crosses with population size (K) varying from 1000 to 50,000 when the probability of an individual cross succeeding (P1) is constant 0.005 or a constant 0.2. The probability of a plant succeeding (P2) is a constant 0.001 in both cases.

more efficient than more, as exemplified by the comparison of 10 versus 200 crosses (Figure 4.3).

An S-shaped decline gives intermediate values between those from a linear or exponential decline. We consider the S-shaped decline to be the most likely because there will usually be a few parents about which a great deal is known (e.g. released varieties); more about which much is known (e.g. lines in advanced trials); and many more about which little is known (e.g. newly produced lines, and lines tested in only one year). We consider the exponential decline to be the next most likely scenario, and the linear decline unlikely.

The advantages of fewer crosses has been under-estimated, as models have failed to take into account the trade-off in costs between making more crosses versus growing more plants. For example: although, given a linear decline and high K, more crosses will yield a very slightly higher probability of success (Figure 4.3), it is much more cost-effective to make fewer crosses.

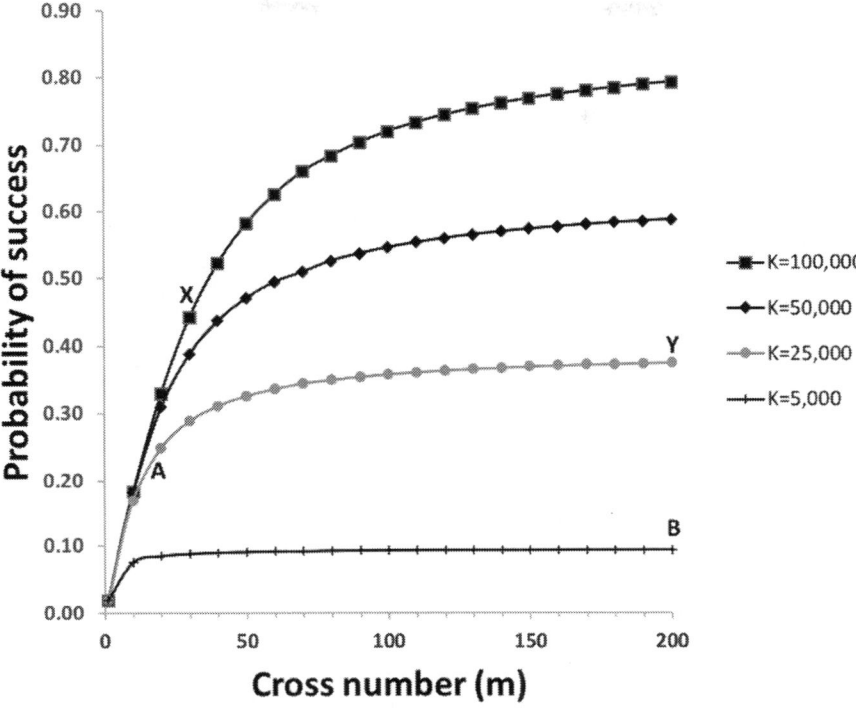

Figure 4.2 Overall probability of success with four population sizes.

Source: Adapted from Witcombe *et al.*, 2013.

Note: Effect on the overall probability of success of increasing the number of crosses with four total population sizes (K). P1 (probability of cross succeeding) = 0.02 and P2 (probability of a plant succeeding) = 0.001 in all cases. 10 crosses with K = 25,000 (point A) is more likely to succeed than 200 crosses with K = 5,000 (point B). Similarly 30 crosses with K = 100,000 (point X) is more likely to succeed than 200 crosses with K = 25,000 (point Y).

P1 was assumed to start at 0.4 on the basis of our experiences without COB (client-oriented breeding) programmes. As more resources are spent on evaluating parents and more time is spent on choosing crosses the starting level of *P1* increases. A most effective way of increasing *P1* to a high level is to make one or more backcrosses to a variety that is popular with farmers because of agronomic and quality superiority. Selection is then made for the traits of the recurrent parent and for those targeted to come from the complementary parent. This approach is commonly used in successful few–cross, marker-assisted selection breeding programmes (see below). The advantage of making fewer crosses continues when *P1* falls below 0.4. For the S-shaped decline, 200 crosses become superior to 10 only when *P1* falls to 0.075 (when *K* = 50,000), while 10 crosses are always superior to 200 for the exponential

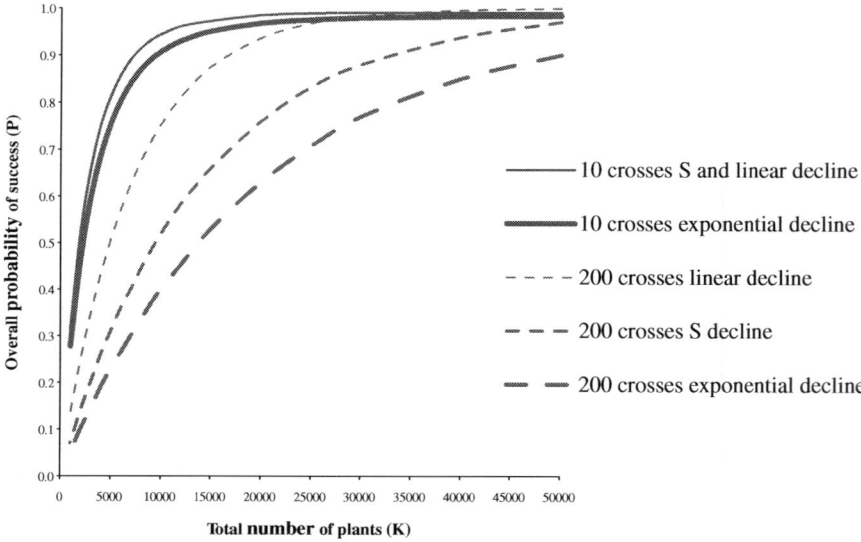

Figure 4.3 Overall probability of success using either 10 or 200 crosses with varying population sizes.

Source: Witcombe et al., 2013.

Note: The overall probability of a programme succeeding using either 10 or 200 crosses with varying population sizes (K varying from 1,000 to 50,000). The probability of an individual cross succeeding is initially 0.4 and declines in either an S-shape (S), linearly (L), or exponentially (E) to 0.05. The results of using 10 crosses with an S-shape or a linear decline are almost identical, so only the mean of the two is shown. The probability of plant succeeding within a cross (P2) is a constant 0.001.

decline no matter how small is *P1*. For the linear decline (which we deem unlikely) the initial value of *P1* must fall to about 0.2 before 200 crosses have any advantage over 10 (and then only for when $K > 30\,000$).

The contradiction between number and judgement

There is an inherent problem with making many crosses, due to the acknow-ledged need to select parents carefully. PBI Cambridge (1990) reported in their wheat-breeding programme that parents for cross-pollination were chosen with great care; further, that some 1,500 crosses were made per year. However, it is difficult to envisage how meticulous consideration could be given to all these cross combinations.

Social construct for high cross-number breeding

Why, then, do breeders make so many crosses? Cleveland (2001) argues that plant breeders decide what they do from objective evidence but also from a social construction they make on the basis of institutional and social settings. He discusses the non-objective dichotomy between viewpoints on yield stability: either breeding for yield in favourable environments provides a spill over of higher yield in more marginal ones; or that, in order to target marginal environments, it is necessary to breed for specific adaptation by carrying out selection in them. Baranski argues how Indian scientists created a social construct in favour of the first of these viewpoints:

> Many people have criticized the Green Revolution for its unequal spread of benefits, but none of these critiques address wide adaptation – the core tenant held by Indian agricultural scientists to justify their focus on highly productive land while ignoring marginal or rainfed agriculture.
>
> (Baranski 2015, 41)

The contrasting viewpoints on wide adaptation and specific adaptation are constructed by considering different ranges of environments. Those supporting spillover from high-yield environments fail to include the extreme low-yielding environments of the supporters of specific adaptation.

Making many crosses would also appear to be a *social* construction rather than an objective one. First, Green Revolution varieties such as those in wheat (see, e.g. Hanson *et al.*, 1982; Maredia and Byerlee, 1999) were bred by making very many crosses. Although this is theoretically not the most efficient method (see above) high-volume-cross breeding programmes proved very successful. However, success per se provides no evidence on how efficient a method is relative to the alternatives. This vital consideration is ignored when breeders unquestioningly follow a many cross-strategy simply because it has been shown to work, not because it has been shown to be the most efficient. Green Revolution breeding programmes succeeded, not through evidence-based choice of method, but because they were well-resourced and long-term. They were effectively large-population, recurrent selection programmes: in a long-term breeding programme with many crosses each year, the parents used for crossing will inevitably include lines produced by the breeding programme itself.

The social construct of many crosses is reinforced by intuition – doing a lot of the most difficult thing in the breeding process ought to be better than doing only a little. Making crosses is difficult and is simple to monitor, so it readily becomes elevated above other components in the breeding process. For example, the International Rice Research Institute (IRRI) marked its 100,000th cross by a public ceremony and a press release. However, it did not mark, or may not have measured, other possible milestones, such as the cumulative area of plants grown, or the number of F_4 lines. Indeed, in most or all developing-country breeding programmes, the skill and diligence of the breeder is judged in terms of the number of crosses made.

Institutional factors support the social construction. Changing to a few-cross method can be interpreted as the breeder escaping the burden of making many crosses – and managements are reluctant to lose what they see as an ideal measure of how hard the breeder is working. In any case, institutional change is difficult in plant breeding. McGuire (2008) reviewed the social construction of the technology adopted by breeders in the Ethiopian sorghum breeding programme (emphasis on hybrids for high-input agriculture, when most sorghum is grown in marginal areas). Early choices become 'locked in', reinforced by technical constraints and established breeding routines, so changes to more participatory research such as PPB are resisted. McGuire concludes that, if new institutional approaches are to be introduced, then the social, technical and political factors that reinforce existing pathways will have to be challenged.

Evidence for a few-cross strategy

There is evidence that a few-cross strategy can be effective when attempted (Witcombe and Virk, 2001). In a programme in Nepal, led by the late Dr Bhuwon Sthapit, four out of only eight crosses for improving a local landrace produced released or pre-released varieties. At the West Africa Rice Development Association (WARDA) during the crossing programme between *Oryza sativa* and O. *glaberrima*, emphasis was placed on selecting the parents for crosses because the crosses were difficult to make (Jones *et al.*, 1997). Only eight parents of *glaberrima* and five of *sativa* were chosen for their combination of traits, and only seven crosses set seed. All seven 'New Rice for Africa' (NERICA) varieties that were released in 2000 (WARDA, 2006) were from just one of these crosses – a success rate of 14 per cent, as against only 0.5 per cent in most rice breeding (Witcombe *et al.*, 2013).

In a 15-year rice-breeding programme in Nepal with varietal testing in India and Nepal, we paid great attention in choosing the parents, and we made few crosses. Six crosses had sufficient time to be evaluated on the basis of adoption and official release of varieties from them (Witcombe *et al.*, 2013). Four of these six produced significantly higher-yielding varieties, on-farm as well as on-station, as the best available alternatives in three rice ecosystems.

With maize, the equivalent of a few-cross approach is to make only a single composite population. We tested this in western and eastern India by starting two independent programmes. In each region, one population was made from white and yellow grain types; both populations have produced a officially released variety (Witcombe *et al.*, 2003; Virk *et al.*, 2005).

The advanced technology of marker-assisted selection (MAS) has made breeders turn to a few-cross strategy that uses principles of participatory research by identifying varieties popular with farmers. These are then improved using MAS for traits that are important to farmers such as submergence tolerance, drought tolerance and disease resistance. One example is the success of Swarna-Sub1 from a cross between a widely grown lowland variety Swarna and a submergence-tolerance donor Indian landrace FR13A having the *SUB1* gene (Mackill *et al.*, 2012). Later, this gene was introgressed into four highly popular ('mega') rice varieties of South Asia commonly grown in flood prone areas where submergence is common, of which Swarna–Sub1 in India, Bangladesh and Nepal, and BR11–Sub1 in Bangladesh have been particularly successful (Singh *et al.*, 2013). The gene controlling tolerance (*SUB1*) resulted in Sub1 varieties with yield advantages of 1 to over $3\,t\,ha^{-1}$ following submergence; these varieties reached over 3.8 million farmers in Asia within three years of official release (Ismail *et al.*, 2013). Other examples, currently in yield trials in Nepal (Ram Baran Yadaw pers. comm.), include drought tolerance QTLs (quantitative trait loci) (Yadaw *et al.*, 2013) introduced into popular Nepalese variety Sabitri, and the addition of bacterial leaf blight (BLB) resistance to Swarna Sub1, and to the popular Indonesian variety Ciherang Sub1.

Case studies

Our COB programmes have produced farmer-preferred varieties in rice in India and Nepal. All methods used very few crosses and large populations of progeny. Selection was always delayed until after the F_2, as that generation has the lowest between-plant heritability.

Upland rice in eastern India

The success of the upland rice varieties in India, Ashoka 200F and Ashoka 228 (Virk *et al.*, 2003) has been reviewed by Witcombe and Yadavendra (2014). The impact of these varieties has been very well studied (Table 4.1).

All the surveys revealed that a very high proportion of farmers who had been given seed continued to grow the variety (usually over 80 per cent, sometimes nearly all) also several years later. Those farmers that dis-adopted often did so because they had been given seed despite not having suitable upland. Farmers distributed the seed to others, in their own villages and to other villages. The adopting farmers always grew the varieties on a high proportion of their upland (always over 85 per cent). Areas grown per household were often small (these are resource-poor, land-poor farmers), but when farmers also grew the variety in more favourable medium land (upland is the least favourable rice growing environment and lowland the most), as was the case in Chhattisgarh, then the area averaged about 1 ha per household.

These results represent a qualitative change in varietal adoption – the difference between success or failure, rather than incremental improvement. Previously farmers rarely grew modern varieties: nearly all the uplands were devoted to landraces. Participatory research produced the very first significant success for the uplands, after decades of conventional plant breeding research.

Farm surveys are not straightforward, however. Lakra *et al.* (2012) describe the results of IRRI surveys conducted in Jharkhand, May–December 2007. The Ashoka varieties were not reported among modern varieties (MVs).

Table 4.1 Adoption surveys of Ashoka 200F and Ashoka 228 upland rice varieties, various Indian States, 2004–2012

Year	Type	Number	Where	Reference
2004	Household interviews	151	West Bengal, Jharkhand, Odisha	Virk and Witcombe (2007). Witcombe *et al.* (2011)
2004	Whole-village surveys	17	West Bengal, Jharkhand, Odisha	Mottram (2004)
2004	Household interviews	165	Gujarat, Madhya Pradesh and Rajasthan	Yadavendra and Witcombe (2013)
2008	Whole-village group discussion followed by household-level questionnaires	20 villages 296 households	Rajasthan, Odisha, West Bengal and Jharkhand	Conroy *et al.* (2008)
2012	Group discussion followed by household-level questionnaires	9 villages 180 households	Odisha, Jharkhand and Chhattisgarh	Witcombe and Yadavendra (2014)

However, because local dialect was used, researchers could not differentiate whether a variety was the same or different from one reported by the farmers in other areas. The landraces Gora Dhan or Goda (= Gora) were very popular; we know that farmers have also called Ashoka varieties 'Gora' or 'Ashoka Gora'. In the same region, Sinah and Xaxa (2014) identified 26 upland rice varieties, most commonly called Gora Dhan. In the true uplands, these Gora varieties are known by 19 different names, including Arsunga Gora, Kala Gora, Karanga Gora, and Lal Gora. The Ashoka varieties can easily get lost amongst those with 'Gora' in their name.

Birsa Dhan 108, the only MV in the uplands similar in maturity to the Ashoka varieties, was grown on only 1.4 per cent of the upland area (Lakra *et al.*, 2012). At the time of the survey this variety was recently released so seed could only have been distributed in limited amounts. It was probably present in a few villages where it was promoted by Birsa Agricultural University. Of the MVs reported in the uplands, two of the three most popular – Vandana and Anjali – are better suited to medium land as they mature 15 days later (90 to 95 days compared with 80 days for the Ashokas).

Hence, reliable conclusions about the Ashoka varieties cannot be drawn from the IRRI survey because of naming ambiguity, the lack of data categorized into district and village, and the surveyed upland area, including medium land.

A subsequent limited survey with focus group discussions in 2016 (Virk, 2016) was conducted in Singari village of Angada block of Ranchi district in Jharkhand, where the last seed of Ashoka and PY84 was sold in 2013. Both varieties were still grown, however. A sample of harvested seed from farmer Sarju Bedia was tested by molecular markers and was found to be Ashoka 200F, which he claimed to be PY84 – a typical case of farmer misidentification of similar varieties.

The Ashoka adoption studies were repeated over several years, covered many states, districts and villages, and were conducted by different institutions and personnel. In the literature, we have not found other studies on adoption that are as well replicated and repeated. Our surveys have shown, beyond doubt, that once farmers have access to seed, they choose to grow the Ashoka varieties on significant portions of their available land. However, despite this evidence, institutional support for the distribution of these varieties was not sustained beyond 2012. The Rockefeller Foundation recognized the value of these varieties and provided substantial funds for their dissemination from 2005 to 2009, and the Department for International Development also supported their dissemination (2008 to 2012). However, despite formal approaches, dissemination failed to gain support from any other international funders. The Bill and Melinda Gates Foundation funded projects in eastern India to disseminate new rice varieties (Yamano *et al.*, 2013). These were led by IRRI, who promoted IRRI-bred varieties and varieties released by formal partners in the project, to the exclusion of all other varieties. This can be described as the 'not-invented-here' problem that, unfortunately, limits the choices available to farmers. It happens because repeat funding is more likely for the lead institute when all successful project varieties have been bred in-house or by its partners.

Rice in Nepal

We have bred varieties for low- and high-altitude regions. Witcombe *et al.* (2016) reviewed the results of the dissemination and uptake of COB varieties for the low-altitude regions of Nepal. Here we consider in some detail the COB breeding programmes for high-altitude rice in which the late Dr Bhuwon Sthapit played a pivotal role. We do not include his significant work on the improvement of rice landraces such as Jetho Bhudo (Gyawali *et al.*, 2010) but concentrate on the breeding of high-altitude rice in the Kaski region of central Nepal.

Using a few-cross, participatory approach, the research he led produced two farmer-accepted, cold-tolerant rice varieties – Machhapuchhre-3 (M-3) and Machhapuchhre-9 (M-9) (Sthapit *et al.*, 1996). Both were derived from the cross Fuji 102/Chhomrong Dhan. Later a third variety was produced from Dr Sthapit's PhD programme from a cross he made at Bangor University in Wales, between IR36 and Chhomrong-1L. In 1992, 230 F3 progenies were grown at Yampaphant, Nepal. One of these lines was named as Lumle 2 after further inbreeding and selection. The subsequent adoption of these three varieties was studied in villages in Kaski district (Joshi *et al.*, 2001; Joshi and Witcombe, 2003; Steele *et al.*, 2009; Witcombe *et al.*, 2011). The adoption of these varieties was high. For example, in half of eight study villages in 2004, all the farmers were growing at least one of the varieties. In each

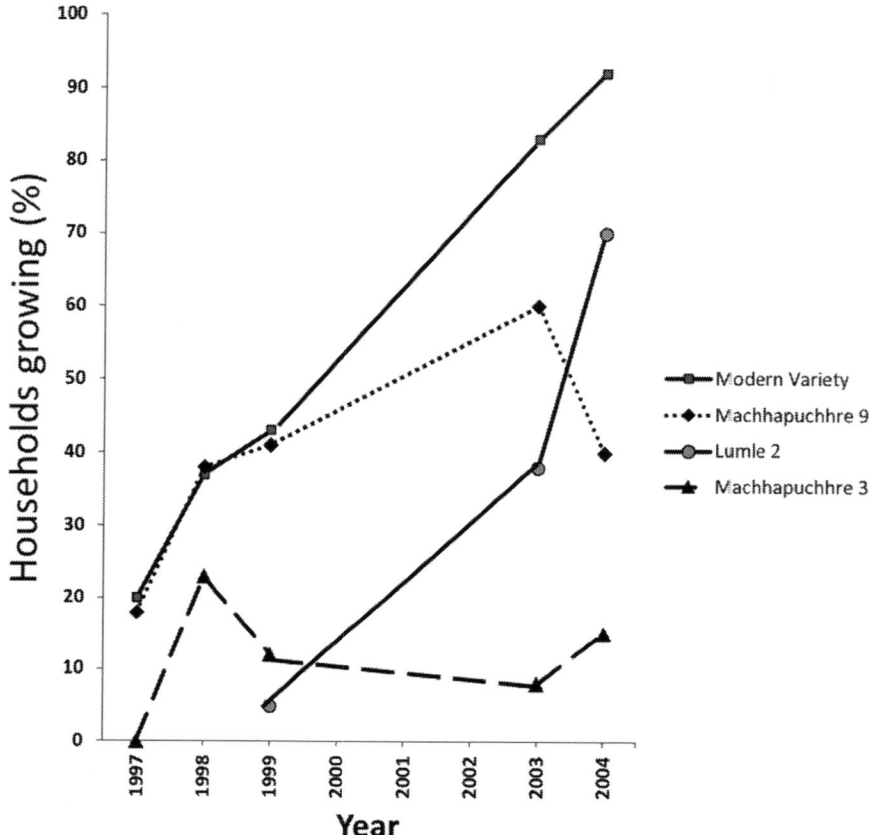

Figure 4.4 Adoption of three new modern varieties in Marangche village, Nepal, 1997–2004.

Source: Adapted from Witcombe et al., 2011.

village, the area occupied by them varied from 37 per cent to 83 per cent of the rice land. In the mid-hill village of Marangche, we were able to monitor varietal change; the popular Machhapuchhre varieties were largely replaced once Lumle 2 was introduced.

Institutionalization

India

There are two interrelated aspects to institutionalization – institutionalization of client-oriented breeding (COB) methods (Witcombe *et al.*, 2005), and the institutionalization of the varieties in the formal seed-supply system. The two are highly interdependent, as successful COB varieties will help to convince breeders to change because the methods have proven effective.

Under certain circumstances, constraints to institutionalization in the formal system may be absent. For the maize variety GM-6 bred by participatory methods (Witcombe *et al.*, 2003), formal linkages proved effective. The popularity of GM-6 was established through project-based participatory seed supply. This created sufficient demand for the Gujarat State Seed Corporation to undertake profitable production of the variety, facilitated by breeder seed supplied by Gujarat Agricultural University. Large quantities of seed were produced and GM-6 became the most-produced variety in Gujarat (Witcombe and Yadavendra, 2014).

However, the seed supply of the Ashoka varieties was constrained by the economics of seed production. In the irrigated off-season, it is less economic to grow a short-duration upland variety than a longer-duration lowland variety. Hence, upland varieties require seed production subsidies greater than those for lowland varieties – but these are not forthcoming. The promotion of varieties from COB released after the Ashoka varieties, such as PY84, Barkhe 3010 and Sugandha 1, have been even more constrained because they came at the end of a project-supported breeding programme and lacked the longer-term project support for seed multiplication and dissemination that the Ashoka varieties had received.

It is unsurprising that the institutionalization of the breeding methods in eastern India is weak, as there has been a lack of strong institutional support for the formal dissemination of successful varieties produced by COB. Institutional support for seed supply is lacking because varietal release is the endpoint for a breeder, whereas promoting the adoption of new varieties is left for the extension system. However, our institutional linkages were mainly with breeders, not with extension workers.

In the State Agricultural University system in India, there are many constraints that prevent change in fundamental methods in breeding. This was discussed above in connection with the continuation of social constructs in plant breeding.

Nepal

The breeding of high-altitude rice proved very successful. The breeding programme could start with a clean slate, as the area under rice was too small to justify a large formal breeding programme. Participatory methods made the breeding for these environments economical. Moreover, the adoption studies have shown that farmers clearly have effective networks for seed distribution in a region where difficult communications prevent a viable formal seed-supply system.

We have not reviewed the low-altitude COB. Here it should be noted that the challenges involved in breeding new lowland varieties in Nepal are of a different order than, for example, those entailed in breeding high-altitude rice or upland varieties for India. A COB programme for the Nepal lowlands will have to compete with the introduction and testing of varieties from many countries in Asia, and with public- and private-sector organizations from India, as many varieties grown in Nepal are introduced from India.

Our project-based breeding programmes had limited resources to compete with these programmes. We had no access to disease-screening nurseries, and no resources to conduct replicated yield trials on research stations to add to the results of PVS trials. Thus, it must be seen as a positive outcome for COB that two 'non-specialist' high-yielding varieties were released from the programme, gaining high acceptance with farmers in 2011 (Witcombe *et al.*, 2016). However, COB methods in these environments are more applicable to the breeding of specialized varieties, such as aromatic ones (released variety Sunaulo Sugandha) or high-grain-value Masuli-type varieties (e.g. Madhyam Dhan 7042, which is proposed for release).

Conclusions

COB programmes in India and Nepal have succeeded in breeding varieties that have been formally released, and some of these have been widely grown. However, ultimate success has been limited by the lack of formal uptake beyond that of project-based support. When supported by formal seed production, COB approaches can be outstandingly successful, as described for GM-6 maize variety above (Witcombe and Yadavendra, 2014).

COB can contribute greatly to the efficiency of breeding programmes. However, it has not been mainstreamed, despite the positive evidence from many crops and institutions. Administrators, policymakers and researchers still need to be convinced. Impetus for change could be provided by integrating into all plant breeding programmes the seed production and scaling-up of varieties, and surveys of their adoption. If the outcomes from formal and COB programmes are followed through to these stages, more convincing evidence might be produced. This can happen only if there is an international thrust by the funding agencies of mainstream institutions. However, development funders tend to follow the latest fashions in development theory: participatory methods in plant breeding risk being labelled as 'old hat'.

Apart from COB in general, the uptake of specific components can be considered, but no substantial efforts have been made to quantify this. The most widely adopted individual component of COB is PVS. As to making fewer crosses, there are anecdotal reports that breeders have heeded the evidence and changed their methods, but no formal evaluations. Given the almost-universal high-volume crossing method in the CGIAR (The Consultative Group on International Agricultural Research) system, it may take many years to achieve wider acceptance of this approach, despite its sound theoretical basis. However, one rapidly developing area of plant breeding is marker-assisted selection; for reasons of cost, breeders typically use very few crosses by having a popular variety as the recurrent parent. As more evidence emerges on the success of a marker–assisted, few-cross approach, this may become more widely accepted for mainstream breeding that relies solely on phenotypic selection.

References

Baranski MR, 2015. Wide adaptation of Green Revolution wheat: International roots and the Indian context of a new plant breeding ideal, 1960–1970. *Studies in History and Philosophy of Biological and Biomedical Sciences* 50, 41–50.

Cleveland DA, 2001. Is plant breeding science objective truth or social construction? The case of yield stability. *Agriculture and Human Values* 18 (3), 251–270.

Gyawali S, Sthapit BR, Bhandari B, Bajracharya J, Shrestha PK, Upadhyay MP, Jarvis DI, 2010. Participatory crop improvement and formal release of Jethobudho rice landrace in Nepal. *Euphytica* 176, 59–78.

Hanson H, Borlaug NE, Anderson RG, 1982. *Wheat in the Third World.* Boulder, CO: Westview Press.

Ismail AM, Singh US, Singh S, Dar MH, Mackill DJ, 2013. The contribution of submergence-tolerant (Sub1) rice varieties to food security in flood-prone rainfed lowland areas in Asia. *Field Crops Research* 152, 83–93.

Jones MP, Dingkuhn M, Aluko GK, Semon M, 1997. Interspecific Oryza sativa L. X O. glaberrima Steud. progenies in upland rice improvement. *Euphytica* 94, 237–246.

Joshi A, Witcombe JR, 1996. Farmer participatory crop improvement. II. Farmer participatory varietal selection in India. *Experimental Agriculture* 32, 461–477.

Joshi KD, Witcombe JR, 2003. The impact of participatory plant breeding (PPB) on landrace diversity: A case study for high-altitude rice in Nepal. *Euphytica* 134, 117–125.

Joshi KD, Sthapit BR, Witcombe JR, 2001. How narrowly adapted are the products of decentralised breeding? The spread of rice varieties from a participatory plant breeding programme in Nepal. *Euphytica* 122, 589–587.

Lakra V, Rahman ANM, Jaim WMH, Paris T, Hossain M, Singh RP, 2012. Diversity, spatial distribution, and the process of adoption of improved rice varieties in Jharkhand, India. In Hossain M, Jaim WMH, Paris TR, Hardy B, (eds) *Adoption and Diffusion of Modern Rice Varieties in Bangladesh and Eastern India*, 59–73. Los Baños (Philippines): International Rice Research Institute.

Mackill DJ, Ismail AM, Singh US, Labios RV, Paris TR, 2012. Development and rapid adoption of submergence-tolerant (Sub1) rice varieties. *Advances in Agronomy* 115, 299–352.

McGuire SJ, 2008. Path-dependency in plant breeding: challenges facing participatory reforms in the Ethiopian Sorghum Improvement Program. *Agricultural Systems* 96, 139–149. doi:10.1016/j.agsy.2007.07.003.

Maredia MK, Byerlee D, (eds) 1999. *The Global Wheat Improvement System: Prospects for Enhancing Efficiency in the Presence of Spillovers.* CIMMYT Research Report No. 5. International Centre for Wheat and Maize Improvement, Mexico, D.F.

PBI Cambridge, 1990. *Cereals: A Guide to Varieties.* Plant Breeding International, Trumpington, Cambridge.

Sinah H, Xaxa M, 2014. Traditional paddy varieties of Jharkhand and conservation priority. *Jharkhand Journal of Social Development* 8 (1 and 2), 1–13. ISSN 0974 651x.

Singh US, Dhar MH, Singh S, Zaidi NW, Bari MA, Mackill DJ, Collard BCY, Singh UN, Singh JP, Reddy JN, Singh RK, Ismail AM, 2013. Field performance, dissemination, impact and tracking of submergence tolerant (Sub1) rice varieties in South Asia. *SABRAO Journal of Breeding and Genetics* 45 (1), 112–131.

Steele, KA, Gyawali S, Joshi KD, Shrestha P, Sthapit BR, Witcombe JR, 2009. Has the introduction of modern rice varieties changed rice genetic diversity in a high-altitude region of Nepal? *Field Crops Research* 113, 24–30.

Sthapit BR, Joshi KD, Witcombe JR, 1996. Farmer participatory crop improvement. III. Farmer participatory plant breeding in Nepal. *Experimental Agriculture* 32, 479–496.

Virk DS, 2016. Visit Report for the project: Better rooting rice for drought prone Eastern India. ESRC Impact Acceleration Account (IAA), Global Challenges Research Fund IMPACT Grants. Bangor University, UK.

Virk DS, Chakraborty M, Ghosh J, Prasad SC, Witcombe JR, 2005. Increasing the client orientation of maize breeding using farmer participation in eastern India. *Experimental Agriculture* 41, 413–426.

Virk DS, Singh DN, Kumar R, Prasad SC, Gangwar JS, Witcombe JR, 2003. Collaborative and consultative participatory plant breeding or rice for the rainfed uplands of eastern India. *Euphytica* 132, 95–108.

WARDA (Africa Rice Centre), 2006. NERICA for Africa-Upland. Cotonou, Benin. www.warda.org/warda/uplandnerica.asp.

Witcombe JR, Virk DS, 2001. Number of crosses and population size for participatory and classical plant breeding. *Euphytica* 122, 451–462.

Witcombe JR, Yadavendra JP, 2014. How much evidence is needed before client-oriented breeding (COB) is institutionalised? Evidence from rice and maize in India. *Field Crops Research* 167, 143–152.

Witcombe JR, Joshi A, Goyal SN, 2003. Participatory plant breeding in maize: A case study from Gujarat, India. *Euphytica* 130, 413–422.

Witcombe JR, Joshi A, Joshi KD, Sthapit BR, 1996. Farmer participatory crop improvement. I: Varietal selection and breeding methods and their impact on bio-diversity. *Experimental Agriculture* 32, 445–460.

Witcombe JR, Joshi KD, Virk DS, Sthapit BR, 2011. Impact of introduction of modern varieties on crop diversity. In Lenné J, Wood, D, (eds) *Agrobiodiversity Management for Food Security: A Critical Review*, 87–98. Wallingford, UK: CABI.

Witcombe JR, Gyawali S, Subedi M, Virk DS, Joshi KD, 2013. Plant breeding can be made more efficient by having fewer, better crosses. *BMC Plant Biology* 13, 22 doi:10.1186/1471-2229-13-22

Witcombe JR, Khadka K, Puri RR, Khanal NP, Sapkota A, Joshi, KD. 2016. Adoption of rice varieties. 2. Accelerating up take. *Experimental Agriculture* 53, 627–643.

Witcombe JR, Joshi KD, Gyawali S, Musa AM, Johansen C, Virk DS, Sthapit BR, 2005 Participatory plant breeding is better described as highly client-oriented plant breeding. I. Four indicators of client-orientation in plant breeding. *Experimental Agriculture* 41, 299–319.

Yadaw RB, Dixit S, Raman A, Mishra KK, Vikram P, Swamy BPM, Cruz MTS, Maturan PT, Pandey M, Kumar A, 2013. A QTL for high grain yield under lowland drought in the background of popular rice variety Sabitri from Nepal. *Field Crops Research* 144, 281–287.

Yamano T, Malabayabas M, Dar M, 2013. *Stress-Tolerant Rice in Eastern India: Development and Distribution*. STRASA Economic Briefs, No. 1. Manila: International Rice Research Institute.

Yonezawa K, Yamagata H, 1978. On the number and size of cross combinations in a breeding programme of self-fertilizing crops. *Euphytica* 27, 113–116.

5 Twenty years of participatory varietal selection at AfricaRice

Lessons from farmer involvement in variety development

Nicholas Tyack, Baboucarr Manneh,
Abdourasmane Konate, Theodore Kessy,
Seydou Alexis Traore and Moussa Sie

Introduction

The Africa Rice Center (AfricaRice), founded as the West Africa Rice Development Association (WARDA) in 1971, is a CGIAR research organization and pan-African intergovernmental association working for poverty alleviation and improved food security in Africa through rice-related research, development and partnership activities. AfricaRice's membership is composed of 27 countries covering the West, Central, East and North African regions. The organization works extensively with the national agricultural research systems (NARS) of these countries, as well as with academic and advanced research institutions, extension services, farmers' organizations, NGOs and donors.

Since 1996, AfricaRice and its partners have focused on a participatory approach to rice breeding, drawing lessons from the poor results of the Green Revolution in tropical Africa (attributed in part to insufficient consultation between scientists and farmers). The organization had developed a set of new interspecific rice varieties bred by crossing higher-yielding Asian rice (*Oryza sativa*) and more locally-adapted African rice (*Oryza glaberrima*), and had chosen to undertake widespread Participatory Varietal Selection (PVS) trials to enable rapid multi-location evaluation and disseminate the new varieties to smallholder rice farmers in member countries. WARDA (the former name of AfricaRice) participatory varietal selection (PVS) trials started in Côte d'Ivoire in 1996 (Gridley and Sie, 2008), with other early adopters (Benin, Ghana, Guinea Conakry and Togo) launching programmes the following year. By 2000, trials had been initiated in all 17 member countries, involving 4,000 farmers at 105 sites. This early testing of 58 NERICA varieties was essential for identifying which characteristics and varieties were most desired by farmers (Gridley *et al.* 2002). After extensive testing and diffusion of more than 25 New Rice for Africa (NERICA) varieties (WARDA/FAO/SAA, 2008), AfricaRice has continued to work with African NARS, using the PVS approach to evaluate and subsequently release more rice varieties for upland, rainfed lowland and irrigated lowland production systems (AfricaRice, 2010a).

This chapter presents AfricaRice's experiences in implementing PVS, a key component of its breeding and dissemination activities over the past two decades. We first explain the methodological aspects, and then present examples of PVS activities conducted in Burkina Faso and elsewhere to show how PVS has been implemented in practice. Finally, we discuss the benefits of the PVS breeding methodology in contrast with conventional approaches to breeding and evaluation, and offer some conclusions.

The participatory varietal selection process at AfricaRice

Participatory varietal selection is an approach to breeding that involves both on-station and on-farm testing of fixed breeding lines, with farmer involvement. Whereas conventional breeding is a centralized and sequential process whereby a breeder crosses and evaluates germplasm under controlled conditions with elite materials – with farmers' involvement often restricted to limited on-farm trials – participatory breeding aims to involve multiple stakeholders, particularly farmers, all the way from the conceptualization of varieties and setting breeding objectives, up to varietal release and dissemination (Ceccarelli and Grando, 2007; Joshi and Witcombe, 1996; Morris and Bellon, 2004; Witcombe *et al.*, 1996). PVS aims to involve farmers as full partners and to ensure that breeding activities meet the needs of farmers and consumers (Dorward *et al.*, 2007). This was chosen as the preferred approach to varietal evaluation: it minimizes the time needed for varieties to reach farmers (and hastens the varietal release process), involves inclusion of gender perception in varietal selection, and provides direct feedback to breeders during the breeding process. PVS brings together scientists, NGOs, extension agents and farmers, and allows farmers to select material adapted to their local fields and socio-economic conditions – which in turn contributes to the development of an effective agricultural technology transfer process.

PVS is conducted during the evaluation phase of breeding materials; it involves planting advanced breeding lines provided by plant breeders alongside farmers' local varieties, which are used as 'checks'. Evaluation of breeding lines is conducted in farmers' fields and on-station, with farmers and scientists working together to determine the agronomic, technological and post-harvest properties of the varieties in question. Grain quality is also tested by various persons – members of the general public, processors, millers, restaurant owners, traders, and village leaders – to determine the milling capacity, texture, cooking time, swelling capacity, aroma and taste of the different varieties.

In addition to the PVS-Research approach, AfricaRice also supports PVS-Extension programmes, in which PVS-research programmes are complemented through efforts to reach locations and areas where farmers had not previously been exposed to AfricaRice's work and varieties (Gridley and Sie, 2008).

The participatory varietal selection cycle at AfricaRice is a three-year process, including seed production during the dry season. Briefly put, a PVS cycle consists of the following steps (AfricaRice, 2010b):

- In the **first year**, an observational nursery is established on a farmer's plot or at a research station. Farmers are invited to visit the trial at different growth stages and select the best and worst lines, giving reasons for their choice.
- In the **second year**, the selected lines are tested by scientists in 'mother and baby trials', a more general CGIAR concept related to participatory breeding. Mother trials contain all selected lines; these trials are conducted at multiple locations (with or without farmer collaboration). Baby trials are carried out by farmers using their own agronomic practices; each trial contains a subset of lines randomly selected from the mother trials, which are tested against the farmer's own variety. Also here, farmers are invited at different growth stages to make selections and give their views on the lines being evaluated. At this stage, the varietal release committee of the ministry of agriculture of the country in which the trial is being conducted joins the evaluators to determine the Value for Cultivation and Utilization (VCU) and the Distinctness, Uniformity and Stability (DUS) criteria, which together constitute the two evaluation conditions which must be fulfilled prior to varietal release and registration in the official national catalogue of varieties.
- In the **third year**, farmers repeat the 'mother trial' and 'baby trials', to analyse the impact of annual weather variability, to confirm the varietal characteristics and to explore possible extension to other farmers. Observations of experimenting and visiting farmers are taken into account at this stage.

These steps are shown in Figure 5.1. While these steps represent the ideal, and are conducted in most countries, sometimes financial constraints restrict the application of the whole programme. Recently, funding from AfricaRice through the STRASA programme and other breeding projects under the auspices of the Breeding Task Force has helped to fund the full programme of PVS in several countries as described here.

PVS in Burkina Faso

As an example of relatively recent participatory varietal selection activities undertaken by AfricaRice, we present a case study from Burkina Faso, where AfricaRice worked with its national partner, the Institut de l'Environnement et de Recherches Agricoles (INERA). Both partners managed several PVS trials

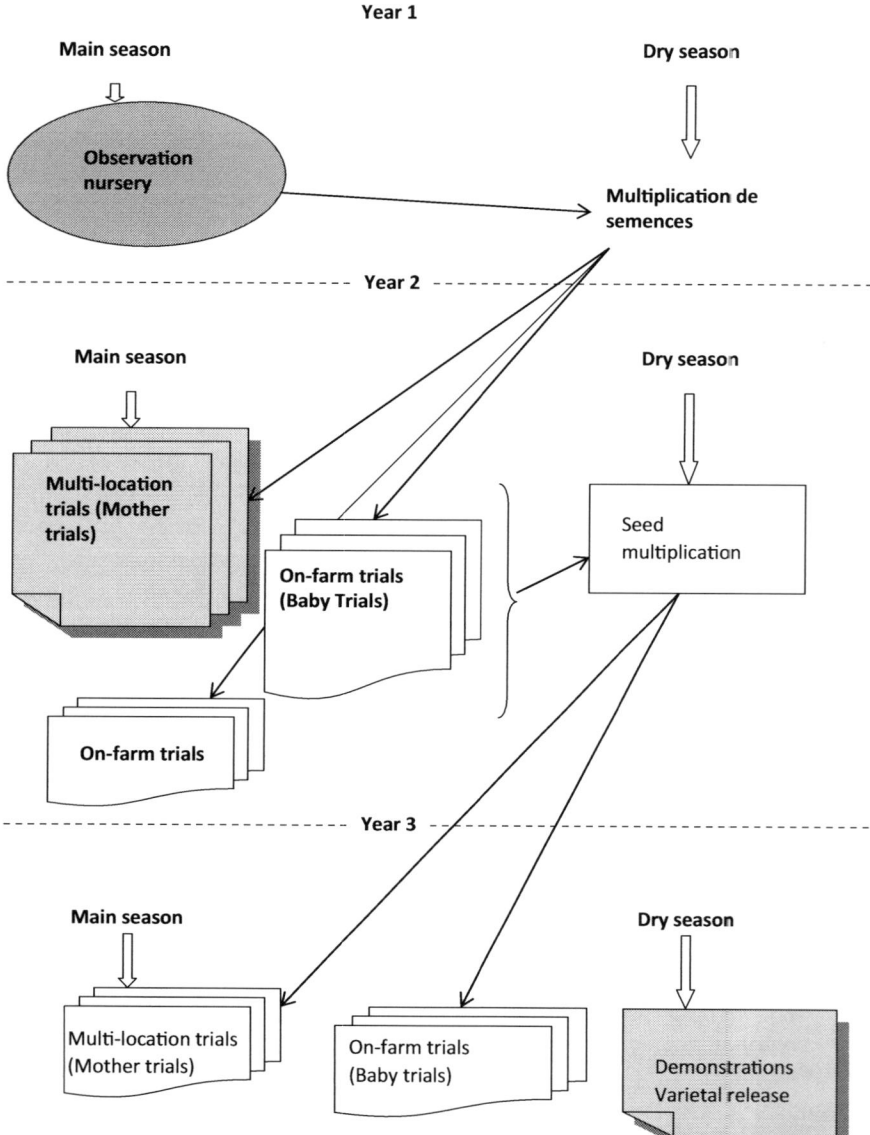

Figure 5.1 Participatory varietal selection cycle: outline.
Source: Africa Rice Center (AfricaRice), 2010b.

in 2011/2012 in Burkina Faso, under the STRASA (Stress-Tolerant Rice for Africa and South Asia) project, funded by the Bill and Melinda Gates Foundation to develop and disseminate rice varieties tolerant to abiotic stresses to millions of farmers living in unfavourable rice-growing lands in sub-Saharan Africa and South Asia. The Burkina Faso case study consists of two components: first, a PVS programme designed to evaluate varieties for tolerance to iron toxicity in lowland rice-growing environments; and, second, evaluation for drought tolerance in rainfed upland rice-growing environments.

The overarching goal of the PVS programme in the African component of the STRASA project was to develop and disseminate rice varieties with drought tolerance (for upland rice varieties), submergence, salinity, iron toxicity (for lowland rice varieties) and low-temperature stress, while maintaining high grain quality. In addition, the project had the more specific goals of allowing rice producers to identify and select varieties that best matched their preferences and growing constraints, to accelerate the adoption and diffusion of new varieties, and enable the rice breeders to take sufficient account of the characteristics of rice varieties desired by producers and consumers in their programme of varietal creation (Ismail, 2018; Manneh *et al.*, 2014).

Breeding for iron tolerance

The PVS trials for tolerance to iron toxicity stress were carried out in three locations: the Kou Valley, Banfora and Niéna Dionkélé. The evaluated materials included nine 'check' varieties popular in the test areas as well as eight new varieties chosen as a result of earlier PVS tests in 2009 and 2010. The experimental set-up in all three sites was the same, and featured a randomized complete block design with three replications for each variety planted.

At the maturity stage of the rice crop, at each site, farmers were invited to visit the trials, walk around the fields and select what they considered the three best and three worst varieties. The selection of participating farmers was made in close collaboration with extension agencies to determine who would be available to conduct the test, with special focus on inviting a balanced group of male and female farmers. Sixty-six farmers participated at Niéna Dionkélé, 49 at the Kou Valley, and 43 at Banfora. Farmers were asked to list the reasons for their choices, and rank them in terms of priority from high to low. Varietal traits identified by farmers included grain quality, tillering capacity, yield, panicle weight, tolerance to iron toxicity, disease resistance, grain appearance, panicle size, and plant vigour. Despite the number of traits assessed, and several differences in ranking for certain traits, WAT 1046-B-43-2-2-2, which also exhibited tolerance to iron toxicity and consistently had yields greater than local check varieties in earlier field trials (Table 5.1 and Table 5.2), was selected by farmers as the preferred variety in two of the three sites. These data were used to help the Breeding Task Force to select and

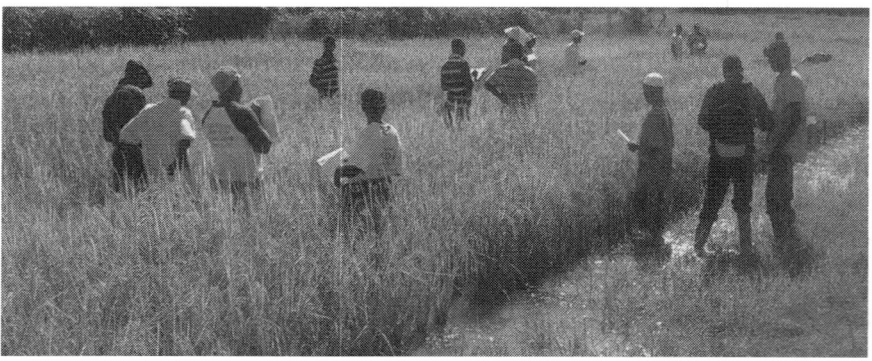

Figure 5.2 Farmer visits to evaluate rice varieties grown at Niéna Dionkélé, Burkina Faso.
Source: A. Konate.

Table 5.1 WAT 1046-B-43-2-2-2 yields compared to local check varieties, field trials
conducted in iron-toxic lowland soils, Banfora, Kou Valley and Niena
Dionkele, Burkina Faso, 2010/2011

Year	Country	Site	Grain yield (kg/ha)			Yield advantage over best check (%)
			WAT 1046-B-43-2-2-2	NL19 (Local Check2)	FKR19 (Local Check1)	
Year 2011	**Burkina**	Kou Valley	1,240	820	520	51
		Banfora	6,280	5,460	4,180	15
		Niena Dionkele	6,240	3,920	3,500	59
Year 2010	**Burkina**	Niena Dionkele	4,278	–	4,027	6
		Kou Valley	966	682		42

Table 5.2 Farmers who selected WAT 1046-B-43-2-2-2 compared to local checks (NERICA-L-19/SIK 9-164-5-1-3), PVS visits at three rainfed lowland trial sites in Burkina Faso, 2011

Selection criterion	WAT 1046-B-43-2-2-2			Local checks			Total selections
	Banfora	Kou Valley	Niena Dionkele	Banfora	Kou Valley	Niena Dionkele	
Grain yield	33	33	14	25	25	50	180
Grain characteristics	66	20	29	37	37	66	255
Plant height	55	20	40	12		8	135
Iron toxicity tolerance	33	41	18		31		123
Panicle characteristics	22	33	47	12	12	55	181
Total selections	209	147	148	86	105	179	

propose varieties as candidates for release through the ARICA (Advanced Rice for Africa) programme.

Breeding for drought tolerance

Like iron toxicity, the subject of the first component of the STRASA PVS programme, drought is a significant abiotic constraint to rice growers in Burkina Faso. With over 75 per cent of the rice-growing area in Burkina Faso being rainfed (Diagne et al., 2013), a major challenge facing rice producers is drought stress. Most of the popular varieties are sensitive to water stress arising from rainfall deficit, which causes frequent problems for rainfed rice production. INERA and AfricaRice collaborated to evaluate several local varieties and breeding materials for drought tolerance at four locations in Burkina Faso: Boucle du Mouhoun, the Cascades, the Hauts Bassins and the South West. These sites were not included in the varietal evaluation for iron toxicity tolerance.

PVS trials were conducted at the four locations mentioned above. Guided visits were conducted with producers at the maturity stage at each site, in order to determine what farmers considered to be the three best and three worst varieties. Each visit involved the following steps:

1   The lead researcher introduced the objective of the exercise and emphasized its purpose, giving participants an hour to visit the plot.
2   Farmers were divided into groups by gender, with each individual in each group evaluating varieties separately, using a card assigned to each.
3   Each producer was then asked to choose which varieties he or she would like to grow on the farm and why.
4   Farmers were asked to state which varieties they considered to be performing poorly, and why.
5   Producers were asked to list the most important traits for their decision-making process in choosing varieties to plant on their farms.
6   Researchers and extension agents collected socio-economic information for each producer.
7   A synthesis meeting was held at the end of the session, typically lasting some 30–45 minutes.
8   Finally, a follow-up session was scheduled to summarize the conclusions of the experiment, combine the PVS data with agronomic trial data and then submit all the data to the varietal release committee.

In addition to evaluations of the drought-tolerance of different rice varieties, culinary tests were also conducted to determine how desirable the varieties were to consumers. Two kilogrammes of each variety of rice were prepared, each being assigned a code number to ensure anonymity. The tasting panel was made up of men and women from different socio-professional backgrounds. The cooked rice varieties – all served without

Figure 5.3 Visiting PVS trials of rainfed rice varieties, Boucle du Mouhoun, Burkina Faso.
Source: A. Konate and A. Traore.

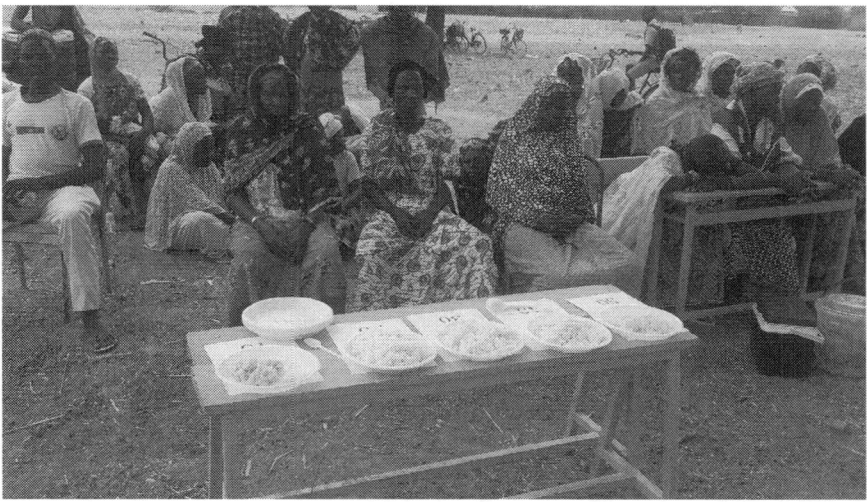

Figure 5.4 Example of culinary tests of rice varieties included in PVS trials.
Source: A. Konate.

sauces – were assessed for colour, texture, stickiness, aroma and taste. Three varieties scored the best in the tasting tests, NERICA 8, FKR 45N, and WABC165, which also scored well for agronomic traits in the PVS trials.

Using the data from the PVS and agronomic trials, NERICA 8, WABC165 and WAT 1046-B-43-2-2-2 were released in Burkina Faso as FKR 55, FKR 61 and FKR 66 in 2013, respectively, and were registered in the West African varietal catalogue in 2014 (CEDEAO, UEMOA, CILSS, 2016). Subsequently, the STRASA project in Burkina Faso supported the production of seeds of the highest-performing varieties identified through the PVS, by means of collaboration between INERA and AfricaRice to support the dissemination of new rice varieties and to increase producer access to the improved varieties.

Current practice and ARICA development through PVS

AfricaRice has continued to utilize PVS in its varietal development efforts. Participatory methods have been central in its efforts to develop 'Advanced Rice for Africa' (ARICA) varieties, a next generation of high-performing rice varieties carrying forward the earlier advances achieved with NERICA. These varieties are evaluated through the Africa-wide Breeding Task Force, consisting of national and international rice breeders from AfricaRice, IRRI and 30 African countries (Kumashiro *et al.*, 2013).

In 2015, AfricaRice helped to conduct PVS evaluations in Tanzania in collaboration with KATRIN Agricultural Research Institute through the Breeding Task Force, focusing on rainfed lowland agro-environments. Four trials were undertaken, at Ifakara, Dakawa, Matombo and SUA (Sokoine University of Agriculture), with the final round involving farmers growing the trial varieties in their rice fields alongside their own varieties and using their own agricultural practices. Data were collected on various traits, including grain yield, grain weight, number of panicles, panicle length, plant height, days to 50 per cent flowering and disease susceptibility, and phenotypic acceptability (see Tables 5.3 and 5.4).

Based on results from agronomic performance in PAT, FAT, and PVS across the two locations of KATRIN and Dakawa, the entries CT 19558-2-17-4P-3-1-1-M, FAROX 521-357-H1 and CT 21407-9P-5P-4SR-1 were selected for further evaluation and potential release for rainfed lowland environments in Tanzania.

Today PVS is widely used in sub-Saharan Africa to evaluate and release rice varieties that are adapted to the needs of local rice farmers and consumers – from The Gambia and Senegal in West Africa to Ethiopia, Madagascar, Mozambique and Uganda in Eastern and Southern Africa. Most of AfricaRice's collaborative trials with NARS follow the Participatory Environmental Trial (PET) scheme. As a result of this collaborative work, AfricaRice and its national research partners have released more than 100 varieties in the

Table 5.3 Agronomic performance of a set of rainfed lowland rice breeding lines, Dakawa, Tanzania, 2015

Entry No.	Designation	Days to flower (days)	Plant height (cm)	Grain yield (t/ha)
1	CT 19558-2-17-4P-3-1-1-M	77a★	111a	7.3ab
2	WAC 13-WAT21-2-1	77a	168b	6.4a
3	FAROX 521-101-H1	77a	178b	8.9ab
4	WAC 13-TGR 5	81ab	174b	6.9ab
5	FAROX 521-357-H1	76a	112a	8.9ab
6	FAROX 521-176-H1	82abc	112a	8.0ab
7	CT 21407-9P-5P-4SR-1	83abc	121a	9.2b
8	CT 18838-1-1-2-1SR-2P	86bc	107a	6.5a
9	WAB 1573-22-B-B-FKR 4-2-WAC 1-TGR 3-WAT9-1	89c	123a	7.2ab
10	WAB 1529-7-B-B-FKR 4-WAC 1-2-TGR 2-WAT7-1	90c	123a	8.4ab
11	TXD 306 (SARO 5)	83abc	121a	7.5ab
Grand mean		82.03	132	7.75
F. probability		<0.001	<0.001	0.003★
LSD (α = 0.05)		4.6	9.6	1.44
SE±		2.7	5.64	0.84
CV (%)		3.3	4.3	10.9

Note
Means separation by Tukey's Test at P< 0.05; ★ – Figures followed by same letters are not significantly different (p>0.05).

Table 5.4 PVS results of a set of rainfed lowland rice breeding lines, Dakawa, Tanzania, 2015

Entry No.	Designation	Male (N = 16)		Female (N = 36)		Total (N = 52)		Preference ranking		Remarks
		+ve	−ve	+ve	−ve	+ve	−ve	Best	Worst	
1	CT 19558-2-17-4P-3-1-1-M	6	2	15	6	21	8	3	4	
2	WAC 13-WAT21-2-1	0	6	1	24	1	30		1	Tall & lodging
3	FAROX 521-101-H1	1	8	0	17	1	25		2	Tall & lodging
4	WAC 13-TGR 5	0	4	3	4	3	8			Tall & lodging
5	FAROX 521-357-H1	1	2	0	1	1	3			
6	FAROX 521-176-H1	0	5	4	7	4	12		3	
7	CT 21407-9P-5P-4SR-1	2	0	15	2	17	2	4		
8	CT 18838-1-1-2-1SR-2P	8	0	15	2	23	2	2		
9	WAB 1573-22-B-B-FKR 4-2-WAC 1-TGR 3-WAT9-1	0	1	1	2	1	3			
10	WAB 1529-7-B-B-FKR 4-WAC 1-2-TGR 2-WAT7-1	1	1	3	5	4	6			
11	TXD 306 (SARO 5)	11	1	17	3	28	4	1		

past 10 years using the PVS approach through the Breeding Task Force, thereby helping to raise yields and farmer incomes (Agboh-Noameshie *et al.*, 2007) and reduce poverty (Kijima *et al.*, 2008).

Benefits and challenges of participatory varietal selection

The primary goal of PVS is to involve farmers in the breeding process. It helps to ensure that varieties are created that match the preferences of farmers and consumers by determining their varietal selection preferences and criteria more precisely, so that these can be included as breeding objectives. PVS can also strengthen farmers' autonomy and increase their freedom to select varieties that match their needs, as well as helping to strengthen and empower often-neglected groups by assisting rural communities in maintaining and accessing genetic diversity and contributing to the creation of new rice varieties that are more resilient and higher yielding. Further, PVS promotes incorporation of the preferences of both genders in varietal development – an important point, as male and female farmers may value traits such as yield, plant height or seedling vigour differently (Gridley and Sie, 2008).

Participatory varietal selection also offers several practical benefits. It helps to shorten the period between varietal development and varietal adoption by farmers (five years for PVS vs. seven years with conventional breeding methods) and helps to streamline the process of varietal registration by involving national representatives in the breeding process. In addition, by ensuring that farmers are exposed early on to new varieties and that new varietal releases match the needs of cultivators, it helps to increase the dissemination of new varieties in farmers' fields and raise adoption rates. For example, Dorward *et al.* (2007) found that PVS efforts in Ghana had helped to speed up the spread of PVS rice varieties through informal seed systems. Further, since consumer preferences are taken into consideration, rice produced using the new varieties is more competitive in rural and urban markets. That enables farmers to sell their produce more easily, in turn improving their livelihoods with the proceeds.

However, there are some substantial challenges to PVS programmes. First, the success of any PVS programme depends heavily on the availability of large quantities of high-quality seeds (or plants) for selection across many farmers. This also implies that the institution(s) carrying out PVS must be able to produce or procure pure seeds of many different varieties. In addition, sustainable funding is necessary, to maintain momentum and to expand activities to include community-based seed systems. AfricaRice's use of PVS in its breeding activities has indeed provided noteworthy benefits – but running PVS programmes in many countries across multiple agro-environmental systems is no simple task.

Conclusions

Ever since AfricaRice began implementing participatory varietal selection (PVS) in 1996, it has been a pioneer within the Consultative Group on International Agricultural Research (CGIAR) system of mainstreaming participatory plant breeding methods. While most CGIAR and national plant breeding programmes tend to rely on conventional breeding methods, AfricaRice has continued to champion the decentralized, PVS methodology in which farmers are treated as full partners in the breeding process. This has allowed the research centre to match the objectives of its breeders with the preferences of farmers; to develop new, locally adapted breeding materials; and produce and distribute new varieties to farmers more quickly.

AfricaRice's experience over the past 20 years has shown that implementing PVS in its breeding programmes has helped the organization to achieve its goal of empowering farmers by integrating them into the breeding process, and has also had substantial practical benefits – shortening the time required to develop new varieties, as well as increasing adoption rates and streamlining varietal registration on the national and regional level. PVS seems set to continue as a key element of AfricaRice's efforts to develop and release new ARICA varieties in the coming decade.

References

Africa Rice Center (AfricaRice) 2010a. *New Breeding Directions at AfricaRice: Beyond NERICA*. Cotonou, Benin: 24 pp.

Africa Rice Center (AfricaRice), 2010b. *Participatory Varietal Selection of Rice – The Technician's Manual*. Cotonou, Benin: 110 pp.

Agboh-Noameshie AR, Kinkingninhoun-Medagbe FM, Diagne A. 2007. Gendered impact of NERICA adoption on farmers' production and income in Central Benin. *AAAE Conference Proceedings* 189–191.

Ceccarelli S, Grando S, 2007. Decentralized-participatory plant breeding: An example of demand driven research. *Euphytica* 155, 349–360.

CEDEAO, UEMOA, CILSS, 2016. *Catalogue Régional des Espèces et Variétés Végétales*. www.israsaintlouis.sn/images/documents/Regional_Catalogue.pdf.

Diagne A, Eyram A-A, Futakuchi K, Wopereis M, 2013. Estimation of cultivated area, number of farming households and yield for major rice-growing environments in Africa. In Wopereis M, Johnson DE, Ahmadi N, Tollens E, Jalloh, A, (eds) *Realizing Africa's Rice Promise*. Boston, MA: CAB International.

Dorward P, Craufurd P, Marfo K, Dogbe W, Bam R, 2007. Improving participatory varietal selection processes: Participatory varietal selection and the role of informal seed diffusion mechanisms for upland rice in Ghana. *Euphytica* 155, 315–327.

Gridley HE, Jones MP, Wopereis-Pura M, 2002. Development of new rice for Africa (NERICA) and participatory varietal selection. Breeding rainfed rice for drought-prone environments: Integrating conventional and participatory plant breeding in South and Southeast Asia. In Witcombe R, Parr LB, Atlin GN, (eds) *Proceedings of a DFID Plant Sciences Research Programme/IRRI Conference*, IRRI, Los Baños, Laguna, Philippines, 12–15 March 2002, 23–28. Department for International

Development (DFID) Plant Sciences Research Programme, Centre for Arid Zone Studies (CAZS)/International Rice Research Institute (IRRI), Bangor and Manila.

Gridley H, Sie M, 2008. The role of PVS. In Somado EA, Guei RG, Keya SO, (eds) *NERICA®: the New Rice for Africa – a Compendium*, 44–48. Cotonou, Benin: Africa Rice Center (WARDA), Rome: FAO, Tokyo: Sasakawa Africa Association.

Ismail A, 2018. *Stress-Tolerant Rice for Africa and South Asia.* http://strasa.irri.org/.

Joshi A, Witcombe JR, 1996. Farmer participatory crop improvement II. Participatory varietal selection, a case study in India. *Experimental Agriculture* 32, 461–477.

Kijima Y, Otsuka K, Sserunkuuma D, 2008. Assessing the impact of NERICA on income and poverty in central and western Uganda. *Agricultural Economics* 38, 327–337.

Kumashiro T, Futakuchi K, Sie M, Ndjiondjop M-N, Wopereis M, 2013. A continent-wide, product-oriented approach to rice breeding in Africa. In Wopereis, MCS, Johnson, DE, Ahmadi, N, Tolles, E, Jalloh, A, (eds) *Realizing Africa's Rice Promise*, 69–78. CABI, Oxfordshire, UK.

Manneh B, Singh U, Singh RK, 2014. *STRASA-Africa: Stress-Tolerant Rice for Africa and South Asia*. www.africarice.org/strasa-africa/.

Morris ML, Bellon MR, 2004. Participatory plant breeding research: opportunities and challenges for the international crop improvement system. *Euphytica*, 136, 21–35.

Witcombe JR, Joshi A, Joshi KD, Sthapit BR. 1996. Farmer participatory crop improvement. I. Varietal selection and breeding methods and their impact on biodiversity. *Experimental Agriculture* 32, 445–460.

6 Native maize in Mexico

Participatory breeding and connections to culinary markets

Martha C. Willcox, Fernando Castillo-Gonzalez,
Flavio Aragón-Cuevas and
F. Humberto-Castro Garcia

Introduction

Mesoamerica is the centre of origin of maize, with greatest diversity found in Mexico – 59 identified native races of maize (CONABIO, 2012; Sanchez-Gonzalez et al., 2000). With the many colour and texture variants within these native races, this is diversity on a scale un-imaginable in any other maize-growing countries.[1] Most of Mexico's 59 races feature several different texture variants as well as several colour variants, creating hundreds of recognizable racial variants. Farmers often cross-pollinate distinct types of maize intentionally; wind pollination disperses pollen between different types of maize in close proximity, creating vast diversity of thousands of types of maize found throughout Mexico.

Farmer-saved seed of traditional landraces is still the major seed source for planting maize in Mexico. Estimates range from 60 per cent to 80 per cent of the hectares planted to maize being sown with farmer-saved unimproved maize. The exact number of hectares planted to certified hybrid maize can be known through sales tracking – about 15 per cent of maize hectares (Donnet et al., 2012). Landrace maize hectares are extrapolated from this figure, but there are also sales of open-pollinated varieties, multiline synthetics, and there are famers who save seeds from hybrids, which they save and replant. These types of maize plantings may represent perhaps 20–25 per cent of the area sown to maize. Bellon et al. (2018) estimated the enormous contribution in conservation of diversity of these traditional landrace farmers based solely on hectares that produced less than three tons of grain per hectare. That study documented the role of these traditional landrace farmers not only in conserving maize diversity but also as part of the evolutionary process of creating diversity and as being responsible for production of about half of the maize grain produced per year in Mexico. Traditional seed saving is clearly the dominant method in Mexico. Most hectares are planted to traditional landraces sown for home consumption, with any excess grain reaching mainly local markets (occasionally regional ones). The tortilla industry prefers hybrid maize because of its uniformity.

Hybrid maize accounts for half of the maize production in Mexico; it has higher yields per hectare and is grown in very fertile areas, with irrigation and high agrochemical inputs.

Community-based maize landrace improvement

The community-based landrace improvement approach recognizes that maize, as an open-pollinated crop, is dependent on the interaction between differing genetic variants (alleles) to provide the vigour known as heterosis. This in turn requires continually maintaining sufficient genetic variation in the number of alleles in a population of maize plants.

In many communities in Mexico, often two or three races of maize are grown, each of which may have several colour variants (Gonzalez-Gonzalez, 2007). These distinct races and colour variants are frequently conserved and selected as distinct entities because of their various micro-environmental adaptations ('the red does better on lighter soils'); different races mature at different times, short season and long season, thereby spreading the risk of loss from drought across different months, and ensuring at least some harvest (Soleri and Cleveland, 2001). Moreover, there are usually differences in culinary uses, with preferences for specific types of maize for specific dishes. In addition comes the variation in farmer-held samples also within the same race and colour variant in the same village (Gonzalez-Gonzalez, 2007; Herrera-Cabrera, 2013; Muñozcano-Ruiz, 2011). Each farmer-held sample has its own selection history – based on the farmer's criteria, total number of ears selected each cycle, history of trading maize with farmers nearby or far away – making it to a greater or lesser extent different from other farmer-held samples in the same community (Herrera-Cabrera *et al.*, 2004; Louette, 1997; Louette and Smale, 1996).

This existing maize diversity must be assessed and understood in order to improve the maize system within a community. Selection should focus on most productive and healthiest farmer samples for each type of maize grown there. Improvement should focus on traits that are important to the community to improve, and the selection method should not diminish any valued trait, grain type in particular. Examples of characteristics that communities wish to improve may include root strength to decrease plant lodging, decrease in barren plants, or decrease in ear rots. In order to improve a trait there must be selectable variation within samples in the community; otherwise, a donor source must be identified, but this must be similar enough to the material being improved that it does not significantly change the climatic adaptation or grain type of the local material.

Methods

Several breeding strategies are employed under the banner of 'participatory breeding' in Mexico. Each is described briefly below.

The three-step method[2]

This is a holistic approach to community-based participatory breeding (Esquivel-Esquivel *et al.*, 2009, 2011; Gonzalez-Gonzalez, 2007; Herrera-Cabrera, 2000; 2004, 2013; Muñozcano, 2011; Romero-Peñaloza, 2002; Zambrano-Zambrano, 2013). It respects the sovereignty of the farmer germplasm base and uses knowledge of population genetics for improvement of the community's maize based on local preferences. This method has been documented through publications of students of Fernando Castillo at the Postgraduate College at Montecillo, cited below.

1   *Evaluation of existing diversity in the community and selection of best landrace variants (farmer samples).* Samples are planted of all available germplasm in the community: first, seed samples are requested of all variants grown by each individual farmer; then the variants from all farmers within the community are planted in replicated trials. Measurements are taken on morphological traits, as well as agronomic traits for each farmer variant. Analysis of these data allows researchers to group maize races, or colour/texture variants within races, on the basis of their morphological differences. These groupings usually correspond to the farmer's seed selection groups, which are planted either in separate fields or at different times so that there is no-cross pollination, in order to maintain the distinct types. Within each grouping (blue, white, cream), the farmer-held samples that are most productive based on their grain weight are determined within each type grown in the village. Selecting the top 20 per cent most productive farmer-held samples in the village and replacing the lower-yielding ones can give an immediate yield increase at the community level. Herrera Cabrera *et al.* (2013) have documented a 15 per cent yield increase on the community level through selection of the top 20 per cent of each type of maize grown in the community.

2   *Selection for specific traits using mass selection (ear rot, ear height).* Selection of specific traits using mass selection is conducted within the 20 per cent most productive farmer-held samples. Mass selection can emphasize direct selection for traits that are important to improve in this material. In multiple populations of Chalqueño maize, Zambrano-Zambrano (2013) demonstrated significant decreases in ear rot and ear height over cycles of mass selection, as well as significant increases in yield (Figure 6.1).

3   *Yield enhancement of landrace variants through heterotic complementation by crossing with similar type of maize (grain, plant and environmental adaptation) from a geographically distant area.* The materials selected from steps 1 (evaluation of diversity) and 2 (mass selection) can then be crossed with another landrace of the same grain type, and environmental adaptation, but from a different state or geographic region. The underlying rationale here is that,

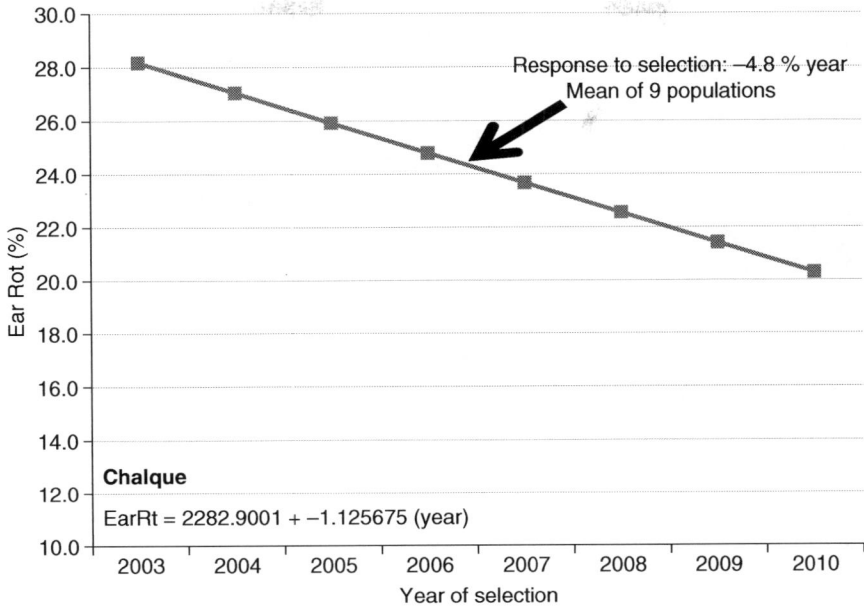

Figure 6.1 Response to seven cycles of mass selection to decrease ear rot (average of nine populations of Chalqueño maize).

Source: Zambrano-Zambrano, 2013.

although the materials have similar characteristics, their geographic separation due to drift and/or selection will have caused sufficient genetic changes for the two materials to exhibit heterosis, with differing allelic patterns (Figure 6.2). This step has been shown to increase yield by 20 per cent (Esquivel-Esquivel *et al.*, 2009, 2011; Romero-Peñaloza, 2002; Romero-Peñaloza *et al.*, 2002).

Mass selection alone

Mass selection alone can be effective if the trait and selection methods are well considered. The trait should be selectable in the field; it should show sufficient genetic variation and be of importance to the local community. Negative selection can be accomplished by de-tasseling plants with undesirable characteristics before pollen shed – principally barren plants with no ear-shoot formation, lodged or leaning plants, weak or foliar diseased plants. Positive selection can be made for such traits as ear health, ear placement, and presence of multiple ears at harvest. Two caveats: (1) A minimum of 200 ears should be selected for next-generation seed (Crossa, 1989), selected

Crosses between similar landrace variants from different regions
10–15% yield gain

- Landrace variants from different regions, expected genetic divergence.
- Maizes from SE state of México crossed with same race from other regions.

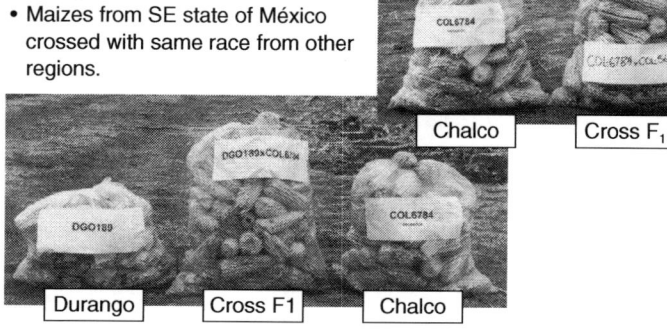

Figure 6.2 Three-step method: Step 3.

Source: Romero-Peñaloza *et al.*, 2002.

Note: Increased grain yield is demonstrated when similar landrace variants from distant areas have diverged sufficiently to display heterosis.

from a field of at least 10,000 to 20,000 plants, giving a selection intensity of about 1 per cent–2 per cent. (2) Selection must occur in plants that are in competition with other plants (plants with no extra space around them due to missing neighbour plants) which can appear more vigorous due to more water and nutrients.

This method has been taught in many regions of Mexico and elsewhere in Latin America to farmers as well as local technicians. It is the simplest technique for farmers to do in their own fields. However, since farmers often have limited time, there should be prior consideration and consultation with the community, to make sure the trait being selected is of high importance to the farmers, and that there is selectable variation within the maize population for that trait. It is common, when asking farmers about prior experiences and training in participatory selection, to find that they have been trained with specific emphasis on selection for lower plant height, aimed at decreasing lodging. Technicians who train farmers have often taught selection for a specific pre-selected trait, without analysis of the actual needs of the local community.

There is a moderate positive correlation between plant height and lodging, between 18 per cent before flowering and 41 per cent after flowering (Larsson *et al.*, 2017). Selecting for shorter plants will have some correlated impact on reducing lodging – but there is a contrary correlation with lower yield when selecting for shorter plants. In addition, in Mexico, maize plants are the main source of animal fodder, usually as dried stover (leaves and stalks). That means that using farmer's time in selecting for decreased plant height as opposed to direct selection against lodging is not justified, because of the potential decrease in yield and loss of animal fodder. Moreover, populations of maize that have experienced some inbreeding due to reduced population size will be shorter due to inbreeding depression (Klenke *et al.*, 1988): in such cases, selection for shorter plants may further increase endogamy, thereby decreasing overall yield. Direct selection for a trait is usually more effective than selection for a correlated trait. Simply tugging on the lower part of the stalk to see if the plant is well anchored, or selecting for erect plants, is simple to do when walking a field during seed selection.

Limited backcrossing (retrocruza limitada)

In this method, after evaluating the diversity of the community, the best farmer samples are crossed with improved material (hybrid or inbred line) and then backcrossed with the selected farmer sample to form a limited backcross population. Selections can then be made to increase uniformity for the traits of interest, using family formation, or individual plant selection. This method is employed almost exclusively by professional breeders, with some farmer input at harvest time, for opinions used in selection. The critical step is identifying the donor parent to improve a trait deemed very important by farmers but for which there is no selectable variation within the farmer samples. Simply crossing a landrace with a high-yielding hybrid will not always give the best results, and may result in a grain type that is less appreciated than the local landrace.

This methodology, taught in universities in Mexico for several decades by Dr Fidel Marquez of the University of Chapingo, has had lasting impact on breeders of native maize landraces (Marquez Sanchez *et al.*, 2000). It was a response to the need to improve yields in native maize landraces in an era when distribution and sales of improved hybrids was limited. The assumption was that the selection by professional breeders in the formation of hybrids would accumulate sufficient favourable alleles that any hybrid would be an improvement of a landrace in terms of yield. Grain type or characteristics were not of primary consideration, though backcrossing to the landrace partially recovers grain type and local adaptation. However, breeders often focus on stringent selection for the hybrid ideotype, of short plant height with low ear placement. Although the theoretical proportion of the donor genome in a first backcross generation, such as is used in the limited backcross method is 25 per cent, such specific selection for the donor ideotype increases the percentage of the donor genome, which may also cause other traits such as grain type to be more similar to the hybrid. Experiments conducted in Oaxaca in 2017 and 2018 compared the backcross-1 generation of local landraces where hybrids were used as donor parents to the original landraces. The preliminary results indicate statistically significant changes in yield only in cases where the original farmer samples were low-yielding. With higher-yielding farmer samples, the major effect seemed to involve homogenization of the population as regards plant and ear heights. In Mexico, this method has been used extensively by the national agricultural research programme (INIFAP) as a rapid way to produce varieties sufficiently homogeneous to be registered with the national seed system.

Breeder-family formation and farmer selection

Breeder-family formation can be effective when selecting for more complex traits, including quality and yield. Both half-sibling and full sibling families can be employed in improving landraces using controlled pollination, usually conducted by or under the direction of a professional plant breeder. Ear-to-row selection is a type of half-sib family structure using open pollination and controlling the maternal parent by planting a row or plot from each ear, which is then evaluated as a unit. That allows the best maternal parents to be selected and then recombined. Farmers can participate in selection of ear-rows that can be recombined to form improved local populations. This method is perhaps the most feasible of the family structures for use in farmers' fields.

Connecting farmers with high-end markets

Professional chefs have been showing greater interest in native Mexican maize types for their culinary properties. That has begun to change our ways of thinking about landrace improvement. Consumers now want gourmet or authentic Mexican cuisine, and that requires authentic ingredients, principally maize in the form of tortillas. This culinary trend started in Mexico in highly acclaimed restaurants like Pujol,[3] and has spread to hundreds of restaurants throughout the United States as well as in Europe. The export of maize from small farmers in Mexico (Figure 6.3) has expanded exponentially. Instead of yield, the focus is on authenticity of flavours and textures, where unimproved landraces do better than improved ones. The top end of this culinary market also values the preservation of local and farmer identity, and appreciates the history of original local landraces within a community or an individual family.

Currently maize landraces that are low in yield but have exceptional colour and flavour profiles commanding prices triple, quadruple, even quintuple the price for hybrid maize grain. Here, participatory breeding efforts concentrate on aspects of plant improvement quite distinct from the Limited Backcross described above: the focus is on conservation of local landraces, and selection within a farmer population or recombination among the best local farmer populations. Several non-breeding objectives have emerged, such as organizing farmers' cooperatives and creating a certification system for native landraces. Both these efforts involve attempts to benefit the very small-scale, usually marginalized, farmers who are the primary conservers of the diversity

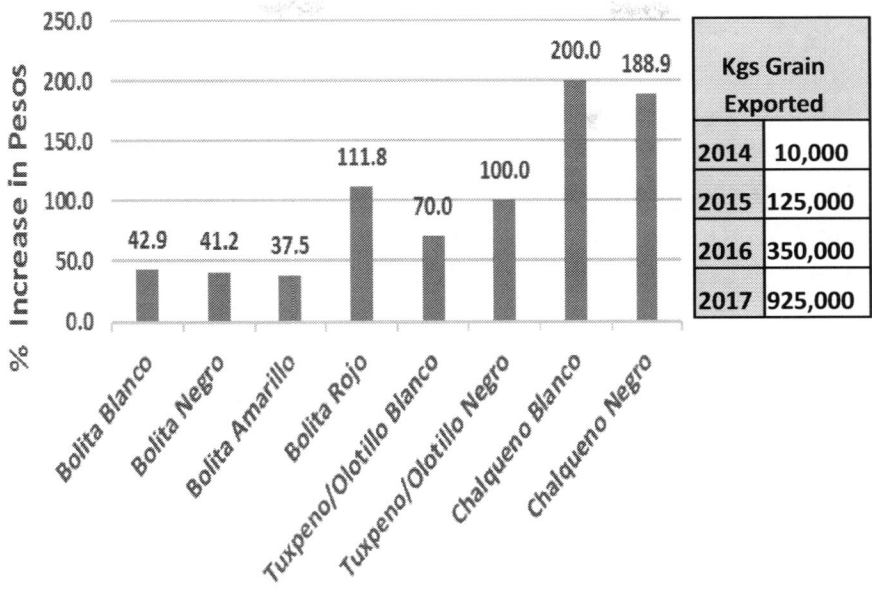

Figure 6.3 Prices paid to native maize farmers, and native maize (kgs) exported, 2014–2017.

Source: Martha Willcox.

and quality of these landraces. Farmers with only very modest landholdings must also live off their harvests, so any excess quantities of grain are too small individually to move the grain to buyers or for buyers to make individual bank transfers to each farmer. Organizing farmers into cooperatives can distribute costs within the community, making small-farmer access to markets economically feasible.

As maize is a wind-pollinated crop, conservation of traditional varieties depends on maintaining populations – not only of single farmers, but of the original native maize by the majority of farmers within a community – in order to provide a buffer from cross-pollination by hybrids or other types of maize. Community-level conservation also provides sufficient population size and genetic diversity to prevent inbreeding depression and to maintain vigour and yield through the genetic diversity of the community's maize population. In Mexico, there still exist communities where the majority of farmers grow the native landrace that was inherited from the ancestors of today's farmers. These communities consist almost exclusively of smallholders, frequently indigenous and/or women farmers who have a cultural and culinary history with the native maize, which has evolved in their community. The fact that these farmers consume their own maize on a daily basis has a positive impact on the conservation of specific culinary properties and flavours. Such small communities are found throughout Mexico, conserving specific adaptations and tolerances to adverse environmental conditions and forming genetic reserves still evolving as the climate changes.

Final remarks

Mexico is rich in maize diversity. It is also rich in scientific expertise of breeders and geneticists who have spent entire careers studying native maize and working with small farmers. There is a wealth of accumulated knowledge. However, these experts often work in various governmental agencies or universities which do not always collaborate formally, so interchange of information and evaluation of different techniques is lacking. Conducting experiments in farmers' fields with the precision needed for statistical analysis is complicated, making it difficult to recommend which methods are best for improving yield and/or other agronomic traits. Many of the methods taught to students, technicians and farmers are based on the historical training of those heading the specific project in question. The three-step method developed by Fernando Castillo is backed by a body of scientific experimentation and statistical analysis, but it is not widely known because the information has been presented in a wide range of publications and student theses, not collected in the format of an extension manual.

The expanding interest in 'authentic' Mexican maize has changed the needs of native landrace farmers. Instead of altering their maize to try to find a market niche, they are becoming involved in conserving their family's excess grain in pristine condition for high-end markets, and finding ways to produce greater excess grain, whether by spending more time/money on fertilizing and cultivation or by planting abandoned plots. The increase in prices that farmers receive for native maize has motivated large-scale commercial hybrid-maize producers to search for sources of native maize to plant commercially. That is giving rise to new concerns. To protect this market niche, a group of native maize experts has formed a non–profit association to certify landrace maize from the small producers who have been the traditional guardians of these native maize races, in order to set them apart in the culinary market. This is a nascent effort, but has promise for both maize conservation and improving the livelihoods of small-scale marginalized farmers. The products of participatory or breeder selection of native maize have often been registered as varieties with the national seed certification service (Servicio Nacional de Inspección y Certificación de Semillas – SNICS) by the institution with which the breeder scientist involved in the project has been connected. By registering the native landrace as an improved variety, the intention is to protect it from possible future misappropriation by large international seed companies. However, such institutional registration of native maize does not generally include the farmer/community from which the original maize was improved as co-registrants, even if selection was farmer-participatory. The intention behind improving these landraces was to benefit the communities that have preserved them, by providing the additional expertise of professional breeders. However, such varietal registration of improved landraces may enable the institutions holding the registration to increase and sell these materials to large-scale farmers seeking to enter the emerging culinary market – with no benefits accruing to the originating farmers. Until recently, this has been a moot point, as these materials were registered and kept in germplasm banks. Today, however, with an expanding market for native maize, many large-scale farmers are trying to buy native maize seed. If the institutions holding the registrations license these materials within Mexico, that would keep the benefit of the germplasm within the country – but without necessarily protecting the rights or benefits of the original farmers. This brings to the fore the acute need for better information and legal structures that can protect the interests of small-scale custodian farmers.

Notes

1 A 'race' of maize can be defined as populations of maize with enough similarities to enable recognition as a related group based on their plant morphology, growth period, growth habit, environmental adaptation, ear formation and grain type.
2 Created by Dr Fernando Castillo Gonzalez, Professor of Genetics, Colegio de Posgraduados, Montecillo, Mexico.
3 Pujol has been named by the *Wall Street Journal* as the best in Mexico City; it was ranked seventeenth best restaurant in the world by *Restaurant* magazine in 2013.

References

Bellon MR, Mastretta-Yanes A, Ponce-Mendoza A, Ortiz-Santamaria D, Oliveros-Galindo O, Perales H, Acevedo F, Sarukhan J, 2018. Evolutionary and food supply implications of ongoing maize domestication by Mexican campesinos. *Proceedings of the Royal Society B*, 285 (August). http://dx.doi.org/10.1098/rspb.2018.1049.

CONABIO. 2012. Maíces y Razas de Maíces. Portal Biodiversidad Mexicana de la Comisión Nacional para el Conocimiento y Uso de la Biodiversidad (CONABIO) www.biodiversidad.gob.mx/usos/maices/razas2012.html.

Crossa J, 1989. Methodologies for estimating the sample size required for genetic conservation of outbreeding crops. *Theoretical and Applied Genetics* 77, 153–161.

Donnet L, López D, Arista J, Carrión F, Hernández V, González A, 2012. *El potencial de mercado de semillas mejoradas de maíz en México*. International Wheat and Maize Improvement Center (CIMMYT), Mexico, DF.

Esquivel-Esquivel G, Castillo-González F, Hernández-Casillas JM, Santacruz-Varela A, García de los Santos G, Acosta Gallegos JA, Ramírez-Hernández A, 2009. Aptitud combinatoria y heterosis en etapas tempranas del desarrollo del maíz. *Revista Fitotecnia Mexicana* 32 (4), 311–318.

Esquivel-Esquivel G, Castillo-González F, Hernández-Casillas JM, Santacruz-Varela A, García de los Santos G, Acosta-Gallegos JA, Ramírez-Hernández A, 2011. Heterosis en maíz del altiplano de México con diferente grado de divergencia genética. *Revista Mexicana de Ciencia Agrícolas* 2 (3), 331–344.

Gonzalez-Gonzalez M, 2007 Diversidad del maíz: potencial agronómico y perspectivas para su conservación y desarrollo in situ, en el sureste del Estado de México, PhD thesis, Postgraduate College, Montecillo, Texcoco, Estado de México.

Herrera-Cabrera BE, Castillo-González G, Sánchez JJ, Ortega Paczka R, Goodman MM, 2000. Caracteres morfológicos para valorar la diversidad entre poblaciones de maíz en una región: caso la raza Chalqueño. *Revista Fitotecnia Mexicana* 23, 335–354.

Herrera-Cabrera BE, Castillo-González F, Ortega-Pazkca RA, Delgado-Alvarado A, 2013. Poblaciones superiores de la diversidad de maíz en la región oriental del Estado de México. *Revista Fitotecnia Mexicana* 36 (1), 33–43.

Herrera-Cabrera B, Castillo-González F, Sánchez-Gonzáles J, Hernández JM, Ortega-Pazkca RA, Goodman MM. 2004. Diversidad del maíz Chalqueño. *Agrociencia* 38 (2), 191–206.

Klenke JR, Russell WA, Guthrie WD, Smith OS, 1988. Inbreeding depression and gene frequency changes for agronomic traits in corn synthetic selected for resistance to European corn borer. *Journal of Agricultural Entomology* 5 (1), 225–233.

Larsson SJ, Peiffer JA, Edwards JW, Ersoz ES, Flint-Garcia S, Holland JB, McMullen MD, Tuinstra MR, Romay MC, Buckler ES, 2017. *Genetic Analysis of Lodging in Diverse Maize Hybrids.* www.biorxiv.org/content/biorxiv/early/2017/09/07/185769.full.pdf.

Louette, D. 1997. Seed exchange among farmers and gene flow among maize varieties in traditional agricultural systems. In Serratos JA, Willcox MC, Castillo-González F, (eds) *Gene Flow Among Maize Landraces, Improved Maize Varieties, and Teosinte: Implications for Transgenic Maize*, 56–66. CIMMYT, México, D.F.

Louette D, Smale M, 1996. Genetic diversity and maize seed management in a traditional Mexican community: Implications for in situ conservation of maize. NRG papers 96–03. México, DF: CIMMYT.

Marquez-Sanchez F, Barrera Gutiérrez E, Carrera Valtierra J, Sahagún Castellanos L, 2000. *Retrocruza limitada para el mejoramiento genético de maices criollos*, Universidad Autónoma de Chapingo, México, DF. ISBN: 968-884-681-3.

Muñozcano-Ruiz M, 2011. Diversidad genética del maíz, perspectivas para su conservación y desarrollo en una comunidad mixteca de Oaxaca: Santa María Tataltepec. MSc thesis, Postgraduate College, Texcoco, Estado de México.

Romero-Peñaloza Jorge, 2002. Diversidad genética y heterosis en cruzas de poblaciones nativas de maíz de la raza Chalqueño en los Valles Altos de México. DC Thesis, Postgraduate College, IREGEP-Genética. Montecillo, Estado de México.

Romero-Peñaloza J, Castillo-González F, Ortega-Paczka R, 2002. Cruzas de poblaciones nativas de maíz de la raza Chalqueño: II. Grupos genéticos, divergencia genética y heterosis. *Revista Fitotecnia Mexicana* 25, 107–115.

Sanchez-Gonzalez JJ, Goodman MM, Stuber CW, 2000. Isozymatic and morphological diversity in the races of maize of Mexico. *Economic Botany* 54 (1), 43–59.

Soleri D, Cleveland DA, 2001. Farmers genetic perceptions regarding their crop populations: An example with maize in the Central Valleys of Oaxaca, Mexico. *Economic Botany* 55, 106–128.

Zambrano-Zambrano EE, 2013. Valoración del mejoramiento genético participativo in situ en poblaciones de maíz *(Zea mays L.)* criollo en el sureste del Estado de México. MSc thesis, Postgraduate College, Texcoco, Estado de México.

7 Pushing back against bureaucracy

Farmers' role in decentralizing plant breeding and seed production in Honduras

*Marvin Gomez, Juan Carlos Rosas,
Sally Humphries, Jose Jimenez, Merida Barahona,
Carlos Avila, Paola Orellana and Fredy Sierra*

Introduction

El Otro Sendero, written in 1989 by the Peruvian economist Hernando de Soto, focuses on how cumbersome bureaucratic legal systems in lower-income countries typically exclude micro-, small- and medium-scale entrepreneurs from the formal economy. Forced to work 'extra-legally' within the informal sector, these entrepreneurs have difficulties expanding their businesses because their assets are not recognized within the law. The protagonists of de Soto's study were small-scale entrepreneurs within Peru's *urban* informal sector, but they might just as well have been Honduran *rural* farmer-researchers and seed producers, who have been excluded from formal recognition because what they produce – seed – cannot be registered or certified within the formal system. In this chapter, we examine Honduran farmer-researchers and their partner NGOs who, with the support of formal-sector breeders, have actively collaborated for nearly two decades in participatory plant breeding to improve access to genetic resources for the country's hillside farmers. In working to decentralize breeding across the landscape, they too have sought to change seed legislation in Honduras to make their product legal.

Introducing the concept of farmer participatory research in Honduras

In 1993, the International Centre for Tropical Agriculture (CIAT), a member of the Consultative Group on International Agricultural Research (CGIAR), brought its farmer participatory research programme to Honduras. The programme of Local Agricultural Research Committees (CIALs), developed by Jacqueline Ashby and her team in CIAT's Hillside Programme (Ashby *et al.*,

1995, 2000), was launched in two hillside communities on the north coast of Honduras (Humphries *et al.*, 2000). CIAT subsequently trained a dozen Honduran organizations in the CIAL methodology. Today, two of those organizations, La Fundación para la Investigación Participativa con Agricultores de Honduras (FIPAH) and El Programa de Reconstrucción Rural (PRR) support 153[1] CIALs across 8 of the country's 18 departments.[2] Other local NGOs have adopted this methodology at various points, but it requires budgetary stability and a commitment to research – not always a feature of those organizations.

The CIAL methodology teaches farmers to manage simple randomized replication trials designed to test new materials against a local, sometimes also a national, control. These trials are conducted on a very small scale so that even poor farmers with limited land resources may participate. Whereas research is conducted by the group, often over several years, any seed production ensuing from the research is generally carried out individually, although more than 20 CIALs have chosen to undertake seed production as a group endeavour. Farmers are initially selected by their communities for leadership roles, but anyone may join a CIAL as a member; over time, natural leaders emerge and typically take on these roles. CIALs are not for everyone, however. They require a commitment to work collaboratively and to dedicate time to collective endeavours, an eagerness to learn new ideas, as well as other motivating factors. CIAL members describe themselves as *futuristas* who seek to bring about changes in their lives. They compare themselves with others whom they describe as *conformistas*, whose fatalistic outlook, they argue, condemns them to poverty, without hope of progress. For the findings of evaluative research on CIAL membership (see Classen *et al.*, 2008; Humphries *et al.*, 2012).

At the outset, the CIALs were dominated by men since agriculture in Honduras is generally perceived as a male undertaking. While women may participate in certain agricultural tasks (like weeding and harvesting beans), this normally takes place under male supervision and women often dislike agricultural activities under these conditions, preferring to limit themselves to gender-prescribed roles that keep them tied to the home (see Roquas, 2002; Humphries *et al.*, 2012). Thus, unsurprisingly, women were not selected by their communities to participate in the early CIALs. To encourage the participation of women, the NGOs sought to establish women-only CIALs, often in the same communities as the de facto men-only CIALs. Gradually, as women gained recognized skills and greater self-confidence in agriculture, the CIALs evolved into mixed-gender groups. Women have now come to play active roles alongside men in all aspects of research as well as in leadership. FIPAH, in particular, has played a proactive role in including women; this has led to the expectation that all new CIALs should be mixed-gender (Humphries *et al.*, 2015). Today, women make up 50.3 per cent of the membership and occupy 67 per cent of the leadership positions in CIAL, and related organizations (savings and credit and micro-business) that make up the CIAL associations supported by FIPAH (FIPAH, USC Canada Annual Report, July 2018, unpublished).

Honduran CIALs are organized regionally and nationally into CIAL associations (ASOCIALs) that bring members together at local and national events to share knowledge generated by the local research teams. This involves a wide array of research that goes beyond participatory plant breeding and varietal selection to include diversified forms of agroforestry and carbon sequestration measurement, post-harvest crop research, and other aspects. Some regional CIAL associations and associated seed committees are also beginning to play a role in service provision to members, as through the distribution of seed kits prior to planting, and the marketing of members' seeds and produce. Individual CIALs within the regional networks are attended by farmer facilitators: these are local farmers who are employed part-time by the NGOs to support and monitor the experiments, and to transmit the findings back to the NGOs. Agronomists working with FIPAH and PRR are closely aligned with scientists at the Pan American Agricultural School (Escuela Agrícola Panamericana, EAP-Zamorano), who have provided germplasm to the CIALs for experimentation for the past 25 years, and have shared in the results of farmers' research at many hillside locations. In Honduras, where there is no public sector extension, it is this link between the CIALs and their partner NGOs, with plant-breeders at EAP-Zamorano, that has provided a *nexus for continuous innovation in agriculture through participatory plant breeding.*

Background

When the CIAL concept was introduced in Honduras by CIAT in the 1990s, and subsequently taken up by FIPAH (then known as Investigación Participativa en Centroamerica, IPCA) and PRR, it quickly became apparent that the priority of hillside communities was finding a solution to food insecurity, epitomized by the widespread use of the term *los junios*, to describe the hungry period starting in June and lasting until the bean harvest in August and the September/October maize harvest. The concept of *los junios* encapsulated the stark reality that the local staples were inadequate to sustain households throughout the entire year. Hillside families had to borrow from wealthier households, leading to a downward spiral of indebtedness, as well as to scaling back on consumption, and/or looking for outside work opportunities.

The first experiments of the early CIALs sought to find a solution to *los junios* by testing varieties of maize and beans available through the formal research system, against farmers' local varieties ('*criollos*'), to see if low crop yields could be remedied through the use of previously released materials. The results from this period showed that the *criollos*, while low-yielding, proved less of an obstacle to food security than conventionally produced materials. In multiple trials (1996–2000) on hillside plots, farmers' varieties out-yielded modern, public sector varieties approximately four out of six times in the case of beans, and five out of six times in the case of maize

(FIPAH field documents, cited in Humphries *et al.*, 2005). This led the CIALs, NGOs and EAP-Zamorano to seek a solution to food insecurity in the Honduran hillsides through participatory plant breeding (PPB) and participatory varietal selection (PVS). For simplicity, in the following both PVS and PPB are referred to as 'PPB', despite the difference between the two (see Almekinders and Hardon, 2006, for discussion, and Berg and Westengen, this volume).

Participatory bean breeding in Yorito

EAP-Zamorano joined with PRR and FIPAH in supporting PPB in two regions, located primarily in the departments of Santa Barbara and Yoro in northern Honduras. Here we document the process of innovation in bean varieties through PPB in Yorito, Yoro, where FIPAH has worked continuously since the late 1990s. PRR has led a similar process of innovation in Santa Barbara. Both regions are mountainous, with steep slopes and poor soils, characterized as suitable for forestry, while poverty among the local people is endemic. In the municipality of Yorito, gang violence and insecurity in recent years have compounded the challenges. Notwithstanding multiple obstacles, FIPAH and PRR-supported CIALs have engaged extensively in PPB, developing 23 PPB bean varieties since 2000 (13 by PRR, 10 by FIPAH). Most of these materials have been released locally, at the municipal level. To date, only one PPB bean variety, developed in parallel with the formal system, has been released at the national level.

The first step in seeking to improve farmers' access to well-adapted bean germplasm was to identify the ideal traits sought by farmers – women and men – in their beans. This was first accomplished through a focus group in 2000, made up of CIAL members (17 women and 20 men) from upland locations in Yorito. The men tended to focus on agronomic traits, whereas the women were generally more concerned with culinary characteristics. However, as women have become more involved in agriculture through the CIALs, their interest has expanded to include agronomic and marketable traits, while men have increasingly shown an interest in how food quality and taste relate to selection and breeding. The important point here is that the involvement of both women and men through mixed CIALs has been essential for ensuring that PPB selections are broadly informed.

Three narrative accounts of PPB in small red beans

The process of varietal development is summarized below through three bean narratives. These provide one example of localized PPB in early generations (Macuzalito), and two examples of varietal selection in advanced generations (Amilcar 58 and Pueblo Viejo 16) as part of larger regional trials.[3] In the first case, CIAL members, supported by FIPAH, determined the criteria for improving their local landrace. Members drew up a list of the characteristics they liked, and those that they wanted to improve. This information, along with local seed, was passed on to Zamorano. In the two other cases, the CIALs participated in the evaluation of advanced lines as part of broader regional trials led by Zamorano. The decision for one option over another depends on many factors, including the capacity of specific CIALs to manage genetic materials in early generations, the expressed germplasm needs of specific CIALs, and the capacity of FIPAH and Zamorano to support whichever process is selected.

Box 7.1 Macuzalito

Macuzalito was the first PPB bean released in Honduras. It was developed in the upland area around Yorito from a widely used local trailing landrace, Concha Rosada, with the support of Zamorano at the request of CIAL members and FIPAH. The improved landrace has gained broad acceptance among CIAL and non-CIAL members alike.

At Zamorano, scientists selected five elite lines[4] for crossing with Concha Rosada, aiming to improve disease resistance, yield, and architecture while retaining the desirable traits of this landrace, including its early maturity. Poor farmers, and especially women, value fast-maturing varieties because they shorten the hungry period (*los junios*), and provide food early. Earliness, farmers explain, allows the variety to 'escape poor weather' like drought or heavy rains, depending on the growing season. However, they also recognize that this benefit is offset by lower yields than those provided by later-maturing varieties. Thus, farmers typically include both early- and late-maturing varieties in their seed portfolios.

The development of Macuzalito involved 53 members (30 men and 23 women) from 4 CIALs over 4 years. Originally, the process had been conceived by EAP–Zamorano as being centralized in one upland community; however, to ensure adaptation to local conditions, the four participating CIALs decided to decentralize it, selecting materials among (F3) lines from 120 families for planting in their own communities, lying at between 1350 and 1650 metres above sea level (masl). The ten best results from the individual community trials subsequently underwent replicate trials (F6) in the four communities, along with five materials selected on-station by EAP–Zamorano, plus the local check. At that stage, farmers selected four lines for multiplication (F7), followed by verification trials from which they selected one line, Macuzalito, for local varietal release. In fact, none of the lines selected by EAP–Zamorano featured among those identified by farmers, reflecting the very different environmental conditions of the two participant groups, as well as differences in preference criteria used by farmers compared to those typically employed by the scientific community. Macuzalito, named after the highest point in the municipality, had the best traits on average of the finalist materials: including good yields (but not the highest), moderate maturity; medium disease tolerance and good commercial value. It was released in the municipal government seat of Yorito in 2004.

In 2014, Macuzalito was among six PPB materials included in a Triadic Comparison of Technologies (TRICOT) trial developed by Bioversity International (Steinke and van Etten, 2016), involving Bioversity International, Zamorano, FIPAH and PRR in 49 communities in four regions of Honduras. The PPB materials, including Macuzalito, outperformed the formal sector check, Amadeus 77, on most criteria; neither altitude nor zone explained any differences, except on the criterion of vigour.

In spring 2018, Macuzalito was excluded from national multi-locational validation trials due to lack of seed availability. The goal of the trials was selection of varieties for national release. However, a sister-line of Macuzalito was released nationally in El Salvador in 2012 by the government agency, Centro Nacional de Tecnología Agropecuaria y Forestal (CENTA), in collaboration with Zamorano, as CENTA CPC.[5]

Box 7.2 Amilcar 58

Amilcar 58 was developed from seed[6] originally distributed as part of a set of 58 materials contained in a Central American Adaptation Trial (Vivero de Adaptación Centroamericano (VIDAC)). The CIALs have regularly participated in VIDAC trials as a means of ensuring that the evaluations of hillside farmers are included in the results. Prior to the formation of the CIALs, it had been the larger farmers, mostly in the major bean-producing, foothill regions, whose criteria and local conditions determined which beans were selected for national release. It was precisely for that reason that the early CIALs had rejected formal-sector materials in their own trials. Amilcar 58 is derived from a simple cross between the formal-sector variety Amadeus 77 and the local variety Cincuenteño, followed by one backcross to Amadeus 77 to increase yield and improve plant architecture.

Among the groups receiving the VIDAC trial was CIAL La Esperanza. The community of La Esperanza is situated at 1350 masl, atop a mountain ridge with 360-degree views across valleys to other mountain ranges. It enjoys moderate humidity year-round, allowing for crop production even when other communities are experiencing drought. The 14-member CIAL team is among the most motivated and innovative of any within the CIAL association. Some members have larger-than-average properties (up to 10 ha) at sites located at various elevations (800–1600 masl) around the community. This allows for the testing of multiple bean lines and the production of diverse seed varieties over three seasons (spring, autumn and summer). These conditions have undoubtedly contributed to the leadership position that this CIAL holds in relationship to PPB and seed production in Yorito.

Between 2006 and 2008, CIAL La Esperanza members engaged in successive rounds of selection of lines from the VIDAC trial, selecting four materials in the fifth round: two lines for early maturity, and two lines for lateness. CIAL members subsequently selected a variety from each of these two groups. The one from the early maturing group was named Esperanceño, after the community. Esperanceño is still used by farmers in some communities because of its early maturity and disease tolerance. However, its dark red grain colour makes it unsuitable for the market and limits its appeal. By contrast, the variety subsequently called Amilcar 58, which was developed from the late-maturing group, has gained acceptance across the country because of its marketability and culinary appeal.

What became known as the seed Amilcar had been identified earlier in the trial by the wife of CIAL member, Amilcar. At the various stages of a trial, extra seed is typically shared out among members and is eaten and/or planted in farmers' fields. Amilcar's wife identified the excellent culinary properties of this late-maturing line and made sure that her husband got some of it in a fifth round of testing, to plant on their land. Amilcar selected seed from the line over a few cycles, improving the seed quality. During this period, the CIAL and farmer facilitator typically monitor seed distributed to individual farmers, comparing how the various lines are progressing. In 2010, Amilcar's seed was selected over the other late-maturing line.

Subsequently, a new bean, Amilcar 58, a golden mosaic virus–resistant selection derived from Amilcar, was identified by scientists at EAP-Zamorano and returned to the CIAL for seed multiplication and dissemination. The process at Zamorano involved single-plant selection of the advanced breeding line IBC 308–24 for resistance to bean golden yellow mosaic virus (BGYMV) using molecular markers and phenotypic selection. The virus–resistant variety has made possible broad dissemination of Amilcar beyond Yorito into lowland areas where disease pressure from BGYMV is high. It is highly valued because of its disease resistance and tolerance of heavy rainfall, along with its commercial and culinary value.[7]

In 2018, Amilcar 58 was one of 10 bean varieties (eight improved and two controls) included in ongoing national validation trials at more than 200 locations in the most important bean-producing regions of the country.

Amilcar 58 was released in Yorito in 2018. Members of CIAL La Esperanza have begun to produce seed equivalent to the basic seed official category of this variety, instead of receiving it from Zamorano, as in the past. If CIAL production of basic seed is recognized, that will represent a considerable step forward in ongoing discussions on seed production, as discussed below. Currently, Amilcar 58 grain commands higher prices than the most popular landrace, Rosado, in the local Yorito market, due to better conservation of its grain qualities under storage.

Box 7.3 Pueblo Viejo 16

The bean Pueblo Viejo 16 is named after an early, women-only CIAL. Nearly 20 years after it was originally founded, CIAL Pueblo Viejo has 10 women members and just one man. It has remained mainly a women's group because most of the men in Pueblo Viejo have worked as day labourers, either in local coffee production, or in crop agriculture outside the community. Since existing bean materials were poorly adapted to the mainly coffee-growing environment of Pueblo Viejo, production of beans was generally restricted to the use of rented plots at lower elevations in the autumn *postrera* planting, or by seeking access to higher elevation plots in the early spring *primera* planting. Women had little role in agriculture, except occasionally supporting their husbands in weeding or pulling up beans outside the community.

For local women whose primary responsibilities are child and home care, being able to engage in agricultural activities means working at a location close to home. It was with an eye to improving food security for their families that women CIAL members began to engage in bean PPB. Starting in 2000, the CIAL was among the first to begin work in PPB. The bean variety, which members named Cayetana 85 after the oldest CIAL member, that emerged from this early research was never released, mainly due to its poor marketability, although the members themselves saved seed and regularly planted it in their community. Regardless, in this initial period of research, CIAL women started to become recognized in the community for their new agricultural skills. This represented a fundamental shift in thinking about women's roles, as is evident in the following quote from the CIAL coordinator, recorded by Classen in 2005:

> I was planting beans in my own field and this man saw me and he told me, 'You are doing bad seeding'. He was laughing at me. I asked him why. He said, 'Because the moon is not in the right position. You will lose all the beans'. I said, 'I'm fine. This is my experiment. Leave me; I know what I'm doing. I am an organized woman'. He had eight farm laborers with him and they too all laughed and joked and said that I would lose everything. 'Don't worry yourself. I will not lose anything. Do what you want, plant in whatever moon you want', I said. I continued planting for three days and they continued laughing at me … Now, you should see how my field looks. I saw him recently and he said, 'It is true, you've learned'. 'Why do you say that?', I asked. He responded, 'I have been watching your crop and you have an excellent yield'. Today, his yield is lower than mine. I told him I learned this in the CIAL. Together we take out loans and run our experiments and we learn a lot. Before I only made tortillas and now, with the group, I am confident in all I do. Working in the group there is love and support and alone there is nothing.
> (Association of CIALs of Honduras and Classen, 2008: 66).

With increasing access to new PPB bean varieties, Pueblo Viejo farmers have begun to embrace bean production, planting on land previously dominated by coffee. This motivated the CIAL to reengage in PPB. In 2014, they began collaborative research with CIAL La Esperanza, evaluating advanced lines of small red and biofortified black beans from an AGROSALUD trial at two sites: in the zone of CIAL La Esperanza (at various bean-growing locations), and another at La Montañita, a site lying above Pueblo Viejo at 1520 masl. Over a two-year period, the two CIALs identified three promising lines (MIB 397–72, BFS 24, BFS 10) that showed good adaptation in both sites over the different cycles. In particular, members of Pueblo Viejo CIAL selected BFS-24 for its high commercial and culinary value, as well as for its architecture. The lines BFS-24 and MIB-397–72 were both passed to a validation trial along with other materials (SJC 730–79, ALS 0532–6) that were added as part of a COVA (Comprobación de Variedades) trial in which four lowland and intermediate location CIALs had previously been involved (2015–2016). These materials, along with local and national controls, were evaluated at three altitudes (high, intermediate, low). In the final validation, the two lines, MIB 397–72 (Honduras Nutritivo) and BFS 24 demonstrated the highest yields across the three altitudinal locations, as well as having excellent culinary and commercial value.

The variety Pueblo Viejo 16, derived from the line BFS 24, was released locally in 2016. CIAL members from both Pueblo Viejo and La Esperanza are currently producing seed of this variety. However, seed production by CIAL Pueblo Viejo depends on land being available to rent. Demand for Pueblo Viejo 16 seed demonstrates the evolution of CIAL women's evaluative capacities in the areas of commercial and agronomic traits, in addition to culinary traits, compared to the earlier period of PPB.[8]

Learning from bean narratives

These narratives point to several recurring themes. Perhaps most significant is the length of time that farmers are willing to commit to developing a new variety. Even when farmers receive advanced lines (F6), it may still take three or four years of selection before they are released locally. This requires a long-term commitment and a *futurista* vision on the part of the farmers. Many of these *futuristas* come from the community of La Esperanza, where CIAL researchers are also *individual* seed producers. As discussed above, CIAL members in this community have ideal bean-growing conditions, with a range of elevations available to them for testing materials and for producing seed in three seasons: growing seed in the 'late autumn (*postrera*) planting' between February and May during the dry season, allowing for the availability of fresh seed prior to the spring (*primera*) planting, when many CIAL members choose to grow beans, and also during the *primera*, in time for the *postrera* planting, the most important bean-growing season in the country, when demand from other farmers is greatest. Thus, for these farmers, a successfully developed PPB variety not only aids local food security but also promises farmer-breeders a stream of future income from seed sales. This link between PPB commitment and seed production is a critical one. In the Yorito area, approximately 100 farmers have been trained in seed production and between 20 and 40 farmers engage in it annually. As mentioned, whereas most farmers engage in seed production on an individual basis, some CIALs produce seed as a group, even if this necessitates renting land. However, only 16 CIAL seed producers have access to drip irrigation systems; without this, seed production over the dry season is often a risky undertaking. While there is sufficient local capacity to produce an estimated 3000 quintals (136.2 US tons) of seed, this would require seed producers to have access to markets beyond Yorito, or else run the risk of falling prices from local oversupply. Reaching more distant market requires that local seed be registered and certified – which the government has resisted up to now. We return to this theme below.

The relationship between EAP-Zamorano and the two NGOs (FIPAH, PRR) has been instrumental in providing CIALs with materials to test on an ongoing basis and to isolate disease-free lines (e.g. Amilcar 58) when these are available. Bean-breeders at EAP-Zamorano have long been supported by the Bean Research Programme through the Bean/Cowpea and Dry Grain Pulses Collaborative Research Support Programme (CRSP), and, more recently, by the Feed the Future Legume Innovation Laboratory, both of which are/have been funded primarily by USAID in partnership with US universities. Since EAP-Zamorano breeders partially develop their lines from germplasm received from CIAT's Bean Programme, this relationship serves to link the CGIAR to hillside farmers, exactly in the manner that Jacqueline Ashby and her team envisioned when they developed the CIAL programme back in 1990. Without this link, formal-sector scientists would lack a mechanism for getting germplasm to farmers and, importantly, for receiving feedback from farmers' trials. It is this feedback that enables the continuous development and improvement of materials to meet the needs of the country's hillside farmers. The Honduran government has not provided sufficient resources to develop and evaluate materials effectively; moreover, it has had no outreach pro-gramme since the early 1990s, when national structural adjustment pro-grammes were first enacted. The latter resulted in the closure of agricultural extension entirely, severely curtailing public agricultural research. Both FIPAH and PRR, which act as boundary spanners between formal breeders and farmer-breeders, are supported by foreign charities, primarily by USC Canada and by World Accord, also located in Canada. In addition, the Development Fund (Utviklingsfondet), a Norwegian NGO, has provided long-term support, particularly to PPB. The process is heavily dependent on funding from outside Honduras, since the government does not allocate ade-quate resources to the agricultural sector to include poor farmers in its pro-gramming. The decades-long reliability of foreign assistance has facilitated skills development in local NGOs and CIALs to support the long-term com-mitment to PPB. However, such long-term support cannot be expected to continue indefinitely. Ultimately, a sustainable system will require either con-tinuous public sector support, or that seed producers find the process suffi-ciently profitable that they can underwrite the cost of linkages to EAP-Zamorano themselves.

The market for small red beans

Honduras has faced recurring deficits in the availability of small red beans since 2000 – most recently in 2014/15, when a two-season El Niño-related drought forced the country to import beans from Ethiopia to meet national demand.[9] While in some years Honduras has had surplus beans for export, between 2009 and 2014 the value of imports has generally been double the value of exports, while dependency on imports has averaged around 14 per cent, ranging from 2 per cent to 16 per cent between 2006 and 2013 (Secretaría de Agricultura y Ganadería, 2015: 11–13). As beans are the primary source of protein, iron and other minerals for most of the Honduran population, bean deficits are a major concern for national food security. For the private sector, deficits represent a lost opportunity, given the market for small red beans in neighbouring countries, as well as among the Central American population in the United States.

Demand for an increase in the production of small red beans, for domestic consumption and export purposes, has helped to foster the development of a National Bean Chain designed to increase values along the chain and to stimulate production. It is supported by committees from six bean producing-departments,[10] by bean-growers from those departments, and by the public, private and non-profit sectors. Agrobolsa is a private sector organization aimed at providing commercialization services to its members – agricultural producers, cooperatives, farmers' unions, agro-industrial producers, research institutes, and service and financial agents.[11] It has been pushing hard to get the bean platform functioning, not least by seeking to provide the platform with legal status through an Asociación de Beneficio Mutuo.

The bean platform is envisaged to provide contracts to growers to produce beans at fixed prices, provide credit, support for seed production, market opportunities, policy advocacy and sectoral development.[12] It would also serve the government by making it possible to purchase beans for the country's strategic reserves and the daily rural school snack (*merienda*)[13] through the national bean platform, rather than relying on the open market, as at present. Only after satisfying national demand would the bean platform be available to serve the export market. Agrobolsa sees opportunities for expansion in the export market, given existing external demand. The drawback is the lack of domestic supply. And this is where PPB seed comes in.

Notwithstanding severe budgetary restrictions that limit its capacity for research, DICTA (Dirección de Ciencia y Tecnología Agropecuaria), the department in charge of the government's agricultural research programme, has continued to play a dominant role governing the release of beans, as well as maize. Most importantly, the department ensures that all nationally released food-grain materials are first subjected to trials conducted across the major growing regions of the country. Since 1996, working in conjunction with EAP-Zamorano, DICTA has authorized the release of seven types of small red beans[14] and two of small black beans.[15] Only one of these varieties, Don Rey, involved the CIALs: indeed, its name comes from the local CIAL facilitator.[16] In general, NGOs and farmers' organizations have not been in a position to submit their materials to national trials involving the collection of data from multiple sites across the country, so PPB varieties have been excluded from national release. However, Agrobolsa has recognized the value of PPB materials produced by the CIALs and would like to be able to sell beans from PPB seed. But it cannot do so under current seed legislation, given the requirement for national release of a registered variety before the variety can be promoted as certified seed.[17] And certified seed must undergo a field inspection and laboratory test to ensure seed quality. This has made it difficult for organizations engaged in PPB to get their materials released and to produce certified seed.

In spring 2018, with support from Agrobolsa, and a new minister of agriculture amenable to the position of this organization, the National Bean Chain became involved in validation trials. Some 250 farmers in more than 200 locations across eight departments participated in testing PPB and conventional varieties. These included two varieties from FIPAH-supported CIALs (Cedron and Amilcar 58), three PPB varieties from PRR-supported CIALs (Campechano JR, Paisano PF, and Rojo Fortificado NC), as well as two improved lines (Rojo Tolupan and Rojo Chorti)[18] previously evaluated by the CIALs as part of on-farm validation process led by Zamorano. National release of all, or any, of these varieties will increase the diversity of the bean market. If this strategy is to be successful, however, diversified seed production must also be permitted. DICTA has shown limited capacity to maintain varieties, particularly concerning seed production from released materials. And this shortcoming affects the availability of foundation and registered seed. Instead, bean seed is generally produced by seed-growers under contract with EAP-Zamorano through its Unidad de Granos y Semillas, a unit separate from its Bean Research Programme. Under collaborative projects with FIPAH, PRR, and CIALs, Zamorano's Bean Programme has provided foundation seed of PPB-released bean varieties to CIAL seed-growers for nearly 15 years; however, with funding for PPB activities to support Zamorano's activities diminishing, CIAL seed-growers must purchase foundation seed from EAP-Zamorano. Not only does the current process of seed production serve to delay local production of seed, it also raises the cost. As CIAL-produced seed must be sold as grain, since it is uncertified, this reduces the returns to local seed-growers. Recent discussions through the National Bean Chain, however, indicate that change may be about to occur; indeed, some changes are already occurring with regard to bean seed production.[19] Discussions have focused on *regional* release of bean varieties. Support to accept the idea of *regionally* approved varieties has been forthcoming from a wide range of government agricultural agencies, although DICTA still has to agree to this idea. In addition, the bean platform is proposing a category of *commercial seed* (similar to *quality declared seed*) that would not require field visits for certification: instead, certification would depend only on laboratory proof.[20] If approved, this would allow farmers and their organizations to produce regionally approved, commercial seed without the onerous requirements of national validation trials and field visits. These proposals, under discussion at the time of writing (October 2018) indicate that the prospects for PPB and its corollary, farmers' seed, may be improving in Honduras. Regionalization of bean varieties would allow official use of varieties that have been proven well-adapted to the effects of climate changes at specific locations.

Indeed, since they were specifically selected for adaptation to these changes, this is one of the strongest agronomic characteristics of PPB varieties.

Decentralized PPB seed production could open several doors. Firstly, commercially approved seed would command a higher price and allow seed-growers to access larger product markets. This would encourage more growers to invest in drip irrigation, reducing the risks associated with seed production during the dry season. The availability of fresh seed following the dry season allows hillside farmers to plant beans in the spring, when prices tend to be higher than in the autumn. That should help to improve the incomes of small bean producers as well as those of the seed-growers. Secondly, if the farmers who purchase fresh bean seed could be provided with credit and assured prices, they would be encouraged to invest in post-harvest drying equipment. At present, and despite high-quality seed and careful field management, heavy post-harvest rains imperil grain quality. Such small-scale capital investments would improve the quality of small red beans, allowing farmers to move up the value chain and into higher-value product markets. Already, FIPAH in Yorito has signed a modest contract for the sale of organic beans to a buyer from Tegucigalpa. This buyer, who selected four PPB varieties, is also interested in the associated background narratives to convey to his buyers. The opening of niche markets, nationally and overseas, is one way the CIALs can sell their product higher up the value chain than when farmers must sell to a '*coyote*' (merchant, with negative connotations) at a considerable discount. In summary, certification of PPB seed, leading to decentralized seed growing, in conjunction with services provided through the National Bean Chain, could have a significant impact on the economies of small bean seed-growers, on the farmers they supply, and – importantly – on national food security. Agrobolsa actively supports these changes with a view to finding new niche markets as well as expanding the market for small red beans beyond Honduras.

Conclusions

The model of decentralized PPB bean research and seed generation described here pits the private sector against state dominance over regulation of bean breeding and seed certification. The agricultural public sector in Honduras has seen its influence decline over the decades due to budgetary restraints. DICTA, in particular, has resisted PPB, perhaps fearing further erosion of the public sector and the associated threat to job security, but also due to the loss of power and authority over the regulation of plant breeding and seed production. PPB represents a paradigm shift that offers farmers a new, and decisive, role in agricultural research and plant breeding. In fact, as the bean narratives clearly show, PPB does *not* displace conventional breeding methods. Instead, it enhances their effectiveness through collaboration with the CIALs and NGOs. DICTA – at least those officials with the most influence – has resisted this shift, still seeing farmers' participation as merely that of validating the work carried out by plant breeders. This culminates in extreme sensitivities around the naming of new varieties, where farmers' and breeders' visions of the process are most likely to compete. Given the restrictions that centralized bureaucratic control places on seed diversity and seed availability, it is not unreasonable to hold that DICTA puts greater priority on maintaining its power and authority, than on national food security. This proposition is supported by the claim voiced by the private sector, that bureaucratic bottlenecks are the main impediment to its own drive to expand markets for small red beans, nationally and internationally.

Agrobolsa, the organization spearheading this drive for market opening, finds itself in a curious alliance with the CIALs and their NGO partners. While Agrobolsa represents a range of organizational types from farmer cooperatives to agro-industrial enterprises, its commercial mission suggests that most of its dealings are with large-scale operations, not small-scale farmer-breeders and seed-growers. This presents the CIALs with a challenge from private sector interests that may not be favourable to them in the long run. In the short run, however, Agrobolsa's commercial mission offers an opportunity to strike out against the strictures on the marketplace erected by the state/public sector. Like the extra-legal entrepreneurs of Peru described in *El Otro Sendero* by de Soto (1989), CIAL breeders and seed-growers want to be part of the formal legal system. Seed legislation in Honduras has long worked to exclude them, but now, with the support of the private sector, this could change.

Notes

1 Currently there are 101 CIALs facilitated by FIPAH (13 in collaboration with two other organizations, Red de Institutos Técnicos Comunitarios (RED ITC) and Fundación Helvet (HELVETAS); 62 facilitated by PRR in collaboration with HEIFER; and 2 facilitated by the Asociación de Desarrollo Pespirense (ADEPES). Funding is provided by USC Canada, Helvetas Swiss Intercooperation, World Accord, HEIFER and HEKS.

2 The nine departments are Yoro, Francisco Morazán, Comayagua, Santa Bárbara, Intibucá, Lempira, La Paz, Choluteca, El Paraíso.

3 Parts of the Macuzalito and Amilcar 58 narratives were originally published in 2015 in *Agriculture and Food*; both have been updated since. The Pueblo Viejo 16 narrative is entirely new. It was presented in 2017 by FIPAH at the Annual Meeting of the Programa Cooperativo Centroamericano para el Mejoramiento de Cultivos y Animales (PCCMCA) in El Salvador.

4 The improved breeding lines used for crossing with Concha Rosada, the maternal parent, were SRC 1–1–18, SRC1–2–12 and UPR 9609–2–2 (source: unpublished FIPAH documents).

5 Adapted from Humphries, *et al.* (2015).

6 Amilcar 58 was developed from line IBC 308–24.

7 Adapted from Humphries, *et al.* (2015).

8 Adapted from Barahona *et al.* (2018).

9 (www.fao.org/news/story/en/item/328614/icode).

10 These departments are Yoro, Olancho, El Paraíso, Francisco Morazán, Lempira and Santa Bárbara-Comayagua.

11 www.agrobolsahn.com/.
12 Some of these proposals might raise CAFTA-DR concerns about competitive advantages. Unlike white maize, beans are not protected under CAFTA-DR. However, since small red beans are a naturally differentiated market, with no outside competitors beyond Nicaragua, it is not clear that the introduction of support prices, etc. would produce any pushback.
13 The WFP has recently extended the *merienda* to include some urban schools in poor neighbourhoods.
14 These were Amadeus 77, Cardenal, Carrizalito, DEORHO, Don Rey, Honduras Nutritivo, and Tío Canela 75.
15 Azabache 40 and Lenca Precoz.
16 The bean known as Don Rey was released officially as Paraísito Mejorado 2 (PM2-Don Rey). National trials, organized by Zamorano and DICTA, were held at the same time as local PPB trials. Both processes contributed to the selection and release of the variety now known as Don Rey. The composite official name reflects this concurrence of events.
17 While Honduras initiated a process to join UPOV-91 in 2012, the agreement has never been ratified. Thus, there are no DUS (distinctiveness, uniformity and stability) requirements in effect. Instead, the government's demand for national release of a registered variety has effectively served to prevent the release of farmers' varieties.
18 Rojo Chorti (SJC-730–79) and Tolupan Rojo (ALS-0532–6) developed by EAP-Zamorano, have been evaluated throughout the main bean-producing regions in collaboration with DICTA since 2012. Several CIALs evaluated these lines in 2016 and 2017 in trials distributed by EAP-Zamorano, including the Comprobación de Variedades (COVA), in Intibucá and Lempira, and more recently by Agrobolsa-supported on-farm validation trials.
19 As mentioned in the narrative describing Amilcar 58, in spring 2018, seed-growers in La Esperanza began to produce good-quality seed of this variety, demonstrating a potential shift in seed regulations. Until recently, CIAL members had to obtain foundation seed from EAP-Zamorano. Amilcar 58 is one of the materials tested in nationally approved trials in spring 2018. Assuming the variety meets the requirements for national release, seed producers will have seed ready for sale in time for the autumn planting.
20 The Food and Agricultural Organization's (FAO) concept of Quality Declared Seed has been discussed as an option for the release of farmers' seeds. In Honduras, however, the bean platform employs developed by the Instituto Nacional de Transferencia en Tecnología Agropecuario (INTA) in Costa Rica, to ensure the quality of local seeds.

References

Almekinders C, Hardon J, (eds) 2006. Bringing farmers back into breeding: Experiences with participatory plant breeding and challenges for institutionalisation. *AgroSpecial* 5. Agromisa, Wageningen, The Netherlands.

Ashby J, 1990. Small farmers' participation in the design of technologies. In Altieri A, Hecht S, (eds) *Agroecology and Small Farmer Development*, 245–257. Boca Raton, FL: CRC Press.

Ashby J, Gracias T, Pilar Guerrero M, Quiros CA, Roa JI, Beltran JA, 1995. Institutionalizing farmer participation in adaptive technology testing with the 'CIAL'. *AgREN Network Paper* 57, London: Overseas Development Institute (ODI).

Ashby J, Braun A, Gracias T, Pilar Guerrero M, Hernandez LA, Quiroz CA, Roa JI, 2000. *Investing in Farmers as Researchers*. Cali, Colombia: CIAT.

Association of CIALs of Honduras and Classen L, 2008. *Campesinos científicos:* Farmer philosophies on participatory research. In Fortmann L, (ed) *Participatory Research in Conservation and Rural Livelihoods: Doing Science Together,* 55–69. Oxford: Wiley-Blackwell.

Barahona M, Jimenez J, Herrera JP, Gomez M, 2018. Mujeres liderando el desarrollo de nuevas variedades de frijol para zonas altas de Honduras. Paper presented by FIPAH in 2018 in the Programa Cooperativo Centroamericano para el Mejoramiento de Cultivos y Animales (*PCCMCA*) in el Salvador.

Classen L, Humphries S, Fitzsimons J, Kaaria S, Jimenez J, Sierra F, Gallardo O, 2008. Opening participatory spaces for the most marginal: Learning from collective action in the Honduran Hillsides. *World Development* 36 (11), 2402–2420.

de Soto H, 1989. *El Otro Sendero: la revolución informal/The Other Path.* New York: Basic Books.

Humphries S, Gonzales J, Jimenez J, Sierra F, 2000. Searching for sustainable land use practices in Honduras: Lessons from a programme of participatory research with hillside farmers. *AgREN Network Paper,* 104. London: ODI Agricultural Research and Extension Network.

Humphries S, Gallardo, Jimenez J, Sierra F, with members of the Association of CIALs of Yorito, Sulaco and Victoria, 2005. Linking small farmers to the formal research sector: Lessons from a participatory bean breeding programme in Honduras. *AgREN Network Paper,* 142. London: ODI Agricultural Research and Extension Network.

Humphries S, Classen L, Jimenez J, Gallardo O, Sierra F, Gomez M, 2012. Opening cracks for the transgression of social boundaries: An evaluation of the gender impacts of farmer research teams in Honduras. *World Development* 40 (10), 2078–2095.

Humphries S, Rosas JC, Gomez M, Sierra F, Gallardo O, Avila C, Barahona M, 2015. Synergies at the interface of farmer–scientist partnerships: agricultural innovation through participatory research and plant breeding in Honduras. *Agriculture and Food Security* 4 (1), 1–17. https://doi.org/10.1186/s40066-015-0046-0.

Roquas E, 2002. *Stacked Law: Land, Property and Conflict in Honduras.* Amsterdam: Rozenberg (Thela, Latin America Series).

Secretaria de Agricultura y Ganadería, Gobierno de la Republica de Honduras, 2015. Análisis de coyuntura del cultivo de frijol en Honduras. USDA.

Steinke J, van Etten J, 2016. *Farmer experimentation for climate adaptation with triadic comparisons of technologies (tricot): a methodological guide.* 40pp. www.bioversityinternational. org/e-library/publications/detail/farmer-experimentation-for-climate-adaptation-with-triadic-comparisons-of-technologies-tricota-methodological-guide/.

8 Participatory varietal selection in the Andes

Farmer involvement in selecting potatoes with traits from wild relatives

Maria Scurrah, Raul Ccanto and Merideth Bonierbale

Potatoes originated in the Andes, where they have remained the most important crop – despite the introduction of major world staples like wheat, barley, fava beans, and other crops well-suited to the cool conditions of Andean highland communities. Potatoes play an essential role in food security due to their high productivity, ease of preparation, versatile use, and health and nutrition benefits (high-quality protein and important micronutrients), (Bonierbale *et al.*, 2010). Thousands of traditional varieties handed down from generation to generation, called 'landraces' or 'native varieties', are still grown. The breeding of high-yielding potato varieties started in Peru over 60 years ago by the Peruvian National Programme of that time, located in the Mantaro Valley, but has had little impact on farming communities in marginal lands with low access to the seed of the new varieties (Vernooy, 2003).

Participatory varietal selection (PVS) brings a wealth of experience and knowledge to the breeding process. It is a practical approach for helping farmers to deal with climate change; and facilitates their access to quality seed of new varieties. Genes of wild potato relatives from northern Peru were incorporated into the cultivated gene pool through interspecific crosses, embryo rescue and ploidy manipulation, resulting in high-yielding and late-blight-resistant types. The process, initiated in 2004, was recently intensified thanks to funding for a potato pre-breeding and evaluation project,[1] with the Lima-based International Potato Center (CIP) and Grupo Yanapai among the partners. Grupo Yanapai, an NGO registered in Huancayo in 1982, works with farming communities in the Central Peruvian Highlands, and has served as a bridge between breeders and farming communities. During the 2016 growing season, CIP multiplied seed of 17 of its newly obtained pre-bred selections; Yanapai then planted these with farmers in two communities, to find out what farmers thought of them. Farmers identified certain clones which satisfied their production and household or market needs.

Background

Participatory varietal selection (PVS)

According to Danial (2003), the centralization of conventional plant breeding in Latin America has resulted in varieties that tend to be unsuited for farming communities in marginal areas. New released varieties might not meet the culinary or agronomic preferences of farmers; seed can be expensive and difficult to obtain, and their cultivation may require inputs outside the normal practices of small-scale farmers. As Andean farmers generally use farm-saved seed, the rate of adoption of improved varieties has been low (Danial, 2003). A recent survey found that 60 per cent of the 275,000 ha of the total potato area in Peru was grown with improved varieties (Pradel *et al.*, 2017), but only 4 of the 34 varieties released dominate, and their distribution varies across communities and regions. A study conducted in two rural highland districts of Peru found that the community of Quilcas grew 194 landrace varieties, and 158 landraces were grown in Challabamba. In both districts, farmers grew more landrace varieties than improved ones. In Challabamba, 89.9 per cent of the potatoes grown were native varieties and 10.1 per cent were improved; figures for Quilcas were comparable – 91.2 per cent and 8.8 per cent (Plasencia *et al.*, 2018). These two communities exemplify the complementarity between improved varieties and local landraces where two management systems coexist. There is an increase in the use of improved varieties as new roads reach formerly isolated communities and farmers adapt to markets. Formal variety release regulations may fail to meet the requirements of farmers working in extreme environments not tested by centralized breeding; or breeders may not have similar culinary requirements. Simplifying the farmers' portfolio of varieties could lead to loss of agrobiodiversity – thereby diminishing the capacity of the ecosystem to continue producing genetic resources, as well as affecting sustainability and resilience (Vernooy, 2003). With PVS, however, farmers' choices are instrumental in the late stages of the plant-breeding process, helping to ensure that selected material is well-suited to local conditions – especially the needs of subsistence farmers for high culinary quality, storability and suitability for local management practices. This can be achieved if the farmers have the opportunity to select from genetically variable material with a range of performance traits. The capacity and local knowledge of producers is included in the selection and development of new varieties (Hocdé, 2006). Yield is not the only characteristic that guarantees adoption, and breeders have come to realize the importance of incorporating local preferences into new varieties (Almekinders *et al.*, 2008).

Yanapai has been involved in PVS since 1999, first collaborating with the Peruvian National Potato Programme (INIA), which selected new varieties from the CIP breeding programme. From this work, two varieties were released in 2002: Colparina and Wankita. Farmers were offered a wide range of advanced clones from varietal trials; their selections were then planted locally, as described by de Haan and Bastos (2012). Various tools and methods have since been developed for improved involvement and quantification of farmers' choices and matters related to varietal selection.

From 2007 to 2012, through the CIP-led network Red Latin Papa,[2] Yanapai worked with farmer groups in Junín and Huancavelica. De Haan *et al.* (2017) have developed a training guide to support PVS in potato using the 'mother and baby' trial design adapted from a participatory research approach in maize originally developed by Snapp (1999) in Zimbabwe. This method facilitates and systematizes the procedures and documents the opinions and selections of women and men separately, at four points of the growing period. It lets farmers take individual decisions about their choices in line with local norms, independent of breeders' choices. This process is deployed in contrasting environments, which permits selection according to various local criteria. A further advantage is that farmers have early access to seed, as harvesting is done on their farms, and they can grow these selections several years before the varieties have been officially named, registered and released. Although farmers' selections are not always released as varieties, farmers are free to keep the seed. Through the combination of researcher-led 'mother' trials to which farmers are invited, and farmer-led 'baby' trials exposing materials to local management conditions, breeders obtain data from replicated trials in various environments, in turn enabling them to produce the stringent reports with statistical designs required by the breeding institution and for varietal release or publication. Farmers may plant the 'baby' trial on their farms, under their own management, and are free to choose what they prefer; evaluators are invited to study the plots.

Potato and its wild relatives in the Andes

In 2016, the potato pre-breeding project 'Sustainable use of potato crop wild relatives (CWR) and development of a pre-breeding core collection with key climate change-related traits' (funded through the Crop Trust) brought new funds for work at CIP, which had started in 2000 with the testing of late-blight resistance in wild species endemic to northern Peru. This new project enabled CIP to intensify its pre-breeding work, aimed at improving populations with wild-species ancestry for global use. In 2016 selections from this programme were tested with farmers for acceptability.

Late blight (*Phytophthora infestans*) has now become a serious threat to potatoes at high elevations, because temperatures are rising, and many landraces are susceptible. Unlike in the past, when farmers never experienced late blight in their high-elevation potato crops, it has now become a serious threat. CIP has prioritized breeding for resistance to late blight, but many releases have become susceptible over time, due to the variability of the pathogen. Diversifying sources of resistance can help to mitigate yield losses.

Most wild potato species are diploid (Hijmans *et al.*, 2002; Ochoa, 1973): the plants have two sets of the basic number of 12 chromosomes ($2x = 2n = 24$), whereas the modern cultivated potato, now grown in 100 countries,[3] is tetra-ploid (2n = 4x = 48). However, there are also cultivated diploid (2x), triploid (3x), tetraploid (4x) and pentaploid (5x) potato varieties in the Andes, a phe-nomenon found only in the territory where the crop originated. In order to cross wild species with cultivated ones, and to overcome the 'ploidy barrier', breeders have traditionally duplicated the chromosome number with colchi-cine, a plant extract which causes chromosome duplication and enables such crosses. After backcrossing work in several breeding cycles with commercial cultivars (which took over 20 years on average), a resistant commercial cultivar was achieved, with excellent varieties which also became parental material for virus, nematode and late-blight resistance (Ross, 1966).

The discovery of 2n gametes facilitated the use of wild species (Ortiz 1998; Ortiz *et al.* 2009). With the use of unreduced gametes, especially 2n pollen, which is easier to identify than 2n eggs, direct crosses to tetraploid cultivated females were successful, with the added benefit that hybrid vigour was often obtained. It is with this approach that CIP used the diploid hybrids it obtained from wild species (*S. chiquidenum, S. cajamarquense, S. paucissectum*) to achieve the tetraploid level and obtain the breeding material which, with funding from the CWR project, enabled Yanapai to test these in Andean communities.

The search for new sources of resistance to late blight focused on wild species from northern Peru and Ecuador, where late blight is endemic (Ochoa, 1999). The species *S. Chiquidenum, S. cajamarquense, S. sogarandinum and S. paucissectum* were tested for resistance to *Phythopthora*; then resistant genotypes were crossed with selected clones of the diploid cultivated *S. tuberosum groups Phureja and Stenotomum*, and a diploid clone from the pre-breeding programme. Interestingly, the embryo rescue (ER) technique was necessary for the success of inter-specific crosses. Those ER hybrids which produced tubers were multiplied and exposed to late blight; and resistant selections were further screened for the ability to produce 2n pollen. The resulting resistant, 2n pollen-producing inter-specific hybrids were crossed with advanced clones from the tetraploid breeding programme, carrying addi-tional important traits such as early maturity and virus resistance. The popula-tion resulting from this process was grown in the field and subjected to late-blight pressure, after which a promising, resistant, vigorous and high-yielding subset was identified for PVS. Seventeen clones were multiplied, and delivered to Yanapai for planting in farming communities (Stokstad, 2019). This showed that, through creative bridges and ploidy manipulation tech-niques, pre-breeding material can be streamlined into the main breeding programme.

Selected farming communities

Two contrasting farming communities from the Central Highlands of Peru took part in this study: Pomavilca and Colpar. *Pomavilca* is located in the

Acobamba district in the Huancavelica region, at 3867 masl, and has a population of 1,080. Families have a mixed portfolio of crops and livestock. Both improved and commercial native varieties of potatoes are grown for the Lima and Huancayo markets. In Pomavilca, the improved variety *Yungai* dominates, followed by two commercial native varieties, Group Phureja *Peruanita* and *Amarilla*, which are popular diploid yellow fleshed and mealy potatoes. All three are susceptible to late blight, so farmers turned to the Agronomy Department of the University of Huancavelica in the district capital of Acobamba for help with disease and pest control. Yanapai, which has a collaborative relationship with the University, proposed PVS in connection with the CWR project to the Pomavilca farmers, because of its emphasis on resistance to late blight, and farmers readily agreed.

Colpar is the capital of a district of the same name located in Huancayo Province in Junín Region, at elevations between 3,200 to 4,200 masl, and with a population of around 4,500. The number of inscribed communal farmers has declined over the past decade, from 250 to about 125. Also here, farmers manage a mixed production system of animal husbandry and crops, complemented by income-generating measures such as migrating temporarily to work in mines. Production is mainly for home consumption. Improved potatoes are cultivated in the mid-elevation zone between 3,300 and 3,600 masl, where land ownership is private and cultivation practices are influenced by modern agrochemical agriculture. Landraces are grown on communal land at 3,900 masl, where the traditional sectoral fallow system is still practised, with potatoes returning to the same field every five to nine years and the land reverting to natural pasture during the fallow years. On the communal lands there is limited tilling, with the Andean foot plough 'Chakitacllia' used in order to minimize erosion; the varieties are mixed and crops are fertilized solely with animal manure: indeed, a communal ordinance prohibits the use of external inputs.

In both communities, potato is the staple and the main crop. However, food security is threatened by climate change; increasingly, late-blight damages have seriously affected harvests. One alternative is the use of varieties resistant to late blight.

Materials and methods

Of the 17 late-blight-resistant clones resulting from CIP's work in the potato pre-breeding project, seven were chosen in the 2015/2016 growing season. This was done with farmer involvement in the first year of the CWR project; but due to a seed contamination problem, the trial had to be repeated. Ten were new entries chosen by CIP from the same pre-breeding material. The three commercial check varieties used were *Peruanita* (native), *Amarilla* (native) and *Yungay* (the most widely grown improved variety). Fourteen clones were identical in the two locations, but 19 were planted in Colpar and only 18 in Pomavilca because of insufficient seed (see Table 8.1 and 8.2 for details). For the same reason, no baby trials were planted.

Table 8.1 Clones HER – Pomavilca, Peru, 2016/2017

Order	CIP number	Clones/varieties
1	CIP512002.39	HER-18.39
2	CIP512003.12	HER-27.12
3	CIP512003.13	HER-27.13
4	CIP512003.38	HER-27.38
5	CIP512004.9	HER-42.9
6	CIP512005.4	HER-44.4
7	CIP512006.23	HER-46.23
8	CIP512010.3	HER-45.3
9	CIP512008.20	HER-57.20
10	CIP512009.1	HER-60.1
11	CIP720064	Yungay
12	Peruanita	Peruanita
13	Amarilla crespa	Amarilla crespa
14	CIP512001.4	HER-13.4
15	CIP512010.1	HER-45.1
16	CIP512010.20	HER-45.20
17	CIP512006.17	HER-46.17
18	CIP512007.99	HER-50.99

Source (both tables): Raul Ccanto.

Note
HER: Hybrids with Embryo Rescue.

Table 8.2 Clones HER – Colpar and Quilcas, Peru, 2016/2017

Order	CIP number	Clones/varieties
1	CIP512002.39	HER-18.39
2	CIP512003.12	HER-27.12
3	CIP512003.13	HER-27.13
4	CIP512003.38	HER-27.38
5	CIP512004.9	HER-42.9
6	CIP512005.4	HER-44.4
7	CIP512006.23	HER-46.23
8	CIP512010.3	HER-45.3
9	CIP512008.20	HER-57.20
10	CIP512009.1	HER-60.1
11	CIP720064	Yungay
12	Amarilla crespa	Amarilla crespa
13	Peruanita	Peruanita
14	CIP512005.8	HER-44.8
15	CIP512010.1	HER-45.1
16	CIP512010.20	HER-45.20
17	CIP512006.17	HER-46.17
18	CIP512006.80	HER-46.80
19	CIP512007.99	HER-50.99

Source (both tables): Raul Ccanto.

Note
HER: Hybrids with Embryo Rescue.

The Procedures involve four evaluations per season with groups of farmers (both men and women), invited by community leaders, participate in all four. Seed of selected clones are stored in the community and replanted the following year for a second assessment. The harvest of unselected clones is distributed to participants for eating.

The four evaluations undertaken each season are as follows: (1) appraisals at flowering time, (2) harvest and two post-harvest appraisals, (3) organoleptic evaluation (taste test), preferably on the same day as harvest, and (4) a storage appraisal, conducted in one store in the community every month. Each evaluation consists of three phases. First, participants brainstorm on selection criteria, with discussion of traits deemed important in new varieties. The traits mentioned are noted on cards placed next to ballot boxes. In communities where some people may have difficulty reading, a person stands next to each trait card and reads it aloud for clarity. The second stage is voting. Every participant receives six seeds to vote with: men receive six maize seeds; women are given six fava bean seeds. Each participant then places three seeds in the ballot box to indicate the most important trait chosen. The second most important trait receives two seeds; the trait in third place will be voted for with one seed. When the selection concerns clones at flowering, the focus is on plant type and vigour; at harvest, the focus is on yield, tuber shape and colour, The ballot box is placed on the plant or the pile of tubers, and voting is directed at the phenotype. In the third phase, the top choices are announced, and there is discussion of the rationale behind the selections, including any gender factors.

This three-step procedure is repeated on all four occasions. However, in the organoleptic test the evaluation is written on a form and preferences are individual, without previous discussion. Women who have difficulties in writing receive assistance.

The organoleptic tests are usually conducted on the same day as the harvest, or at least within seven days of the harvest. For this evaluation, five whole, unpeeled, washed tubers of each clone are cooked, as well as check varieties, usually in nylon bags in large pots where all clones are cooked together, in boiling water – a standard way of potato preparation in Andean homes. Once cooked, each clone is placed on a separate plate – some clones whole, as participants will be asked to evaluate 'cooked appearance', and some cut into small portions for each person to taste. It is important to have visible and clear labels for each clone at all times. Participants score the clones in writing, on a 1–5 scale, for three qualities: appearance, texture and flavour. Once all the data are complete in the field book, an analysis programme, the Highly Interactive Data Analysis Platform (HIDAP), is employed.

Results

Farmer participation is important throughout the evaluation process. Pomavilca is a traditional community where the household head is usually male; if a woman is the head of a household she is usually overburdened, which can explain why women were absent from the evaluation at flowering time. During this evaluation, male farmers prioritized 'vigorous plants' (high yield potential), frost tolerance, and 'earliness' as well as resistance to early blight *(Alternaria solani)*. Neither CIPs nor other breeding programmes in Peru are working specifically on early blight, but earliness is an important objective for CIP. However, because early varieties have problems in storage, they are often 'de-selected' after the storage trial. The farmers of Pomavilca selected three clones (CIP512010.1, CIP512.20 and CIP512006.23), along with the check variety Peruanita, which they know and like.

In Quilcas, the majority of the community are smallholder farmers affected by male outmigration, so the women are very active in community affairs. In fact, the majority of participants in the evaluation were women. The most important attributes of new varieties discussed by women at flowering were late-blight resistance, big plants with abundant foliage and large leaves (linked to yield and resilience). For the men, key criteria were thick stems, resistance to late blight, and abundant foliage. Thus, both men and women prioritized vigorous plants with abundan foliage, and disease resistance. That contrasts with the criteria prioritized by men in Pomavilca, where the market determines characteristics like earliness and problems that make potatos unmarketable, such as wart disease. Peruanita, which has abundant foliage and beautiful lilac flowers, was selected in both communities. In Colpar, women and men also selected one clone, but different ones: the women chose CIP512006.17, while the men opted for CIP512010.20 – which also was chosen by men in Pomavilca; the other four selections differed between the communities.

Harvest evaluation

At harvest in Pomavilca, 16 male and 5 female farmers were involved in identifying criteria and setting priorities, and then direct selection of the best clones. The men voted for a slightly flattened tuber shape, modelled on Peru's current main variety Yungay, with resistance to wart disease and superficial eyes, as well as resistance to Andean weevil – all important characteristics in commercial potato cultivation. Women, by contrast, prioritized tolerance to Andean weevil and to late blight, as well as high yield with tubers that shine, associated with suitability for long storage periods. Because of the low number of voters, only weevil resistance featured in general voting priorities for both men and women. The clones selected were the two native check varieties, Peruanita and Amarilla, and a CIP clone (CIP512004.9).

In Quilcas, 17 women and seven men participated. The first five selection criteria were the same for women and men: resistance to late blight, rapid cooking, high production, mealy texture and good flavour. However, the clones selected varied between men and women. Women gave top rating to CIP512010.1, whereas the men favoured CIP512006.17. The native check variety Peruanita received a considerable number of votes from men, even though it was prone to late-blight attack late in the growing season.

Organoleptic evaluation

The results of the organoleptic evaluation in Pomavilca (where 20 women and 20 men took part) and Colpar (17 women and 7 men) showed that the check varieties were generally preferred (both the improved and the native), but one pre-bred clone was selected. In all cases, the preference went to the two native varieties Amarilla and Peruanita, but four of the CIP-bred clones were also highly rated. In general, the flavour and texture of the clones were rated highly, but the skin and flesh colour was not considered as attractive as the native varieties – a frequent problem with material derived from wild species. Different clones were selected in Pomavilca and Colpar, and there were marked gender differences. The conclusion is that enough clones were rated positively in all cases in both communities to make it highly probable that in each community a clone that could lead to a variety that will be identified in one or two more growing seasons, as well as perhaps through independent multi-location trials.

Final remarks

At the end of the season, both communities, working with the Yanapai team, selected five clones. This demonstrated to breeders that excellent progress had been made in the use of potato wild relatives; that pre-breeding, which normally takes many years, can be streamlined; that having selected diploids which produce 2n gametes is a route that can lead to success; and that testing for stresses is essential. Farmers' opinions should give the breeders confidence that the outcome is acceptable to growers who have farms located in marginal and risk-prone environments and who place high demands on cooking quality. Figure 8.1 shows the pedigree of one of these selections; the majority of selections originate from the same cross – a diploid pre-breeding clone CIP59008.29 crossed with the wild species *S. cajamerquense*, used as pollen donor in a cross with the CIP tetraploid breeding line CIP392657.171. Two clones were also selected where the diploid cultivated parent was from group *Stenotomum* and the CWR donor of late-blight resistance was from *S. chiquidenum*. Thus it is important to have wide crosses with differing genetic backgrounds. One further cross of these diploid hybrids with selected tetraploid breeding parents generated the progeny from which excellent clones were selected: this indicates good possibilities for utilizing diploid pre-breeding in tetraploid potato variety development.

This experience has shown that, while agronomic traits are important, cooking quality should not be the last step in a programme, as it is central for acceptance by Andean farmers. Old stresses such as frost, drought, hail and Andean weevil continue to reduce farmers' harvests, and have intensified with climate change; newer diseases such as late blight, early blight and

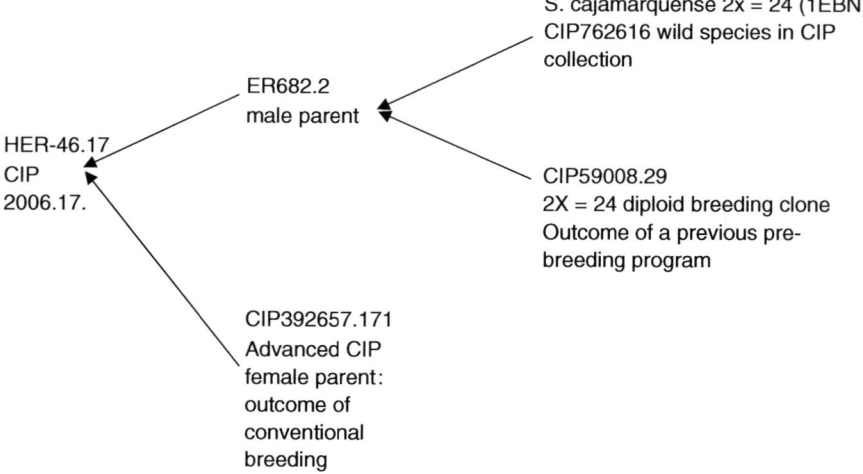

Figure 8.1 Pedigree of the clone CIP512006.17 selected by farmers, two Peruvian communities.

Source: Maria Scurrah.

soil-borne disease such as wart and weevil caused by intensification, shorter rotations, and abandoning sectoral fallow systems (Parsa, 2011; Thurston, 1992) all affect food production. We have not touched on seed production here, as much of the seed used is farmer-saved. However, registering and disseminating new varieties is problematic; and solutions should be found to enable farmers who did not participate in variety trials to access the new selections. The success of varieties selected in farming communities has influenced Peru's varietal release procedures, but comparative yield trials are still required in the final stages. This information is important for policymakers who regulate varietal release in countries where farmers are not permitted to access varieties prior to release.

Further funding for a second phase of the potato pre-breeding project was secured in the course of 2018. The best selected clones are now being tested in two communities in Junín and Huancavelica, in the hope that one of more varieties will be released.

Notes

1 The potato evaluation project is part of the extensive, 10-year project 'Adapting Agriculture to Climate Change: Collecting, Protecting and Preparing Crop Wild Relatives', funded through the Crop Trust. See: www.cwrdiversity.org/partnership/potato-evaluation-project/.

2 This project focused on participatory selection, and a network of eight Peruvian institutions and farmer groups took part. It was also part of wider efforts in countries

such as Colombia and Bolivia, Bhutan, and Bangladesh. For a list of partners see: https://cgspace.cgiar.org/handle/10568/81221.
3 See https://cgspace.cgiar.org/handle/10568/81221.

References

Almekinders C, Hardon J, Guevara F, (eds) 2008. *Un nuevo respeto para los agricultores: Experiencias en Fitomejoramiento Participativo y los desafíos para su institucionalización.* Agromisa Especial 5. Wageningen: Agromisa. www.academia.edu/21900927/Un_nuevo_respeto_para_los_agricultores_Experiencias_en_Fitomejoramiento_Participativo_y_los_desafios_para_su_institucionalización.

Bonierbale M, Burgos Zapata G, Zum Felde T, Sosa P, 2010. Composition nutritionnelle des pommes de terre. *Cahiers de Nutrition et de Diététique* 45 (60) (Supplement 1): S28–S36. DOI: 10.1016/S0007-9960(10)70005-5.

Danial DL, 2003. *Agrobiodiversidad y producción de semilla con el sector informal a través del mejoramiento participativo en la Zona Andina.* Rome: FAO.

de Haan S, Bastos C, 2012. *Catalogo de nuevas variedades de papa: saberes y colores para el gusto peruano.* Red Latin Papa. DOI: 10.4160/978-92-9060-419-8. http://cipotato.org/wp-content/uploads/2013/08/005909.pdf.

de Haan S, Salas E, Fonseca C, Gastelo M, Amaya N, Bastos C, Hualla V, Bonierbale M, 2017. *Selección participativa de variedades de papa (SPV) usando el diseño mamá y bebé: una guía para capacitadores con perspectiva de género.* Lima: Centro Internacional de la Papa.

Hijmans RJ, Spooner DM, Salas AR, Guarino L, de la Cruz J, 2002. *Atlas of Wild Potatoes.* Rome: International Plant Genetic Resources Institute.

Hocdé H, 2006. Fitomejoramiento participativo de cultivos alimenticios en Centro América: panorama, resultados y retos: un punto de vista externo. *Agronomia Mesoamericana* 17 (3), 291–308.

Ochoa C, 1973. El germoplasma de papa en Sudamérica. In French E, (ed) *Prospects for the Potato in the Developing World.* 2nd edn., 68–84. Lima: CIP.

Ochoa C, 1999. *Las papas de Sudamérica.* Lima: CIP.

Ortiz R, 1998. Potato breeding via ploidy manipulations. *Plant Breeding Reviews* 16, 15–86.

Ortiz R, Simon P, Jansky S, Stelly D, 2009. Ploidy manipulation of the gametophyte, endosperm and sporophyte in nature and for crop improvement: A tribute to Professor Stanley J. Peloquin (1921–2008). *Annals of Botany* 104, 795–807 http://doi.org/10.1093/aob/mcp207.

Parsa S, Ccanto R, Rosenheim JA, 2011. Resource concentration dilutes a key pest in indigenous potato agriculture. *Ecological Applications* 21 (2), 539–546. www.esajournals.org/doi/full/10.1890/10-0393.1.

Plasencia F, Juarez H, Polreich S, de Haan S, 2018. Papa en los distritos de Challabamba en Cusco y Quilcas en Junín mediante el uso del Mapeo Participativo. *Revista del Instituto de Investigaciones de la Facultad de Geología, Minas, Metalurgia y Ciencias Geográfica* 21 (41), 17–24.

Ross H, 1966. The use of wild solanum species in German potato breeding of the past and today. *American Potato Journal* 43 (3), 63–80. https://doi.org/10.1007/BF02861579.

Snapp S, 1999. Mother and Baby trials: A novel trial design being tried out in Malawi. *TARGET* (Newsletter of the Soil Fertility Research Network for Maize-based Cropping Systems in Malawi and Zimbabwe) 17. CIMMYT Zimbabwe.

Stokstad E, 2019. The new potato: Breeders see a breakthrough to help farmers face an uncertain future. *Science* 363 (6427), 574–577.

Thurston D, 1992. *Sustainable Practices for Plant Disease Management in Traditional Farming Systems*. Boulder, CO: Westview.

Vernooy R, 2003. *Seeds that Give: Participatory Plant Breeding*. Ottawa: International Development Research Centre (IDRC). Out of print, PDF https://idl–bnc–idrc.dspacedirect.org/bitstream/handle/10625/25864/IDL-25864.pdf?sequence=8&isAllowed=y.

9 A 20-year journey

Participatory breeding of maize in South-West China

Yiching Song, Ronnie Vernooy, Lanqiu Qin, Hexia Xie, Milin Tian and Xin Song

A changing context: challenges facing the new China

As the largest developing country in the world, with a highly diverse culture, economy and ecology, China is experiencing a dramatic transition period in many spheres. Its entry into the World Trade Organization in 2000 marked an important milestone, accelerating the process of economic transition to a fully market-oriented system. However, China remains a largely agricultural country, having 9 per cent of the world's arable land and feeding 22 per cent of the global population. Some 60 per cent of the population relies on farming for food and income. With over 4000 years of farming history, China still has 240 million small farms. The total farming population is about 700 million people and the average farming landholding is only 0.6 ha. (RCRE of MOA, 2017). The majority of farmers today are women. Due to urbanization, the influence of markets and the introduction of information and communication technologies, farming styles and practices are undergoing rapid socio-economic change. Moreover, the impacts of climate change are increasingly felt – for instance, droughts are on the increase in many parts of the country. Ensuring national food security has recently been added to these concerns (for an 'early' warning, see Huang, 2003).

China's rapid and massive transition has led to unbalanced development, with increasing gaps between urban and rural regions, industry and agriculture, the east coast and remote western areas, economic development, and environmental protection. Severe rural poverty, socio-economic inequality, severe environmental degradation and erosion of biodiversity are on the rise (Song and Vernooy, 2010a). China has indeed achieved great success in poverty alleviation, halving the number of people living in extreme poverty and hunger in the course of the three decades preceding 2012 (Xu, 2015). However, this rapid economic growth has come at a very high environmental cost, with severe air, water and soil pollution, ecological degradation and great loss of biodiversity, in addition to the loss of socio-cultural and political rights. About 30 million people remain in extreme poverty (income below US$0.5 a day: State Council Office of Poverty Alleviation, 2007) in 14 poverty-concentrated mountainous areas, especially in western China. Most of those suffering from these conditions are members of ethnic groups living in remote and neglected areas rich in biodiversity but seriously affected by climate change.

Policymakers as well as researchers have become alerted to the impacts of these rapid socio-economic changes on the environment and on vulnerable groups, such as smallholder farmers facing increasing climate change – and the implications for policymaking and the reform of regulations and laws. Public demands for safe, healthy and diversified foods also are increasing due to recent issues of food safety and growing environmental concerns. This has stimulated a drive toward organic, 'green' farming. Local alternatives to promote a 'green transformation' require strong and dynamic farmers' seed-systems.

Feminization and ageing of agriculture

Agriculture in China is undergoing feminization and ageing, bringing a major shift of roles and responsibilities (and burdens) in rural households. With the recent predominantly male outmigration to cities, women, the elderly and children are left behind in rural areas, especially in western China (UNDP, 2003; Song and Zhang, 2004; Song and Vernooy, 2010a). Table 9.1 shows the migration situation in two southwestern provinces, Guangxi and Yunnan, which are home to most of the nation's rural poor mountain ethnic-minority communities. The rising trend of male migration generally marks a move from rural to urban areas, not to overseas destinations. Since the beginning of the twenty-first century, the percentage of migrants in the total labour force in these communities has grown from 42.56 per cent (2002) to 62.09 per cent (2012): a 20 per cent increase in 10 years. Although some young women migrate, men comprise the majority of migrants. Women have come to constitute about 70–80 per cent of the agricultural labour force in the west and southwest (Song et al., 2006; Song and Vernooy, 2010b). These are mainly middle-aged women with limited formal education.

Nationwide, women now comprise about 60 per cent of the agricultural labour force and the majority of the poor in China.[1] They play critical roles in supporting food and nutrition, and improving rural livelihoods, while also performing most of the unpaid care work. However, the income disparity between women and men in rural areas in China has increased: from 79 : 100 in 1990 to 56 : 100 in 2010.[2] Findings from the 2016 study conducted by the

Table 9.1 Migrants in total labour force: women migrants in Guangxi and Yunnan, China (2002–2012)

Year	2002	2007	2012
% migrants, total original labour force	42.56	55.94	62.09
% women, total migrants	38.48	39.84	42.06

Source: Song et al., 2016.

UN Women China and the Ministry of Environment, *Gender Dimensions of Vulnerability to Climate Change in China*,[3] indicate that women farmers are most vulnerable to climate change-induced poverty; moreover, women face persistent systemic biases and challenges in dealing with this. Key gender-based challenges today include the limited capacities of women farmers to build climate resilience, and their restricted access to climate-resilient resources and services. Women farmers often find themselves at the forefront of having to cope with the impacts of climate change, as well as natural resource management (Resurrección, Song and Goodrich, 2018), playing crucial roles in agrobiodiversity conservation and management. They tend to have specific interests in the food quality and nutritional traits of certain crops and crop varieties, as well as a keen eye for pest- and disease-resistance characteristics.

These societal changes put new demands on agricultural development, including plant breeding. Environmental concerns call for greater attention to safeguarding and optimizing the use of agrobiodiversity. The feminization and ageing of agriculture requires social- and gender-sensitive research and policy development.

Environmental degradation and loss of biodiversity and related traditional knowledge

With its long farming history, China is rich in agrobiodiversity and ethnic culture, especially in the southwest. However, in recent decades, the environment and natural resources have been suffering rapid degradation as result of rapid economic development characterized by resource overexploitation and inappropriate interventions. Smallholder farmers and their farming communities find it increasingly difficult to safeguard and enhance agricultural biodiversity. Their traditional knowledge and local practices of experimentation and innovation are under stress. Biological diversity, in particular the landraces in farmers' fields, is disappearing at an accelerating speed (Zhang et al., 2010). This threatens the livelihood and security of the poor, as well as national agricultural sustainability and food security in the long run.

In the past 30 years, the number of traditional varieties of staple food crops (rice, wheat, maize) has declined rapidly: 90 per cent of the landraces have disappeared – rice from 46,000 to 1,000 landraces; wheat from 13,000 to 500; maize from 10,000 to 152 (personal communication, Zhu Youyong, 2017, presented at a FAO workshop in Kunming). Crop landraces in the southwest have decreased rapidly. For example, a recent study in Guangxi and Yunnan (CCAP, 2014) found rapid decline every year since 2000: Guangxi lost the most in the years 2000 and 2006, while Yunnan had the highest loss in 2008 and 2010. The decline is mainly due to local economic developments and agricultural policies.

Recent data from the Guangxi Maize Research Institute (GMRI) indicate the major losses that are occurring in mountain areas of Guangxi; Table 9.2 shows the case regarding maize.

Table 9.2 Maize landraces no longer cultivated in farmers' fields, 1990–2016, eight mountainous counties in Guangxi, China

County	# of landraces in 2016	# of landraces in 1990	Landraces loss (%)
Leye	6	28	79
Xilin	8	15	47
Longshen	5	18	72
Zhiyuan	6	9	33
Shanglin	4	12	67
Nabo	15	82	82
Longlin	26	38	32
Total	82	279	62

Source: Lanqiu Qi (GMRI), personal communication, Nanning, 2018.

The major reason for this loss is the use of hybrids. Market forces backed by government policies and interventions (e.g. seed and input subsidies) 'force' farmers to abandon local varieties and embrace modernization. This shift results in less space and options for farmers' own seed systems. There have been various government subsidies for the production and use of hybrid seed of staple food crops – but no specific support for the conservation and sustainable use of landraces. In Stone Village, in a remote mountain area of Northwest Yunnan, 50 native varieties have disappeared in the course of a mere 10 years, including 13 rice, 10 maize, and 6 beans plus varieties of a few traditional local crops such highland barley, sorghum, oats and highland ramie (CCAP, 2013).

Such severe degradation of agrobiodiversity has meant fewer options for farmers and a narrower genetic base for breeding, threatening farmers' seed systems and the formal seed system alike. A gender-sensitive participatory plant breeding programme is clearly needed, to explore new pathways in this situation.

Failing research and extension

The three major national food crops – maize, rice and wheat – are priority crops of state agricultural research in China. Research on these crops is well organized and has produced good results, but it has been conducted mainly under favourable growing conditions. Less-arable regions, including Guangxi, Yunnan and Guizhou, have not been served well. This is partly due to assumptions prevalent among breeders: that farmers are less knowledgeable than breeders, that selection must be conducted under optimum conditions, that cultivars must be genetically uniform and widely adaptable over large geographic areas, and that landraces and open-pollinated varieties (OPVs, such as those still found in the southwest) must be replaced by high-yielding varieties in order to ensure national food security. Concerns for biodiversity, farmers' diverse livelihoods and their contribution to crop improvement have been largely ignored (Zhang *et al.*, 2010; Song *et al.*, 2009). The narrow focus on hybrid development has not served farmers in mountainous areas affected by such aspects of climate change as temperature rises, greater occurrence of droughts, new and increasing incidence of pests. Not only are most hybrid varieties unable to adapt to conditions in remote mountainous areas, as in Guangxi and other southwestern provinces, they are also susceptible to diseases, pests and drought.

In marginalized areas, farmers' seed systems continue to play a major role in seed supply, while maintaining the diversity essential to sustain the livelihoods of all farmers and indeed the country as a whole. A recent study of six selected provinces (CCAP, 2017) has found considerable regional differentiation in terms of seed use. However, this reality is not known to most agricultural researchers, and such ignorance leads to neglect, and disrespect, of farmers' own seed systems. For example, in 2015, hybrid maize had 100 per cent coverage in Nan County of Heilongjiang province (which is located in the well-developed and modernized northeast of China) whereas the corresponding figure for the same year in Luocheng county in mountainous, marginalized Guangxi province was 38 per cent (CCAP, 2017).

A similar situation exists for wheat: in 2015, Qihe county in Shandong had 100 per cent hybrid wheat, whereas in Stone Village in Yunnan, farmers had never adopted hybrids. Stone Village farmers are satisfied with their local wheat varieties for daily food preparation, and still rely on their own seed systems – improving, exchanging and saving seeds for the next cropping season. They prefer to maintain their seed autonomy (CCAP, 2017).

China's agricultural extension system is indicative of this lack of support. During the country's rapid transformation from a planned economy to a market-oriented one, the extension system became paralyzed and obsolete. In the 1990s, the whole system was on the brink of collapse: no real service delivery took place, few (if any) innovations reached farmers, connections with other rural development agencies were ineffective or non-existent, the capacity of the extension system had not been maintained, and most staff focused on tasks other than serving farmers or contributing to sustainable rural development (Zhang *et al.* 2010). Many local extension stations became shops for the hybrid seed of staple food crops, chemical fertilizers and pesticides. Concern for farmers' own seed systems was mostly absent.

In fact, local varieties and landraces of the staple food crops and other minor food crops such as millet, barley, beans, vegetables are highly relevant to farmers and consumers alike. Farmers continue to take care of the seeds of these crops through cultivation, saving and exchanging in traditional ways. Farmers' seeds systems play a crucial role in achieving food security and sustainable livelihoods, especially in remote areas. These practices represent an excellent entry point for *participatory plant breeding*.

Participatory plant breeding the Chinese way: from maize pilot project in the south-west to national seed platform

The need for bridge-building

At the end of the 1990s, an assessment of the impact on smallholder farming in Southwest China of maize varieties released by the International Maize and Wheat Improvement Centre (CIMMYT) concluded that there had been a systematic separation of the formal seed system and farmers' seed systems (Song, 1998). It found that varieties bred and released by international and national scientific institutions were almost never adopted by smallholder farmers in the remote mountainous regions of the southwest, because they were poorly adapted to local agro-ecological conditions. Researchers involved in the formal seed system, and breeding programmes, seemed unaware of this reality.

The assessment also documented, for the first time in China, the local diversity of maize landraces that had been conserved in the farming communities of the southwest over decades, even centuries. Some 80 per cent of farmers' seed was supplied by their own seed systems, with the remaining 20 per cent purchased commercially from the formal seed-supply system. Farmers' systems relied on self-saved seed and non-monetary seed exchanges among kin, neighbours and friends. Women farmers played a crucial role in seed management.

These findings inspired researchers of the Centre for Chinese Agricultural Policy (CCAP) to establish, in 2000, a participatory plant breeding project, in order to study the usefulness of local varieties in scientific breeding, add value to local varieties and explore the potential of participatory plant breeding (PPB) in China. Such local varieties included farmer-improved, open-pollinated varieties and landraces. The researchers also set out to explore the possibilities of adapting formally released varieties to local conditions through a new way of plant breeding that involves the full participation, knowledge and skills of breeders and farmers. Although this new approach was already being introduced or slowly gaining ground in several countries, including Cuba, Honduras, Mali, Nepal and Syria (for an overview, see Vernooy, 2003), it was totally unknown in China.

A first project to pilot PPB for maize in China and link the formal and farmers' seed systems started in 2000. Although inspired by the work of other participatory plant breeders referred to above, the project developed its own approach in view of the specificities of the situation in China. Project activities initially focused on the southwestern province of Guangxi, with the active collaboration of farmers in six villages, maize breeders from the Guangxi Maize Research Institute (GMRI, the provincial state breeding institute) and the Chinese Academy of Agricultural Sciences (CAAS, which houses the national state breeding institutes), in particular from the Maize

Research Institute under the leadership of China's foremost maize breeder at the time. Funding came from the International Development Research Centre of Canada and the Ford Foundation, with facilitation provided by sociologists and policy researchers of CCAP in Beijing.

In 2008, similar work started in the neighbouring provinces of Yunnan and Guizhou, led by CCAP, the Institute of Crop Science (of the Chinese Academy of Agricultural Sciences) and the Ministry of Agriculture. What started as a small initiative focusing on improving maize varieties in Guangxi grew into a broader effort to revitalize rural development by addressing not only crop production, but also sustainable agriculture, including ecological agriculture, sociocultural development (e.g. supporting farmers' own cultural initiatives), economic empowerment (e.g. improving marketing options) and political aspects (e.g. linking farmers to decision-makers in science, higher education and policy development) (Song and Vernooy, 2010b).

Main project activities now include community-based landrace conservation, participatory variety selection, PPB field trails, local seed registration, community seed banking, community-based seed production and value addition through processing and marketing of processed foods. PPB communities also do circular farming; they practise agro-ecology and have established direct links with consumers to add value to farmers' seeds and related knowledge. This they do by using community-supported agriculture and participatory guarantee systems. For example, in Guangxi, a community-based, women-led seed production cooperative is producing quality PPB seeds for community-supported ecological farming. This farming is linked to urban consumers in the provincial capital of Nanning. Incomes in the PPB village have increased threefold as a result. Similar results have been achieved in Yunnan as well.

Methodology

From the beginning, the participatory action research methodology was adapted to the local context. The work of the entire team, including the farmers, builds on local women farmers' maize breeding experience and expertise developed over many years. The team also draws on and seeks knowledge and expertise from formally trained plant breeders (see Table 9.3). Crop improvements are made through a range of crossing techniques and variety selection processes, which involve field mass selection and line selection undertaken by farmers, with support from breeders. In the experimental fields of the GMRI in Nanning, breeders use more complex methods, including hybrid breeding. The work has involved many parallel activities over several years, using various methods to identify parental materials (through participatory variety selection), improve populations (involving local and formal-system genetic materials) and select further to obtain individual varieties. Trials in villages and at GMRI include both PPB and participatory variety selection. These trials are evaluated by breeders and farmers after each

Table 9.3 Roles of farmers and breeders in the Chinese participatory maize breeding programme: population and hybrid breeding process

Breeding steps	Population breeding		Hybrid breeding	
	Farmers	Breeders	Farmers	Breeders
Defining objectives				
• Evaluating existing varieties on-farm	Y	Y	Y	Y
• Prioritising preferred traits and preferred diversity	Y	Y	Y	Y
Creating genetic variation				
• Collecting, maintaining and/or creating diverse (base) populations	X	X	X	X
• Identifying crossing parents			X	X
• Making crossings:				
for OP breeding	X	X	X	X
for hybrid breeding	X	X	N	Y
producing inbred lines			N	Y
making test crosses			N	Y
improving inbred lines			Y	Y
Selection (including test cross-evaluation)				
• In field (on-station and in multi-locational farmers' fields and kitchens)	Y	Y	Y	Y
• the lab (e.g. disease resistance and quality tests)	N	Y	N	Y
Testing and evaluating varieties	Y	Y	Y	Y
Registration	N	Y	?	Y
Seed production*				
• Parental seed provision	N	Y	N	Y
• Parental seed development	Y	N	Y	N
• On-farm seed production	Y	N	Y	N

Source: developed from resource tables by the Guangxi PPB team, 2018.

Notes

Y = yes; N = no; grey = not applicable; X = farmers and breeders working together.

? = depending on institutional options.

*All activities related to seed production are done by women farmers.

cycle; subsequently, new designs are discussed and agreed jointly. The trials allow for comparisons in terms of locality, approach, objectives and the types of varieties tested (Song and Jiggins, 2003; Song *et al.*, 2006; Song *et al.*, 2009).

New maize varieties

Following discussions among farmers and formal plant breeders, jointly and separately, the field experiments have targeted four types of OPVs and landraces: 'exotic' populations (from abroad), farmers' 'creolized' varieties (developed by breeders but further adapted by farmers, sometimes by crossing them with landraces), farmer-maintained landraces, and formally conserved landraces. So far, more than 180 varieties have been used in trials at the GMRI station and in the villages. Based on 20 years of experimentation, 12 farmer-preferred varieties have been selected and released in the research villages, and have also spread beyond these villages. In addition, five varieties from the International Maize and Wheat Improvement Centre that were showing increasingly poor results have been adapted to local conditions. Another 30 landraces from the PPB villages have been improved by farmers with support from formal breeders. All these varieties exhibit satisfactory agronomic traits, yields and palatability; moreover, they are better adapted to the local environment (Song *et al.*, 2006; CCAP, 2004, 2012; CCAP, 2018b).

The star maize varieties include:

Xin Mo 1 (New Mexico 1), an OPV, derived from a cross between farmer-improved Tuxpeño 1 (as the female line from Wentan village) and Jiahe white (as the male line from Zicheng village) in 2002. Both parents were selected by farmers who were involved in the whole improvement process. Xin Mo 1 is highly drought-resistant; it yields, on average, 15 per cent more than local varieties. It has not been registered, however: in 2003, it failed the registration trials in six provinces. Passing these trials is required by the government before a new variety can be formally recognized and released.

Zhong Mo 1, an OPV, the result of crosses of Xin Mo 1, Suwan 1 and Amarinto 966 (as the male line) in 2004. This variety was developed because PPB farmers wanted to improve Xin Mo 1, which is white, by creating a yellow variety, which would have a higher commercial value. This was the first cross between one parental line and an exotic line. Farmers have been involved from the F1 stage. Zhong Mo 1 has not been formally registered as an approved variety, because it is a PPB variety, the collective result of joint efforts, knowledge and materials – and collective rights are not recognized in the variety regulations.

Zhong Mo 2 was derived from a cross between Xin Mo 1 and Amarinto 9 in 2006. The objective was to produce a yellow variety with improved taste. Farmers and breeders from GMRI worked closely together throughout the whole process. Also this variety has not yet achieved registration.

Guinuo 2006 is a hybrid waxy variety, also called Guangxi Wax 2006. It was produced by GMRI breeders using one line from a PPB project (in Duan county) in 2001. Since 2002, it has been tested, adapted and used for seed production in the PPB villages. Farmers have been involved in testing and adaptation since the F3 stage. It has proven very popular, and has been registered under the name of a GMRI breeder.

Guizuzong, an OPV, has strong lodging resistance and gives higher yields than local varieties. Farmers and breeders from GMRI worked closely together throughout the whole process. Also Guizuzong has not been registered.

Guizhongnuo is a quality sticky OPV bred out together with women farmers in four PPB villages and distributed in neighbouring mountain communities in Yunnan and Guangxi. Table 9.4 summarizes the testing and distribution process of this variety, illustrating the importance and strength of the farmers' seed system.

Seed production: two successful cases

Of the OPV varieties listed above, Guinuo 2006 is the favourite among farmers and local communities, not only for its exceptional taste, but also because of its market potential. The PPB team working with women farmer groups initiated community-based seed production of *Guinuo 2006* in Guzhai village in Guangxi. Seed production and marketing was managed by a women's farmer group in the village. At first, the main difficulty for farmers was their lack of knowledge of hybrid seed production, but with help and technical support from GMRI breeders, they learned the basic skills and knowledge within two years. To manage the process better, farmers have set up a seed production group formally established as a cooperative, to produce seeds each season for their own use and to sell to neighbouring villages. The cooperative also promotes cultural activities such as traditional dance shows.

The PPB research team encouraged farmers and GMRI breeders to establish agreements concerning the exchange of breeding material and seed production methods, to further enhance their collaborative relationship. This type of collaboration is still very new; time and effort by all parties will be required to entrench the practice. It represents novel policymaking in practice and is being followed with interest by the Ministry of Agriculture and the Ministry of Environmental Protection.

Table 9.4 Testing and distribution process of *Guizhongnuo*, a PPB maize variety, in a mountain valley in North-West Yunnan, China

Year	Households # (HH)	Area (mu)*	PPB villages and distribution
2014	2	0.3	SV Started sowing in spring SV: Sharing and distribution to neighbouring villages in autumn through a village clinic
2015	30 HH + 30 HH	2.4	SV and neighbouring villages
2016	30 HH + 40 HH	2.7	A village leader brought 1 kg seed from SV and shared in Wumu and Youmi for testing
2017	SV: 40 HH Wumu: 6 HH Youmi: 5-6 HH Shancha: 2 HH	17.2	Wumu farmer liked it and expanded cultivation to a larger area (10 mu) Youmi: 0.5–0.6 mu Women farmer breeders sent 1 kg to women farmers in Shancha
2018	Stone V: 50 HH Wumu V: 60 HH Youmi: 10 HH Shancha: 6 HH	56.5	Wumu: 50 mu Youmi: 1–1.2 mu Shancha: 1.8 mu
5 years in total	4 PPB villages with 126 PPB HH	Total area 89	Distributed through farmer-to-farmer exchange to 10 neighbouring villages and hundreds of HH

Source: CCAP, 2018.

Note
* 15 mu = 1 ha.

Seed production has experienced some ups and downs, but has continuously generated significant returns for the female farmer seed producers. The women's group has gradually grown into an active women-led farmers' cooperative (also open to men) that conducts circular farming and other value-adding and cultural activities in the region. In 2013, these PPB activities were communicated to other women's groups in Stone Village in Yunnan through farmer-to-farmer exchanges facilitated by the national PPB platform.

Stone Village is a remote mountain village in north-western Yunnan province. Seed production of *Guinuo 2006* took off in the village in 2014. Since then, other PPB activities have been conducted by the women's group, supported by the Guangxi women's group and scientists. As of June 2018, the Stone Village women's group had conserved more than 50 food-crop varieties and improved 10 drought-resistant or quality landraces. The women's group has generated significant revenues from seed production. The group has also started learning ecological and organic farming practices from Guzhai village and plans to register as a women's farmer cooperative. These activities have served to raise women's incomes three times more than 'modern' agriculture has done in the past 10 years. Moreover, the process has enhanced women's community-based collaboration in sharing productive activities, managing natural resources and linking to external information sources, markets and society.

The two cases are examples of women empowerment in practice through self-organization, economically and socially. They have also led to climate-change adaptation and sustainable development in these very remote mountain areas.

Policy and legal challenges

In the early years of the collective efforts, there was no reason to deal with or worry about policy and legal issues. The process was still in an exploratory phase confined to a few villages; no concrete achievements had been made; very few people beyond the team members were aware of what was going on. Policy and legal issues were simply not 'on the agenda'. That changed when the first PPB varieties were produced and the team started to wonder who were the rightful owners of these new varieties produced through a breeding process very different from the conventional approach. Of major concern was how to recognize and reward farmers as rightful breeders. Although the PPB farmers and their communities felt they were entitled to collective rights to the new varieties, such collective rights were not recognized in the Seed Law of the People's Republic of China (adopted in 2000); and are still not in the recently revised Seed Law adopted in 2017. When farmers began seed production of *Guinuo 2006*, another major challenge surfaced: insufficient policy and institutional support for farmers' seed production, distribution and marketing.

The need to protect farmers' rights to seeds

Between March 2016 and October 2017, a participatory assessment of China's seed policies in the last five decades was conducted by a group of chief scientists coordinated by CCAP working with Farmer Seeds Network (FSN) in seven provinces (CCAP, 2018a). The review found that these policies have evolved through four major periods: from being farmer-centred to driven by public research, followed by a move to being market-driven, then focused on seed-industry empowerment and centralization. Further, seed policy has gradually become less supportive of public interest and values, focused increasingly on market values. The review (CCAP, 2018a) observes that these changes have had major negative impacts on farmers' seed systems, related traditional knowledge, and livelihoods. The biodiversity of major grain crops and traditional food crops has decreased rapidly, at field and genetic base levels. Farmers' seed systems have been marginalized, threatened by hybrid technology and mono-cropping. Within the formal seed system (including state research and seed enterprises), the focus has been on hybrid and GM technologies for greater yields and profits. These developments have led to role conflicts and contradictions, resulting in poor-quality hybrid seeds, the narrowing of the genetic base, and general neglect of farmers' interests and changing conditions.

To adjust China's seed policy in support of a green agriculture transformation, reconstruction of public values should be the first priority for the government. Inspiration and inputs for this can be drawn from good practice examples in the country. The PPB initiative with almost 20 years of experience described here and other examples of farmer-led seed conservation and sustainable use have been recommended as good practices and cases. These examples have generated several technological and institutional innovations as well as policy recommendations for further upscaling.

The CCAP review (CCAP, 2018a) presents three major policy proposals:

- support to in-field conservation and sustainable use of farmers' seeds, to protect farmers' interests and ensure seed security;
- building linkages among farmers, state researchers and seed enterprises, for coherent and complementary collaboration;
- use of multiple models and sizes of seed supply in different regions and at different levels, to serve emerging multi-functional agro-ecological farming practices across China and related diversified food systems.

It is positive that the new Seed Law (2016) in China provides some degree of legal protection for farmers' rights and interests. For example, according to Article 26 in the revised seed law, farmers may save, exchange and, on a small scale, sell their seeds. This is crucial for maintaining and enhancing farmers' seed systems and their contribution to national food security and sustainable development. Seeds as a common public resource have existed in China for thousands of years. Farmers have been the main actors in conserving this precious resource, so essential to food security. Farmer seed selection, saving and exchange are at the heart of traditional agriculture systems. These activities contribute to the conservation and evolution of agricultural biodiversity, climate resilient farming and diversified healthy food systems. Ownership and access to seeds are farmers' rights and are vital to achieving a sustainable agriculture transformation.

Reflections and future prospects

In 2013, following a process of reflection on the results and challenges of the PPB process until then led by CCAP researchers, the PPB programme developed into a multi-actor platform, the 'Farmer Seed Network' (FSN) of China. This is a pioneering initiative supporting community-based farmer seed conservation and utilization, and at the same time a multiple-actor platform for linking farmers' voices and field evidence to researchers and policymakers. The network encompasses 36 villages in 10 provinces, and 10 breeding and policy institutes and universities at national and provincial levels.

Twenty years of ongoing and expanding field research in southwest China, combined with strategic policy research at provincial and national levels, have resulted in growing recognition and appreciation of the synergies that can be created between the formal and farmers' seed systems. Given the scope and complexity of the institutional landscape in China, this has been a remarkable achievement.

Integrating PPB activities into local farmers' seed systems in southwest China, by allowing farmers, mainly women, working together with scientists, has emerged as an important institutional innovation for linking formal and farmers' seed systems, empowering women farmers while also enhancing scientific knowledge. Women farmers have successfully conducted OPVs conservation and improvement in their fields, and have been actively involved in various stages of hybrid breeding and seed production, with technical support and benefit sharing from breeders.

In recent years, CCAP, GMRI and CAAS have been joined by other Chinese research institutions to strengthen the efforts pioneered in a few communities in Guangxi. Leading agricultural policy organizations have also become involved, and have begun to incorporate the important results of the field research into relevant policies and laws aimed at creating a more supportive environment for the type of approach piloted by the PPB team. There are now specific policies and mechanisms for recognizing and supporting farmers' saving, exchange and sales of their own seeds in implementing the recently revised seed law, and appropriate access and benefit-sharing (ABS) policies and tools for the ABS Law are under development, to protect farmers' rights, and integrate these into related policies for poverty alleviation, climate-change adaptation and agro-ecology. This should allow more farmers to benefit in the form of recognition of their expertise, improved access to and availability of quality seeds and improved varieties, income generated from seed production and marketing, and the provision of scientific and technical knowhow through collaboration with the formal seed sector.

This has been a long journey of learning for all actors involved, and has contributed significantly to the organizational strengthening of farmer communities and breeding institutes in China. As a result of collective efforts, more than 1,500 farmer-preferred landraces have been safeguarded, 12 PPB varieties have been produced (including 1 maize hybrid), and PPB varieties have achieved yield increases of 15 per cent to 20 per cent over average yields of local varieties, with positive responses to the chief characteristics preferred by farmers: drought resistance, lodging resistance and good eating quality. Through seed production and other value-adding activities, these varieties generate significant value for women's groups.

Women farmers have been at the forefront of these efforts. The women's group cooperatives represent an important rural innovation, offering a path for empowering women for poverty alleviation, food security and climate-change adaptation in the mountain areas of China. PPB has been shown to be a good entry-approach for locally driven adaptation and empowerment processes whereby farmers, led by women, can improve their local varieties, enhance their capacity to deliberate about choices of action, experiment with options, create new practices, and enlarge their network of horizontal relationships with other actors, thereby obtaining more social capital and greater autonomy for collective innovations and the transformation towards equal and sustainable development.

Notes

1 http://politics.people.com.cn/n/2015/0309/c70731-26661075.html.
2 National surveys of women social status in China, conducted by the National Bureau of Statistics and All-China Women's Federation in 1990 and 2010, respectively.
3 http://asiapacific.unwomen.org/en/digital-library/publications/2016/12/gender-dimensions-of-vulnerability-to-climate-change-in-china.

References

CCAP, 2004. *PPB Project Annual Report to IDRC by CCAP PAR team.* Beijing: Centre for Chinese Agricultural Policy.

CCAP, 2012. *First Annual Report on 'Smallholder Innovation for Resilience: Strengthening Innovation Systems for Food Security in the Face of Climate Change' by China case team to IIED.* Beijing: Centre for Chinese Agricultural Policy.

CCAP, 2013. *Second Annual Report on 'Smallholder Innovation for Resilience: Strengthening Innovation Systems for Food Security in the Face of Climate Change' by China Case Team to IIED.* Beijing: Centre for Chinese Agricultural Policy.

CCAP, 2014. *Mid-term Report on 'Smallholder Innovation for Resilience: Strengthening Innovation Systems for Food Security in the Face of Climate Change (SIFOR)' by China case team to IIED.* Beijing: Centre for Chinese Agricultural Policy.

CCAP, 2017. *Final Annual Report of SIFOR project by CCAP Team to IIED.* Beijing: Centre for Chinese Agricultural Policy

CCAP, 2018a. *Seed Policy Review 'Accessing Seeds, Securing Foods and Sustaining Future: A Seed Policy Impact Assessment in China (1949–2016) by Reviewing Team.* Beijing: Centre for Chinese Agricultural Policy.

CCAP, 2018b. *First Annual Report on 'Smallholder Innovation for Resilience: Strengthening Innovation Systems for Food Security in the Face of Climate Change' by China case team to IIED.* Beijing: Centre for Chinese Agricultural Policy.

Huang J, 2003. Input and use of agricultural extension funds. In Chen X, (ed) *The Use and Management of Supportive Agricultural Funds* (in Chinese). Taiyuan, China: Shanxi Economic Press.

RCRE of MOA, 2017. *Annual Report (in Chinese) of the Research Centre of Rural Economic of Ministry of Agriculture.* Beijing, China.

Resurrección BP, Song Y, Goodrich CG, 2018. In the shadows of the Himalayan mountains: Persistent gender and social exclusion in development. In Wester P, Mishra A, Mukherji A, Shrestha AB, (eds) *The Hindu Kush Himalaya Assessment: Mountains, Climate Changes, Sustainability and People.* Kathmandu: ICIMOD. https://tinyurl.com/ybcwllbl.

Song Y, 1998. *'New' Seeds in 'Old' China: Impact Study of CIMMYT's Collaborative Programme on Maize Breeding in Southwest China.* PhD thesis. Wageningen, The Netherlands: Wageningen University and Research Centre.

Song Y, Jiggins J, 2003. Women and maize breeding: The development of new seed systems in a marginal area of southwest China. In Howard PL, (ed) *Women and Plant–Gender Relations in Biodiversity Management and Conservation.* London: Zed Books.

Song Y, Zhang L, 2004. *IFAD Gender Assessment Report for IFAD-supported Poverty Alleviation Projects in China.* Beijing: IFAD China Office Report.

Song Y, Vernooy R, (eds) 2010a. *Seeds and Synergies: Innovating Rural Development in China.* Burton on Dunsmore, UK: Practical Action Publishing/Ottawa: International Development Research Centre. www.idrc.ca/EN/Resources/Publications/openebooks/485-7/index.html.

Song Y, Vernooy R, 2010b. Seeds of empowerment: Action research in the context of the feminization of agriculture in Southwest China, paper for special edition of *Gender, Technology and Development.* http://gtd.sagepub.com/content/current.

Song Y, Zhang S, Vernooy R, 2009. Biodiversity conservation and crop improvement in a fragile agro-ecosystem: insights from Guangxi, China. *Mountain Forum Bulletin,* 9 (2). www.mtnforum.org.

Song Y, Zhang L, Vernooy R, 2006. Empowering women farmers and strengthening the local seed system: Action research in Guangxi, China. In Vernooy, R, (ed) *Social and Gender Analysis in Natural Resource Management: Learning Studies and Lessons from Asia,* 129–154. New Delhi: Sage/Beijing: China Agricultural Press/Ottawa: International Development Research Centre. www.idrc.ca/openebooks/218-X/.

Song Y, Zhang S, Huang K, Qin L, Pan Q, Vernooy R, 2006. Participatory plant breeding in Guangxi, southwest China. In Almekinders C, Hardon J, (eds) *Bringing Farmers Back into Breeding: Experiences with Participatory Plant Breeding and Challenges for Institutionalisation,* 80–86. *Agromisa Special* 5. Wageningen, The Netherlands.

Song Y, Zhang S, Vernooy R, 2009. Biodiversity conservation and crop improvement in a fragile agro-ecosystem: Insights from Guangxi, China. *Mountain Forum Bulletin* 9 (2). www.mtnforum.org.

Song Y, Zhang Y, Song X, Vernooy R, 2016. Access and benefit sharing in participatory plant breeding in Southwest China. *Farming Matters,* special issue, April, 18–24. Available at: www.bioversityinternational.org/e-library/publications/detail/access-and-benefit-sharing-in-participatory-plant-breeding-in-southwest-china/.

State Council Office of Poverty Alleviation, 2007. *Report on National Poverty Alleviation Programme.* Beijing.

UNDP 2003, *Annual Development Report of China.* Beijing: UNDP.

Vernooy R, 2003. *Seeds that Give: Participatory Plant Breeding.* Ottawa: International Development Research Centre.

Xu Z, 2015. *Climate Related Disasters Risk and Analysis for Climate Change Affected Poverty Evaluation.* Working report (in Chinese). Beijing: Oxfam HK China Office.

Zhang S, Li X, Pan G, 2010. *Generic Base Narrowing and Maize Breeding in the Past 20 Years.* PPT report to annual workshop by the national Maize Breeding and Development Programme, Hubei (in Chinese). Hubei, China: Maize Breeding and Development Programme.

10 Evolutionary participatory quinoa breeding for organic agro-ecosystems in the US Pacific Northwest

Julianne Kellogg and Kevin Murphy

Introduction

In response to rapidly increasing consumer demand for quinoa over the past few decades in the USA, farmer interest in – and experimentation with – quinoa production as a regionally novel seed crop has grown considerably in recent years. The field experiments reported in this chapter were designed to test whether (1) yield increased with participatory selection on heterogeneous quinoa populations and (2) yield and agronomic performance of breeding lines selected by farmers surpassed the performance of the commonly grown variety, 'Cherry Vanilla'. A farmer-driven selection index determined the five best-performing breeding lines for each location, and nine unique breeding lines that outperformed Cherry Vanilla were selected. In 2016, the evolutionary participatory breeding (EPB) method ($2090\,kg\,ha^{-1}$) resulted in yields higher than the evolutionary breeding (EB) method ($1718\,kg\,ha^{-1}$). Additionally, two populations that underwent selection by participating farmers [P104 ($2628\,kg\,ha^{-1}$), P107 ($2309\,kg\,ha^{-1}$)] out-yielded the parental variety 'Black' ($1322.17\,kg\,ha^{-1}$). Heterogeneous populations have the plasticity to provide farmers with increased within-field diversity and improved stability over time, while participatory selections made from such populations can lead to the development of uniform varieties adapted to specific agro-environments and farming systems.

After the UN General Assembly declared 2013 as the International Year of Quinoa, the number of countries with farmers growing *Chenopodium quinoa* Willd. has doubled (Bazile *et al.*, 2016a). Recognition that quinoa is suited to cultivation in adverse environmental conditions (Zurita–Silva *et al.*, 2014) has additionally prompted worldwide experimentation with this nutritious crop (Bazile *et al.*, 2016b). In the USA, interest in producing a commercial crop may result in an increase in acreage planted to quinoa. In a recent survey of certified organic farmers in Washington State, 49 per cent of the respondents expressed an interest in growing an organic quinoa crop in the next five years (Detjens, 2016). However, the short history of quinoa production, research and breeding in the USA (Peterson and Murphy, 2015) leaves organic farmers who would like to join the ranks of quinoa innovators with few adapted varieties commercially available from seed companies within the USA. Growing unadapted varieties puts farmers at risk of yield and crop loss. For example, Pacific Northwest (PNW) farmers require early-maturing varieties that can be harvested prior to autumn rains, because quinoa typically lacks seed dormancy. Growing a late-maturing variety could result in partial to entire crop loss.

A successful organic farming system will require varieties bred with adaptations to the soils, inputs, management practices, and abiotic/biotic stresses unique to it (Shelton and Tracy, 2016). In organic systems, such traits often require considerable attention and are not usually present in varieties used for high-input conventional agriculture (Murphy *et al.*, 2007). Breeders can seek to develop varieties with rapid juvenile growth and subsequent early cover and shading of the soil to combat weeds in organic agro-ecosystems (Lammerts van Bueren *et al.*, 2002). Particular plant architecture characteristics that reduce fungal diseases (Lammerts van Bueren *et al.*, 2002) and mechanisms that enhance nutrient-use efficiency are also traits targeted for organic variety development (Wolfe *et al.*, 2008; Lammerts van Bueren *et al.*, 2011). Traits that increase nutrient-use efficiency may include more extensive root systems, increased ability to form *Rhizobia* or mycorrhizal associations, modified root exudate production to alter the rhizosphere, and resistance to root pathogens and pests.

Efficient quinoa cultivar improvement for organic farming systems does not require expedited achievements in phenotypic and genetic uniformity. Breeding to create and maintain genetic diversity over many plant generations can be a goal in itself, as demonstrated by researchers studying a historically significant barley population created by Harlan and Martini (1929) using the composite cross method in plant breeding. This population, referred to as Composite Cross II (CCII) (Suneson and Stevens, 1953), was evaluated for numerous generations with no active selection to determine that, over the span of nine generations, the barley population did not differ in yield from the uniform check variety (Suneson, 1956). In fact, the check variety varied more widely in yield than CCII during the 18-year trial. Allard *et al.* (1992) studied CCII for 45 generations, noting genetic restructuring of the population in response to natural selection in a temporally heterogeneous environment. For example, the frequency of resistance alleles increased in wet years favourable for scald (*Rhynchosporium secalis*) (Allard *et al.*, 1992). Natural selection alone may create gains in a few generations with an inbreeding crop population (Allard, 1999). The evolutionary dynamics of CCII subpopulations grown in differing environments demonstrated that subpopulations shifted toward earlier maturity, shorter stature and higher yield in rainfed xeric environments, whereas in fertile irrigated environments, these subpopulations shifted toward later maturity, larger kernel size and higher yields (Allard, 1999).

The EB method can preserve a high degree of genetic variation in the resulting population, allowing for continued improvement in yield and adaptation (Suneson, 1956). Most agro-ecosystems experience fluctuations in abiotic and biotic stresses. With a mix of genotypes, as in heterogeneous populations, the risk of low yields decreases when the failure of one genotype is compensated by another (Döring *et al.*, 2011). Corte *et al.* (2002) studied segregating dry bean populations for 17 generations under differing environmental conditions and found a mean yield increase of 2.5 per cent per generation over the parent mean. A heterogeneous crop provides a buffer against climatic unpredictability, but it is the EB method that allows natural selection to favour high-yielding genotypes in environments with fluctuating conditions. EB differs from other bulk breeding schemes in the practice of sowing and re-sowing seed from a breeding population for more generations than required to reach homozygosity (Suneson, 1956). After sufficient homozygosity is achieved, pure lines may be selected – but the bulk population should not be discarded at that time, because population yield can continue to improve over time (Suneson and Stevens, 1953). Survival of the populations determines evolutionary fitness (Suneson, 1956) – and survival is positively correlated with agricultural value (Allard and Hanche, 1964). When a breeding programme includes farmers, the participants can contribute farming experience and knowledge of their climate, farming practices, agricultural pests and diseases, as well as possible end-uses and marketing avenues for the crop (Ceccarelli, 2009). In participatory plant breeding (PPB), farmers are involved in various stages throughout the breeding process. This participatory element can be defined as (but is not restricted to) farmer involvement in identifying breeding objectives, making initial crosses, managing research trials on their fields, conducting selection, evaluating varieties, and distributing seed (Ceccarelli, 2009; Morris and Bellon, 2004; Murphy *et al.*, 2016). PPB for organic agro-ecosystems has proven successful in vegetables (Campanelli *et al.*, 2015; Mazourek *et al.*, 2009; Mendum and Glenna, 2010), legumes (Ghaouti *et al.*, 2008), and grains (Dawson *et al.*, 2011; Entz *et al.*, 2015). EPB combines the methods of EB and PPB (Murphy *et al.*, 2005). Breeding populations grown in farmers' fields under organic and low-input conditions are likely to improve as a result of participatory selection, because farmers provide insight into the surrounding environment and how the climate may differ on a yearly basis, and direct researchers' attention to specific traits that improve crop performance under their management practices.

Since 2010, researchers at Washington State University (WSU) have conducted extensive breeding, agronomy and food science trials with quinoa. A key objective in the WSU quinoa research programme has been the development of resilient and regionally adapted varieties across diverse and contrasting environments. Using publicly available quinoa genetic resources and regional organic farmers in the breeding process, a breeding scheme was designed with the objective to develop both heterogeneous quinoa populations and uniform varieties specifically adapted to organic farms in the PNW region of the USA. A *breeding methods trial* comparing the EB and EPB methods was established within the framework of the breeding scheme, to test the hypothesis that participatory selection will improve yields of six bi-parental populations over a relatively short, two-year time-frame. A complementary *pedigree selection trial* evaluated the agronomic performance of single-plant selections across three distinct locations, to test the hypothesis that these breeding lines will outperform a commonly grown variety. In addition, a farmer-driven selection index was developed, to determine the relative importance of traits for selection and the top-performing breeding lines for each of the three locations.

Methods and materials

Plant material and breeding scheme

The parental varieties in this trial were 'Black', 'Cherry Vanilla', 'CO407Dave', 'Kaslaea', 'QQ065' and 'QQ74' (Table 10.1) – commercially available varieties, farmer-maintained varieties and Chilean landraces. Parental varieties were chosen for the initial crossing events based on one or more of the following characteristics from results of several years of preliminary testing in multiple contrasting environments in the PNW: seed yield, day-neutral photoperiod, resistance to downy mildew, heat tolerance, drought tolerance, seed size, plant height, tolerance to lodging, mould and pre-harvest sprouting resistance, and stem, leaf, panicle and seed colour (Murphy *et al.*, 2016). The initial quinoa crossing events performed in 2012 (Peterson *et al.*, 2015) formed six bi-parental populations.[1] The populations were grown out as F_3 seed in western Washington State, without human selection, in 2014. The F_4 seed derived from the six populations was planted in 2015 for use in this study; and the F_5 seed harvested in 2015 from each treatment was planted in 2016 for continued use in this study.

Table 10.1 Source and origin of seed used in initial crossing events, US Pacific Northwest

Parental varieties	Source	Accession	Origin
Cherry Vanilla	Wild Garden Seed	N/A	Philomath, OR, US
CO407Dave	USDA Plant Introduction, Ames, Iowa	PI 596293	Colorado, U
QQ74	USDA Plant Introduction, Ames, Iowa	PI 614886	Maule, Chile
QQ065	USDA Plant Introduction, Ames, Iowa	PI 614880	Los Lagos, Chile
Kaslaea	USDA Plant Introduction, Ames, Iowa	Ames 13745	New Mexico, US
Black	White Mountain Farm	N/A	Mosca, CO, US

Source: Kellogg, 2017.

Experimental design

Breeding methods trial

The six bi-parental quinoa breeding populations were planted as a split-plot RCBD.[2] The six parental varieties used in the crosses forming the populations were included in this study. The parental varieties were planted in each replicate of the breeding methods each year, to enable assessment of the performance of the populations against the parental varieties. Seed of the parental varieties was saved from the 2015 trial and used as the seed for the 2016 trial. The populations underwent two treatments: EB and EPB. The EB treatment experienced only natural selection during the growing season. The EPB treatment involved both natural selection and farmer participatory selection (see Figure 10.1).

Pedigree selection trial

The participatory selection involved farmers who identified and tagged single plants exhibiting traits of interest (*positive selection*) and removed unsatisfactory

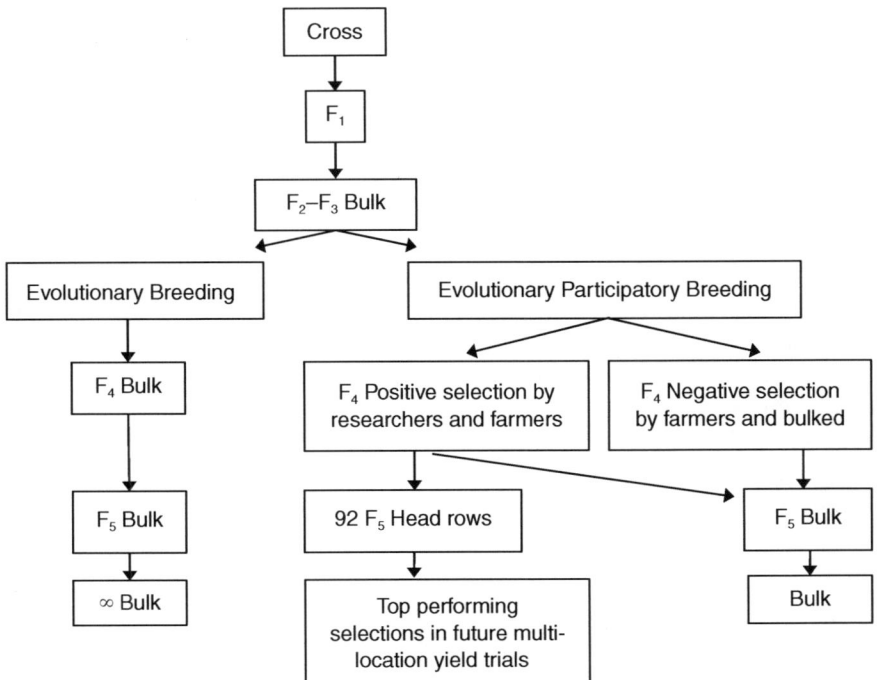

Figure 10.1 Quinoa breeding scheme, US Pacific Northwest.
Source: Kellogg, 2017.

plants (*negative selection*), within each population. The positively selected 213 pedigree selections were hand-harvested just prior to the mechanical harvest of the trial. Only 92 of the selections had an adequate amount of seed for planting in 2016. The pedigree selections trial was designed as a non-replicated augmented RCBD (ARCBD) with 16 breeding lines and three repeated check varieties planted as head rows in each of six blocks. The final block was incomplete, with only 12 breeding lines and three check varieties. The repeated check varieties were Cherry Vanilla, CO407Dave and Kaslaea.

Trial locations and site preparations

Breeding methods trial

In 2015, the breeding methods trial and participatory selection were conducted on a certified organic farm in Chimacum, WA. In 2016, the trial was located on a certified organic farm in Sequim, WA. Farmers prepped the fields; no fertilizer was applied at sowing or during plant growth. Trials were sown, cultivated and harvested by the researchers. The farmers at the Sequim trial sites applied irrigation during emergence, stand establishment and after flowering.[3]

Pedigree selection trial

The pedigree selection trial was planted out in 2016 on three certified organic farms in Chimacum, Quilcene, and Sequim, all in Washington State, and was managed identically to the breeding methods trial.[4]

Participatory methods

Eight farmers from the region and three WSU researchers were invited to engage in participatory selection. On two separate days, participants conducted positive selection by affixing a tag on two preferred plants in each plot of a quinoa population receiving the EPB treatment. One group consisted of three farmers and three researchers operating independently of each other. The second group of participants consisted of five farmers divided up as two pairs and a solo individual to conduct selection. On each day of participatory selection, after positive selection was completed, the farmers – not the researchers – conducted negative selection by removing an unrestricted amount of plants that did not visually appeal to them.

Yield and agronomic evaluations

Breeding methods trial evaluations: F₄ and F₅ populations and parent lines

Yield data for the F_4 populations and parental varieties in 2015 in Chimacum were adjusted by percentage of seedling emergence due to non-uniform seedbed conditions. Yield data for F_5 populations and parental varieties in 2016 in Sequim were adjusted by percentage of mature plants damaged by elk (*Cervus canadensis roosevelti*).

Pedigree selection trial evaluations: F_5 breeding lines

Field data analysed included days after sowing (DAS) for the phenological stages of flowering, seed set, plant height at harvest (cm) and maturity, as per Jacobsen and Stølen (1993). Post-harvest data were collected and statistically analysed on the F_6 seed from the F_5 breeding lines harvested in 2016. The 1,000-seed weight (g) and colour of unprocessed seeds for each breeding line were measured.[5]

Statistical analysis

Breeding methods trial statistical analysis: F_4 and F_5 populations

An analysis of variance (ANOVA) using a linear mixed-effects model fit by residual maximum likelihood was conducted for each year (2015 and 2016) to assess the significance of the breeding method and population factors.[6]

Breeding methods trial statistical analysis: F_4 and F_5 populations and parental varieties

In 2015, yield data[7] of the F_4 EB populations and six parental varieties were analysed to determine differences between populations and parental varieties prior to selection. In 2016, yield data[8] of the F_5 EPB populations and parental varieties were analysed to determine if participatory selection resulted in populations overyielding parental varieties.[9]

Pedigree selection trial statistical analysis: F_5 breeding lines

Non-replicated experiments are often used in early-generation yield trials and progeny evaluations when seed quantity is low, land and other resources are limited, and many genotypes need to be evaluated. In such cases, the ARCBD can be used for spatial adjustment across blocks when there is field variation (Federer, 1956). Adjusted check varieties and breeding lines were analysed as a completely randomized design.

A one-way ANOVA was carried out using a linear model to assess the significance of the effect of the check variety and breeding-line trial entry.[10] Post-hoc tests were conducted for traits when the main effect (trial entries) was found to be statistically significant in an ANOVA. The breeding lines were tested against Cherry Vanilla, because this variety is widely grown along the US West Coast, including western Washington State.[11]

Farmer-driven selection index

Prior to conducting selection in the field, eight participating farmers and two researchers were asked to assign levels of importance to a list of quinoa plant traits developed by WSU researchers. Traits not important to the farmers, their farming practices, their market, or climate were assigned a '1'; traits of necessity or of great interest to the farmer were assigned a '5'. Participants were given the option to add traits not listed and assign levels of importance. Ratings were added up to create cumulative weights for each trait in the selection index. Traits measurable in the field and lab were matched with the traits rated by the participants. In adding up the ratings, the cumulative weights were heavily influenced by farmer preference (Table 10.2).

Table 10.2 List of traits rated by participants prior to conducting selection, US Pacific Northwest

Trait	Importance level	Weight	Traits with analyzed data
Disease resistance	High (41–50)	47	
Early maturity		47	Seed set, harvest DAS
Yield		47	1,000–seed weight (in g)
Lodging resistance		46	
Pest resistance		45	
Post-harvest sprouting resistance	Moderate (31–40)	42	
Panicle architecture		40	
Seed size		40	Seed size (in mm²)
Plant architecture		35	
Leaf colour		33	
Saponin amount		33	
Uniformity	Low (0–30)	29	
Seed colour		27	$L\star$, $a\star$, $b\star$ with weight of 9 for each colour attribute
Heat tolerance		26	Flowering DAS
Height at harvest		26	Height (in cm)
Flavour★		25	

Source: Kellogg, 2017.

Notes
★Not all participants added 'flavour' to their list of traits, giving this trait a low cumulative weight.
Participants rated traits from 0 to 5: '5' for very high importance; '0' being not important. Ratings were added up to create a cumulative weight to apply to the selection index.

Results

For the breeding methods trial, the ANOVA of the F_4 populations grown in 2015 showed that the breeding method did not have a significant effect on yield, but population did (Table 10.3).[12] The ANOVA of the F_5 populations grown in 2016 showed the converse: the breeding method had a significant effect on yield, but population did not (Table 10.3). The EPB method resulted in higher yields than the EB method.[13]

When the populations were compared against the parental varieties, the ANOVA of the F_4 EB populations and parental varieties revealed a significant entry (populations and parental varieties) effect on yield (Table 10.3). Only one population was significantly higher than a parental variety.[14] The ANOVA of the F_5 populations and parental varieties also showed a significant entry effect (Table 10.3): two populations were significantly higher than the parental variety.[15]

For the Chimacum pedigree selection trial, entry (check varieties and breeding lines) had a significant effect on the following response variables: flowering DAS, seed–set DAS, harvest DAS, 1,000-seed weight, seed size, height at harvest, and colour channels 'L^*', 'a^*' and 'b^*'. Out of the 92 breeding lines, 71 were found to differ from Cherry Vanilla for at least one trait.[16] For the Quilcene pedigree selection trial, entry had a significant effect on the following response variables: response variables of flowering DAS, seed–set DAS, harvest DAS, 1,000-seed weight, seed size, height at harvest and colour channels 'L^*', 'a^*', and 'b^*'. Seventy breeding lines were found to be different from Cherry Vanilla for one or more of the traits. For the Sequim pedigree selection trial, entry had a significant effect on the following response variables: flowering DAS, 1,000-seed weight, seed size, height at harvest, and colour channels 'L^*', 'a^*', and 'b^*'. Thirty-two breeding lines were found to be different from Cherry Vanilla for one or more of the traits.

Discussion

Breeding methods trial

Evaluation of F_4 and F_5 populations

In 2015, a yield difference between the breeding methods had been anticipated, with lower yields expected of the EPB populations because negative selection reduced the number of plants in each EPB plot. However, no significant difference between methods was found. It is possible farmers chose to remove only plants that were late-maturing or low-yielding and that would not have contributed to population yield. The yields of two populations, P105 and P106,[17] were exceptionally low in 2015, with yields no different from other populations in both the EB and EPB treatments in 2016 (Table 10.4). P105 and P106 had a high percentage of late-maturing genotypes in

Table 10.3 ANOVA for yield data of F4 and F5 populations and parental varieties, breeding methods trial, US Pacific Northwest

7a. *EB and EPB populations in 2015. Analysis of square root transformed yield data.*

	Df	Sum Sq	Mean Sq	F value	P (>F)
Method	1	109.5	109.49	1.637	0.270
Population	5	3747.9	749.58	11.207	2.985e-05★★★
Method: Population	5	184.4	36.88	0.552	0.736

7b. *EB populations and parental varieties in 2015. Analysis of square root transformed yield data.*

	Df	Sum Sq	Mean Sq	F value	P (>F)
Entry (Populations, Parental varieties)	11	4286.4	389.67	4.905	0.002★★

7c. *EB and EPB populations in 2016. Analysis of raw yield data.*

	Df	Sum Sq	Mean Sq	F value	P (>F)
Method	1	1245161	1245161	6.344	0.019★
Population	5	2053269	410654	2.092	0.101
Method : Population	5	1401197	280239	1.428	0.250

7d. *EPB populations and parental varieties in 2016. Analysis of log transformed yield data.*

	Df	Sum Sq	Mean Sq	F value	P (>F)
Entry (Populations, Parental varieties)	11	1.003	0.091	2.716	0.036★

Source: Kellogg, 2017.

Notes
Type III error with Satterthwaite approximation for degrees of freedom. Significance codes: '★★★' 0.001, '★★' 0.01.

Table 10.4 Mean yield (kgha–1) of F4 and F5 populations and parental varieties, breeding methods trial, US Pacific Northwest

Population	Female parent	Male parent	EB 2015	EPB 2015	EB 2016	EPB 2016	Parental variety	2015	2016
P102	CO407 Dave	QQ74	862.35	1595.36	1717.80	2006.41	Black	375.44	1322.17
P104	Kaslaea	QQ74	1081.57	1368.76	1847.36	2628.43	Cherry Vanilla	11.76	1855.04
P105	QQ065	QQ74	137.68	105.10	1299.93	1676.87	CO407 Dave	1363.12	1960.81
P106	QQ065	Black	77.09	152.93	1738.09	1942.95	Kaslaea	798.56	2422.67
P107	QQ74	Black	335.17	418.34	1419.72	2309.36	QQ065	108.40	1660.95
P108	QQ74	Cherry Vanilla	641.12	1026.81	2283.56	1974.34	QQ74	1203.60	2241.10

Source: Kellogg, 2017.

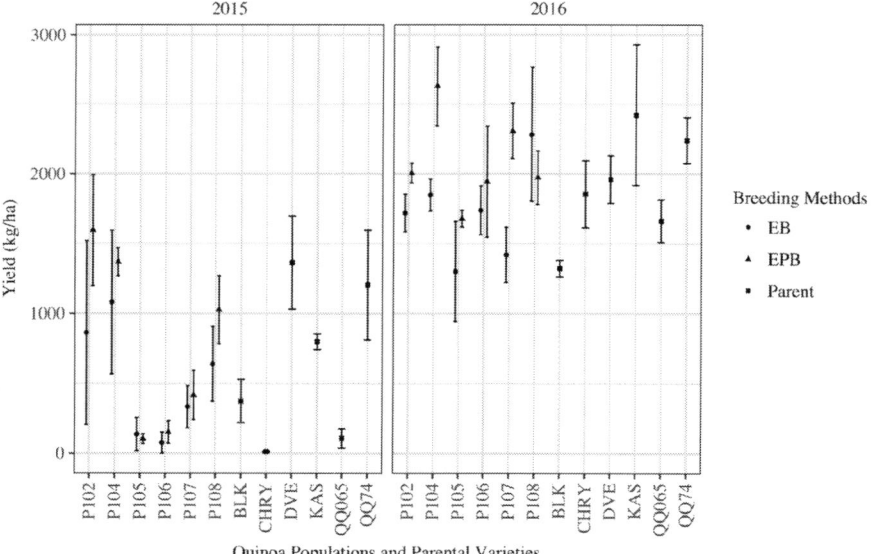

Figure 10.2 Yield (kg ha-1) of quinoa populations treated with EB and EPB methods and parental varieties, US Pacific Northwest.

Source: Kellogg, 2017.

Note: Data represented were not transformed. The mean of three replicates (*point*; ±SE) is reported.

2015. As the trial had been harvested prior to complete senescence of all plots due to inclement weather in 2015, unconscious selection for earlier-maturing genotypes might explain the increase in yield in 2016. Additionally, the soils of the 2016 trial were sandy loam and loam sand with lower water-holding capacity than the gravelly loam soil of the 2015 trial. The amount of precipitation starting on 1 August, when plants began to set seed prior to harvest, was similar in both 2015 and 2016 (12.95 mm and 16.76 mm, respectively). In 2016, the farmer terminated irrigation after flowering and, although precipitation was similar prior to harvest, the soils in Sequim may not have held the additional late-season rainfall that allowed plants to mature earlier than they had in 2015. The 2016 trial also had a higher number of degree-days measuring from 1 August up to harvest (751.25) than the 2015 trial (431.74), and the soils in Sequim tested at lower levels of fertility.

In 2016, an interaction between method and population was expected, as some populations might have received higher negative selection pressure than others. However, neither the interaction nor the population effect proved significant. Analysing the population effect, the high means of P104 and P107 with the EPB method are diluted by the lower means of these populations with the EB method.[18]

Evaluation of F$_4$ and F$_5$ populations and parental varieties

In 2015, the F$_4$ of P104 with no participatory selection applied to it (EB treatment) yielded greater than Cherry Vanilla. However, in 2016, the F$_5$ of P104 subjected to participatory selection (EPB treatment) did not yield greater than Cherry Vanilla: in fact, the F$_5$ of the EPB populations P104 and P107 yielded greater than Black. In 2015, Cherry Vanilla and QQ065 had exceptionally low yields because they had been late in maturing and were harvested prematurely due to impending precipitation. Cherry Vanilla and QQ065 are not completely uniform varieties; therefore, it is possible that, like P105 and P106, the two parental varieties were subject to selection pressures and altered by the early harvest in 2015. In addition, the soil and climatic conditions of the 2016 trial might have caused earlier maturity of these two parental varieties, resulting in higher yields.

In 2016, Black was surpassed in yield by its progeny population P107. Kaslaea and QQ74 also outyielded Black. It is important to note that EB trials in other crops, many more than six generations of natural selection occurred to improve agronomic fitness and outperform the parent varieties. Suneson (1956) originally determined that a minimum of 15 generations of natural selection was needed prior to furthering the breeding project with (1) continued cycles of natural selection to improve population yield, (2) additional hybridization with natural selection, or (3) selection and testing of breeding lines. In some EB projects, heterogeneous populations were not suitable for becoming heterogeneous cultivars because yield was not sufficiently high and they were not stable (Phillips and Wolfe, 2005). In such cases, it was suggested that the populations should be continued for use as breeding populations from which individual lines could be derived. It is for that reason that pedigree selection was conducted on populations in this study.

The heterogeneous quinoa populations selected and evaluated will continue to be grown in Washington State for further evaluation and continued natural selection. The end-use of the heterogeneous populations is less straightforward than with a uniform quinoa variety. The US Plant Variety Protection Act requires registered varieties to be new, distinct, uniform and stable (7 US Code § 2402), prohibiting the registration of a heterogeneous variety resulting from the EPB method. Without variety registration, such heterogeneous varieties are ineligible for seed certification in Washington (WAC 16–302–040) and in other US states. Despite these regulations, heterogeneous varieties from EPB programmes are useful in the unofficial seed sector of farmer seed exchanges and can be sold as non-certified seed. Moreover, a heterogeneous variety can be released as 'germplasm' and added to the National Plant Germplasm System. This was the case with the 'STRKR' barley germplasm developed by Oregon State University (Meints *et al.*, 2015) and eight open-source carrot composite populations developed by the University of Wisconsin-Madison (Luby and Goldman, 2016).

Pedigree selection trial

Evaluation of the F_5 pedigree selections

Unlike typical breeding projects of self-pollinating (inbreeding) crops, some of the quinoa breeding lines were heterogeneous and required pre-harvest rogueing. Selection was conducted in plots of heterogeneous populations sown to a high density for several generations; quinoa outcrossing rates have been documented to range from 0.5 per cent to as high as 17.36 per cent (Gandarillas, 1979; Silvestri and Gil, 2000). Therefore, some of the breeding lines may not represent a F_5, but instead an earlier generation of different yet related parentage.

Using the farmer-driven selection index (Table 10.2), the five breeding lines receiving the highest selection index score in Chimacum and Sequim came from P102 and P107 (Table 10.5). In Quilcene, the top five breeding lines came from P105, P106, and P107. However, breeding lines with high selection index scores were expected of P108 progeny. Participants initially made 213 selections in 2015, but only 92 were of plants that produced 13 g of seed or more. Of these 92 selections, 24 were selected out of P102, 27 out of P104, 2 out of P105, 5 out of P106, 11 out of P107, and 23 out of P108, for entry in the 2016 pedigree selections trial. The largest number of selections made by participants came from P102, P104, and P108, and high-ranking breeding lines were expected from these three populations. These populations displayed average yields higher than P105 and P106 in 2015, the year the selections were made, and probably had plants with larger seed bearing panicles containing larger amounts of seed. However, no breeding lines from P108 outperformed Cherry Vanilla. The parental varieties of P108 are QQ74 and Cherry Vanilla, and the farmer-driven selection index results showed that Cherry Vanilla progeny were not significantly different from the parental variety for the traits measured. This may indicate that the selection index failed to capture the importance of traits for which these breeding lines were initially selected.

Out of the 15 breeding lines selected in for all three locations, 5 breeding lines came from P107.[19] Two of these surpassed Cherry Vanilla at all three locations (WWA15FS3, WWA15FS7), and displayed broad adaptability across the Olympic Peninsula region. These two lines proved short and uniform at all locations. In Chimacum, researchers noted that WWA15FS7 had no *Lygus spp* damage to the seed-bearing panicle, possibly an avoidance mechanism because WWA15FS7 was a short entry surrounded by taller entries. The absence of *Lygus spp* damage would contribute to higher 1,000-seed weights and seed size. WWA15FS65, selected for the Chimacum and Sequim locations, was early to flower in Sequim and uniform in Chimacum. WWA15FS67, selected for Chimacum, was early to flower and mature, with no *Lygus spp* damage to the seed-bearing panicle. WWA15FS78, selected for Quilcene and Sequim, flowered early at both locations. P107 comes from a cross with the USDA accession QQ74 from Maule, Chile, and variety Black from White Mountain Farm in

Table 10.5 Breeding lines, five highest selection index scores, 2016 pedigree selection trial, US Pacific Northwest

CHI

Entry	SIS	PO	PR	RR	NR	SC	PC	PA
WWA15FS3	199	107	F	4	Uniform, short, early to mature	Brown	Orange	No branching
WWA15FS7	216	107	R	2	Early to flower, no *Lygus spp* damage	White	Pink	No branching
WWA15FS13	138	102	FR	0	Early to flower	Red	White Green	No branching
WWA15FS65	140	107	R	2	Uniform	Brown	Orange	No branching
WWA15FS67	216	107	F	4	Early to flower and mature, no *Lygus spp* damage	White	Orange	No branching

QUL

Entry	SIS	PO	PR	RR	NR	SC	PC	PA
WWA15FS3	185	107	F	1	Short, early maturing	Brown	Orange	No branching
WWA15FS7	211	107	R	2	Short	White	Pink	No branching
WWA15FS37	100	106	FR	0	Early to flower	Black	Orange	No branching
WWA15FS78	131	107	F	0	Early to flower	Brown	Orange	No branching
WWA15FS92	147	105	R	2	Early to flower, set seed, and mature	Brown	White	No branching

SEQ

Entry	SIS	PO	PR	RR	NR	SC	PC	PA
WWA15FS3	105	107	F	3	Uniform, short, early maturing	Brown	Orange	No branching
WWA15FS7	122	107	R	3	Uniform, short, early to flower	White	Pink	No branching
WWA15FS14	67	102	F	2	Red foliage	Brown	Red	No branching
WWA15FS65	105	107	R	0	Uniform, early to flower	Brown	Orange	No branching
WWA15FS78	114	107	F	0	Early to flower	Brown	Orange	No branching

Source: Kellogg, 2017.

Notes
Descriptions of the breeding lines include: selection index score (SIS), population of origin (PO), type of participant (farmer (F), researcher (R), or farmer and researcher (FR)) who made initial selection (PR), rating of entry (0–4) made by researchers at each location based on uniformity and uniqueness prior to harvest at each location in 2016 (RR), notable remarks made by researchers in the field (NR), seed colour(SC), panicle colour of mature plants (PC), plant architecture (PA).

Mosca, Colorado. From past variety trial observations, QQ74 is a nearly uniform accession with white seed. Black is a noticeably heterogeneous farmer-maintained variety with mostly black-seeded plants. The EPB F_5 of P107 yielded higher than Black in 2016. This study captures only two years of evaluations of heterogeneous populations and one year towards development of pure lines selected from these populations; however, the performances of P107 yielding higher than one of its parental varieties and the numerous breeding lines from P107 outperforming Cherry Vanilla support Phillips and Wolfe's (2005) conclusion that the development of heterogeneous populations (typically through composite crosses) may be an efficient breeding method that can result in heterogeneous crops and pure lines selected for low-input agro-ecosystems.

P102, a cross with QQ74 and nearly homogeneous variety CO407Dave from Mosca, CO, produced WWA15FS13. This breeding line selected for Chimacum was notable for early flowering. Selected for Sequim was WWA15FS14, with conspicuously large seeds and high 1,000-seed weight. P106, a cross with the heterogeneous USDA accession QQ065 from Los Lagos, Chile, and Black, produced WWA15FS37, which was selected for Quilcene and noted as early-flowering. P105, from a cross with USDA accessions QQ065 and QQ74, produced WWA15FS92, notably early to flower, set seed and mature, and was selected for Quilcene.

Farmer-driven selection index

The farmer-driven selection index was applied to the results of the Dunnett contrasts for each site. The five breeding lines that received the highest selection index score were determined to be the top-performing pedigree selections for each location (Table 10.5). For each trial location, the number of top five performing selections originally selected by farmers was equal to or higher than the number selected by researchers (Table 10.5). For example, as shown in the column in Table 10.5 that identifies the type of participant who initially selected the individual quinoa plant in 2015, three of the five quinoa selections were made by farmers: this demonstrates the importance of farmer participation in this study. With the farmer-driven selection index, the top selections for each location may have been biased towards farmer-selected breeding lines simply because more farmers participated than researchers.

Prior to harvest in 2016, the participating researchers in this study independently rated the breeding lines at each location based on uniformity and uniqueness. The rating was intended to help identify a breeding line preferred for further use in the breeding programme. Researchers gave a rating of '4' as the highest. Table 10.5 shows that only two breeding lines selected using the selection index as top-performing also received a researcher rating of '4'. This may reflect the disparity between researcher preference and the results from the farmer-driven selection index. Ceccarelli *et al.* (2001) conducted experiments in participatory plant breeding with barley to test whether farmers' selection was effective, and whether farmers' and researchers' selection criteria differed. They found that (1) the preconceived idea that farmers are not able to evaluate a large number of plots was disproven, (2) farmers effectively identified some of the highest-yielding entries at the decentralized locations, (3) farmer selection was most effective when farmers conducted selection in groups rather than as individuals, and (4) the farmers used broader selection criteria than the researchers. With the farmer-driven selection index for quinoa, this study has not specifically examined farmer vs. researcher selection criteria, but does shed light on the disparity in preferences. The aim was to develop a collaborative breeding project with farmers as a driving force in selection pressure and breeding-line development. Next steps in quinoa variety development of the breeding lines include entering the breeding lines in multi-location variety trials.

Ranking of traits using the farmer-driven selection index

Participants and researchers rated quinoa traits, and the ratings were added up for each trait. The added value became the trait weight in the selection index; weights ranged from 9 to 47. This helped to distinguish the traits most important to the breeding project. Researchers then identified the traits quantifiable in the field or lab that could achieve the quinoa traits desired by the participants, as follows.

Highly important traits

One of the most important traits, early maturity, was given a weight of 47. Early-maturing quinoa genotypes are important for farmers in western Washington, where there are often early autumn rains in late August and early September. Precipitation or high humidity can cause mature quinoa seeds to sprout, resulting in partial to entire crop loss. Early seed set is measured as the DAS that seed set occurs. Seed-set DAS is included in this selection index as a trait that can confer early maturity, because plants with mature seed (but not dried biomass) can be mechanically swathed or hand-harvested, depending on the scale of the farming operation. Plants can then be dried in the field or in covered storage, allowing an earlier harvest than is possible with a combine.

Jacobsen and Christiansen (2016) observed that white quinoa seeds are lightest in colour if harvested when seeds first mature, typically several weeks before plant stems have dried. This not only improved the quality of the harvest, but also increased the germination rates of the seed. However, they also found that an early harvest resulted in lower seed protein, because protein continued to increase until plant maturity. Early harvest is included as a trait that represents early maturity for farmers who rely on both mature quinoa seed and dried biomass for a successful combine harvest. The yield component trait of 1,000-seed weight is given a weight of 47. Seed weight has been shown to contribute to yield in quinoa (De Santis *et al.*, 2016). However, agronomic practices and environment affect the stability of this trait. Alandia *et al.* (2016) did not find an effect of nitrogen fertilizer or watering conditions on 1,000-seed weight in a pot experiment. With field-grown quinoa, however, Jacobsen and Christiansen (2016) observed increased seed weight as levels of nitrogen fertilizer increased; split applications of nitrogen increased seed weight, but not yield. Miranda *et al.* (2013) found that cold-temperate environments increased 1,000-seed weight.

MODERATELY IMPORTANT TRAITS

Seed size, given a weight of 40, is a trait that does not confer high yields, but does determine commercial quality. Bertero *et al.* (2004) found no association between grain yield and seed size in quinoa, indicating that gains can be made in yield and seed size through simultaneous selection. No additional traits that farmer participants considered moderately important were statistically analysed, due to in-field rogueing prior to harvest.

LESS-IMPORTANT TRAITS

Seed colour, given a weight of 27,[20] determines the commercial quality of white and yellow seeded[21] quinoas and the nutritional value of brown, red, and black seeded[22] quinoas. Tang *et al.* (2015) reported that the pigments of red and black quinoa seeds are betacyanins, shown to have unique health-promoting properties, and are not anthocyanins. Additionally, the darker the quinoa seeds, the higher the phenolic concentration and antioxidant activity.

Heat tolerance was given a weight of 26. In this study, a stress-avoidance mechanism was targeted to confer heat tolerance. One such mechanism is early flowering, which was measured as flowering DAS. High temperatures during flowering can cause flower abortion and pollen sterility (Hinojosa *et al.*, 2018a; Hinojosa *et al.*, 2018b; Jacobsen and Stølen, 1993). In the USA, quinoa seed-yield losses caused by crop exposure to temperatures greater than 35°C during flowering have been recorded in Colorado (Johnson and Croissant, 1985), eastern Washington (Murphy *et al.*, 2016), Minnesota (Oelke *et al.*, 1992) and Utah (Buckland, 2016).

Plant height at harvest, accorded a weight of 26, is an important consideration to farmers who hand-harvest or mechanically harvest the quinoa crop. A tall quinoa plant can be cumbersome to hand-harvest or store for further drying, and can result in increased biomass accumulation in a combine during mechanical harvest. However, quinoa has other uses beyond seed consumption. In some rural Andean communities, a tall quinoa plant is preferable for future use of the dried plant stalks as fuel for home cooking (Nicholas Pichazaca, Mushuk Yuyay Association, personal communication). Diversified small farms may choose to grow quinoa as a multi-purpose crop by harvesting leaves from young plants intended for seed; this will require a plant with greater biomass than one bred solely for seed harvest (Kellogg and Murphy, 2018).

Conclusions

In view of the limited quinoa varieties available to farmers within the USA, this study aimed to develop heterogeneous populations specifically adapted to organic farms within the Pacific Northwest region, and, using seed from individual plants selected by farmers, to develop uniform varieties. In the second year of the study, the EPB method resulted in higher population yields than the EB method. With few instructive local resources detailing how to grow quinoa organically, we relied heavily on collaboration with organic farmers, who proved instrumental in the development of a selection index that established traits important in this breeding project and in management of the trials. Participatory single-plant selection in heterogenous populations resulted in the identification of nine unique breeding lines suited to three distinct locations, using the farmer-driven selection index. Researchers and farmers alike gained critical insights into quinoa production, and many participating farmers grew successful commercial quinoa crops during the 2015 and 2016 trial years.

Notes

1 The six bi-parental quinoa populations created in 2012 were named P102, P104, P105, P106, P107 and P108.
2 Breeding method was the whole plot factor, and population was the sub-plot factor with three replications. Population was nested within breeding method, within replication.
3 Each research plot was $4.88\,m \times 1.22\,m$, into which quinoa seed was planted in three rows spaced at 40.64 cm from centre at a seeding rate of $1.35\,g\,m^{-2}$.
4 Breeding lines were hand-sown as single head rows 4.9 m in length and 40.64 cm from centre at a seeding rate of $4\,g\,row\,m^{-1}$.
5 Seed size (mm^2) was analysed with the free ImageJ software (Schneider *et al.*, 2012). Colour of unprocessed seeds was measured in the CIELAB colour space using a Minolta CR-310 chroma meter.
6 Yield data for the F_4 and F_5 populations were analysed as a split-plot RCBD using the free R software (R Core Team 2016) and the packages 'lme4' (Bates *et al.*, 2015) and 'lmerTest' (Kuznetsova *et al.*, 2016). The t-tests used Satterthwaite approximations for degrees of freedom. Multiple comparisons of yield means were performed using Tukey contrasts for significant effects with the 'multcomp' package (Hothorn *et al.*, 2008). Yield data of the F_4 populations in 2015 were transformed using the square root of the values prior to data analysis, to accommodate model assumptions of normally distributed error and homogeneity of variance. Yield data of the F_5 populations in 2016 were not transformed.
7 Yield data were square-root transformed.
8 Yield data were log-transformed.
9 The R software and packages 'lme4' and 'lmerTest' were used to perform an ANOVA with a linear mixed effects model fit by residual maximum likelihood. The t-tests used Satterthwaite approximations for degrees of freedom. Multiple comparisons of yield means were performed using Tukey contrasts with the 'multcomp' package.
10 Analysis was conducted using R software and the function 'aov'.
11 The R package 'multcomp' was used to conduct multiple comparisons of means, using Dunnett contrasts to test for breeding lines that exceeded the check Cherry Vanilla.

12 Tukey contrasts on the F_4 population data showed that the yields of P102 (1228.86 kg ha^{-1}), P104 (1225.16 kg ha^{-1}), and P108 (833.96 kg ha^{-1}) were greater than the yields of P105 (121.39 kg ha^{-1}) and P106 (115.01 kg ha^{-1}). P102 and P104 yields were also greater than the yield of P107 (376.75 kg ha^{-1}).

13 In 2016, the EPB method yielded 2089.73 kg ha^{-1} and the EB method yielded 1717.77 kg ha^{-1}.

14 Results from Tukey contrasts showed the yield of P104 [1081.57 kg ha^{-1} (EB)] was higher than Cherry Vanilla (11.76 kg ha^{-1}) (Table 10.3; Figure 10.2).

15 Results from Tukey contrasts showed P104 [2628.43 kg ha^{-1} (EPB)] and P107 [2309.36 kg ha^{-1} (EPB)] had higher yields than Black (1322.17 kg ha^{-1}).

16 Dunnett contrasts were used for the analysis.

17 In 2015, P105 yielded 137.68 kg ha^{-1} (EB), 105.10 kg ha^{-1} (EPB) and P106 yielded 77.09 kg ha^{-1} (EB), 152.93 kg ha^{-1} (EPB).

18 P104: 1847.54 kg ha^{-1} (EB), 2628.43 kg ha^{-1} (EPB); P107: 1419.72 kg ha^{-1} (EB), 2309.36 kg ha^{-1} (EPB).

19 P107-derived breeding lines: WWA15FS3, WWA15FS7, WWA15FS65, WWA15FS67, WWA15FS78.

20 The weight given to seed colour was divided up by the colour channels L^*, a^*, and b^*; each channel received a weight of 9.

21 In this study, white and yellow seeds with 'L^*' values in the 40s were observably dark, possibly reducing their commercial viability. Discoloration of white and yellow seeds could be related to time of harvest and secondary fungi (Jacobsen and Christiansen, 2016).

22 Seeds with black, brown and red seed coats have 'L^*' values in the 20s, 30s and low 40s, in this study, with higher values indicating lighter-coloured pericarps. Higher positive values of colour channels 'a^*' and 'b^*' further differentiate the unprocessed seed colours of magenta and yellow, respectively.

References

Alandia G, Jacobsen S-E, Kyvsgaard NC, Condori B, Liu F, 2016. Nitrogen sustains seed yield of quinoa under intermediate drought. *Journal of Agronomy and Crop Science* 202, 281–291. doi:10.1111/jac.12155.

Allard RW, 1999. Evolution during domestication. In Allard RW, *Principles of Plant Breeding*, 2nd edition, 24–35. Hoboken, NJ: Wiley & Sons.

Allard RW, Hansche PE, 1964. Some parameters of population variability and their implications in plant breeding. *Advances in Agronomy* 16, 325–328.

Allard RW, Zhang Q, Saghai Maroof MA, Muona OM, 1992. Evolution of multi-locos genetic structure in an experimental barley population. *Genetics* 131, 957–969.

Bates D, Mächler M, Bolker B, Walker S, 2015. Fitting linear mixed-effects models using lme4. *Journal of Statistical Software* 67 (1), 1–48. doi:10.18637/jss.v067.i01.

Bazile D, Jacobsen S-E, Verniau A, 2016a. The global expansion of quinoa: Trends and limits. *Frontiers in Plant Science* 7, 622. doi: 10.3389/fpls.2016.00622.

Bazile D, Pulvento C, Verniau A, Al-Nusairi MS, Ba D, Breidy J, Hassan L *et al.*, 2016b. Worldwide evaluations of quinoa: Preliminary results from post international year of quinoa FAO projects in nine countries. *Frontiers in Plant Science* 7, 850. doi: 10.3389/fpls.2016.00850.

Bertero HD, de la Vega AJ, Correa G, Jacobsen S-E, Mujica A, 2004. Genotype and genotype-by-environment interaction effects for grain yield and grain size of quinoa (*Chenopodium quinoa* Willd.) as revealed by pattern analysis of international multi-environment trials. *Field Crops Research* 89 (2–3), 299–318. doi:10.1016/j. fcr.2004.02.006.

Buckland K, 2016. Increasing the sustainability of Utah farms by incorporating quinoa as a novel crop and protecting soil health. Dissertation, Utah State University.

Campanelli G, Acciarri N, Campion B, Delvecchio S, Leteo F, Fusari F, Angelini P, Ceccarelli S, 2015. Participatory tomato breeding for organic conditions in Italy. *Euphytica* 204, 179–197. doi:10.1007/s10681-015-1362-y.

Ceccarelli S, 2009. Evolution, plant breeding and biodiversity. *Journal of Agriculture and Environment for International Development* 103 (1/2), 131–145. doi:10.12895/ jaeid.20091/2.28.

Ceccarelli S, Grando S, Bailey E, Amri A, El-Felah M, Nassif F, Rezgui S, Yahyaoui A, 2001. Farmer participation in barley breeding in Syria, Morocco and Tunisia. *Euphytica* 122, 521–536. doi: 10.1023/A:1017570702689.

Corte HR, Ramalhol MAP, Goncalves FMA, Abreu ADFB. 2002. Natural selection for grain yield in dry bean populations bred by the bulk method. *Euphytica* 123, 387–393. doi: 10.1023/A:1015065815131.

Dawson JC, Rivière P, Berthellot J, Mercier F, de Kochko P, Galic N, Pin S, Serpolay E, Thomas M, Giuliano S, Goldringer I, 2011. Collaborative plant breeding for organic agricultural systems in developed countries. *Sustainability* 3 (8), 1206–1223. doi:10.3390/su3081206.

De Santis G, D'Ambrosio T, Rinaldi M, Rascio A, 2016. Heritabilities of morphological and quality traits and interrelationships with yield in quinoa (*Chenopodium quinoa* Willd.) genotypes in the Mediterranean environment. *Journal of Cereal Science* 70, 177–185. doi:10.1016/j.jcs.2016.06.003.

Detjens A, 2016. Experiences and perceptions of quinoa production in the western United States. Thesis, Washington State University.

Döring TF, Knapp S, Kovacs G, Murphy K, Wolfe MS, 2011. Evolutionary plant breeding in cereals – into a new era. *Sustainability* 3, 1944–1971. doi:10.3390/ su3101944.

Entz MH, Kirk AP, Vaisman I, Fox SL, Fetch JM, Hobson D, Jensen HR, Rabinowicz J, 2015. Farmer participation in plant breeding for Canadian organic crop production: Implications for adaptation to climate uncertainty. *Procedia Environmental Sciences* 29, 238–239. doi:10.1016/j.proenv.2015.07.291.

FAO, 2017. *International Treaty on Plant Genetic Resources for Food and Agriculture.* Available at www.fao.org/plant-treaty/en. Accessed 28 February 2017.

Federer WT, 1956. Augmented (or hoonuiaku) designs. *Hawaiian Planters' Record* 55 (2), 191–208.

Gandarillas H, 1979. Mejoramiento genetico. In Tapia ME, (ed) *Quinua y Kaniwa: Cultivos Andinos*, serie libros y materiales educativos, 65–82. Bogota: Instituto Interamericano de Ciencias Agricolas.

Ghaouti L, Vogt-Kaute W, Link W, 2008. Development of locally-adapted faba bean cultivars for organic conditions in Germany through a participatory breeding approach: Participatory breeding of faba bean for organic conditions. *Euphytica* 162 (2), 257–268. doi:10.1007/s10681-007-9603-3.

Harlan HV, Martini ML, 1929. A composite hybrid mixture. *Journal of American Society of Agronomy* 21, 487–490.

Hinojosa L, González JA, Barrios-Masias FH, Fuentes F, Murphy KM, 2018a. Quinoa abiotic stress responses: A review. *Plants* 7, 106. doi:10.3390/ plants7040106.

Hinojosa L, Matanguihan JB, Murphy KM, 2018b. Effect of high temperature on pollen morphology, plant growth and seed yield in quinoa (*Chenopodium quinoa* Willd.). *Journal of Agronomy and Crop Science* (published online 18 September 2018). doi.org/10.1111/jac.12302.

Hothorn T, Bretz F, Westfall P, 2008. Simultaneous inference in general parametric models. *Biometrical Journal* 50 (3), 346–363. doi:10.1002/bimj.200810425.

Jacobsen S-E, Stølen O, 1993. Quinoa – morphology, phenology and prospects for its production as a new crop in Europe. *European Journal of Agronomy* 2 (1), 19–29. doi:10.1016/S1161-0301(14)80148-2.

Jacobsen S-E, Christiansen JL, 2016. Some agronomic strategies for organic quinoa (*Chenopodium quinoa* Willd.). *Journal of Agronomy and Crop Science* 202 (6), 454–463. doi:10.1111/jac.12174.

Johnson DL, Croissant RL, 1985. *Quinoa production in Colorado.* Fort Collins: Colorado State University Extension Service-In-Action No. 112.

Kellogg JA, 2017. Evolutionary participatory quinoa breeding for organic agroecosystems in the Pacific Northwest region of the United States. Thesis, Pullman: Washington State University.

Kellogg JA, Murphy KM, 2018. Grains: Growing quinoa in home gardens. Washington State University Extension Publication FS258e. cru.cahe.wsu.edu/CEPublications/FS258E/FS258E.pdf.

Kuznetsova A, Brockhoff PB, Christensen RHB, 2016. lmerTest: Tests in linear mixed effects models. R package version 2.0–33 https://CRAN.R-project.org/package=lmerTest. Accessed 28 February 2017.

Lammerts van Bueren ET, Struik PC, Jacobsen E, 2002. Ecological concepts in organic farming and their consequences for an organic crop ideotype. *Netherlands Journal of Agricultural Science* 50 (1), 1–26. doi:10.1016/S1573-5214(02)80001-X.

Lammerts van Bueren ET, Jones SS, Tamm L, Murphy KM, Myers JR, Leifert C, Messmer MM, 2011. The need to breed crop varieties suitable for organic farming, using wheat, tomato and broccoli as examples: A review. *NJAS – Wageningen Journal of Life Sciences* 58 (3–4), 193–205. doi:10.1016/j.njas.2010.04.001.

Luby CH, Goldman IL, 2016. Release of eight open source carrot (*Daucus carota* var. sativa) composite populations developed under organic conditions. *HortScience* 51 (4), 448–450.

Mazourek M, Moriarty G, Glos M, Fink M, Kreitinger M, Henderson E, Palmer G, Chicring A, Rumore DL, Kean D, Myers JR, Murphy JF, Kramer C, Jahn M, 2009. 'Peacework': A cucumber mosaic virus-resistant early red bell pepper for organic systems. *HortScience* 44 (5), 1464–1467.

Meints B, Brouwer BO, Brown B, Cuesta-Marcos A, Jones SS, Kolding M, Fisk S et al., 2015. Registration of #STRKR Barley Germplasm. *Journal of Plant Registrations* 9, 388–392. doi:10.3198/jpr2014.09.0066crg.

Mendum R, Glenna LL, 2010. Socioeconomic obstacles to establishing a participatory plant breeding program for organic growers in the United States. *Sustainability* 2, 73–91. doi:10.3390/su2010073.

Miranda M, Vega-Gálvez A, Martínez EA, López J, Marín R, Aranda M, Fuentes F, 2013. Influence of contrasting environments on seed composition of two quinoa genotypes: Nutritional and functional properties. *Chilean Journal of Agricultural Research* 73 (2), 108–116. doi:10.4067/S0718-58392013000200004.

Morris ML, Bellon MR, 2004. Participatory plant breeding research: Opportunities and challenges for the international crop improvement system. *Euphytica* 136 (1), 21–35. doi:10.1023/B:EUPH.0000019509.37769.b1.

Murphy KM, Campbell KG, Lyon SR, Jones SS, 2007. Evidence of varietal adaptation to organic farming systems. *Field Crops Research* 102, 172–177. doi:10.1016/j.fcr.2007.03.011.

Murphy KM, Lammer D, Lyon S, Carter B, Jones SS, 2005. Breeding for organic and low-input farming systems: An evolutionary-participatory breeding method for inbred cereal grains. *Renewable Agriculture and Food Systems* 20 (1), 48–55. doi:10.1079/RAF200486.

Murphy KM, Bazile D, Kellogg J, Rahmanian M, 2016. Development of a worldwide consortium on evolutionary participatory breeding in quinoa. *Frontiers in Plant Science* 7, 608. doi:10.3389/fpls.2016.00608.

Oelke EA, Putnam DH, Teynor TM, Oplinger ES, 1992. Quinoa. In *Alternative Field Crops Manual*. University of Wisconsin and University of Minnesota. Available at http://corn.agronomy.wisc.edu/Crops/Quinoa.aspx. Accessed 28 February 2017.

Peterson A, Murphy KM, 2015. Quinoa in the United States of America and Canada. In Bazile D, Bertero HD, Nieto C, (eds) *State of the Art Report on Quinoa Around the World in 2013*, 549–561. Rome: FAO.

Peterson A, Jacobsen S-E, Bonifacio A, Murphy K, 2015. A crossing method for quinoa. *Sustainability* 7 (3), 3230–3243. doi:10.3390/su7033230.

Phillips SL, Wolfe MS, 2005. Evolutionary plant breeding for low input systems. *Journal of Agricultural Science* 143, 245–254. doi:10.1017/S0021859605005009.

R Core Team, 2016. R: A language and environment for statistical computing. Vienna, Australia: R Foundation for Statistical Computing. www.R-project.org/. Accessed 28 February 2017.

Schneider CA, Rasband WS, Eliceiri KW, 2012. NIH Image to ImageJ: 25 years of image analysis. *Nature Methods* 9 (7), 671–675. doi:10.1038/nmeth.2089.

Shelton AC, Tracy WF, 2016. Participatory plant breeding and organic agriculture: A synergistic model for organic variety development in the United States. *Elementa: Science of the Anthropocene* 4, 143. doi:10.12952/journal.elementa.000143.

Silvestri V, Gil F, 2000. Alogamia en quinua. Tasa en Mendoza (Argentina). Universidad Nacional de Cuyo, *Revista de la facultad de Ciencias Agrarias*, 71–76.

Singmann H, Bolker B, Westfall J, Frederik A, 2016. afex: Analysis of factorial experiments. R package version 0.16–1. https://CRAN.R-project.org/package=afex. Accessed 28 February 2017.

Suneson CA, 1956. An evolutionary plant breeding method. *Agronomy Journal* 48, 188–191.

Suneson CA, Stevens H, 1953. Studies with bulked hybrid populations of barley. Washington, DC: United States Department of Agriculture, *Technical Bulletin* No. 1067.

Tang Y, Li X, Zhang B, Chen P, Liu R, Tsao R, 2015. Characterization of phenolics, betanins and antioxidant activities in seeds of three *Chenopodium quinoa* Willd. genotypes. *Food Chemistry* 166, 380–388. doi: 10.1016/j.foodchem.2014.06.018.

Wolfe MS, Baresel JP, Desclaux F, Goldringer I, Hoad S, Kovacs G, Löschenberger G, Miedaner T, Østergård H, Lammerts van Bueren ET, 2008. Developments in breeding cereals for organic agriculture. *Euphytica* 163 (3), 323–346. doi:10.1007/s10681-008-9690-9.

Zurita-Silva A, Fuentes F, Zamora P, Jacobsen S-E, Schwember AR, 2014. Breeding quinoa (*Chenopodium quinoa* Willd.): potential and perspectives. *Molecular Breeding* 34, 13–30. doi:10.1007/s11032-014-0023-5.

Part III

Overarching concerns and new perspectives

11 Participatory plant breeding

Human development and social reform

Rene Salazar, Gigi Manicad, Anita Dohar and Bert Visser

Introduction

Farmers and their plant genetic resources (PGR) are fully embedded within and influenced by their societies. This chapter seeks to place Participatory Plant Breeding (PPB) in the context of farmers' socio-economic and political realities. PPB cannot address only the technical challenges of managing and improving the local crop and variety portfolio by farmers: it must take into account the socio-economic and political context, if efforts are to be sustainable and meaningful. By doing this, PPB can not only improve the technical capacities of farmers and enhance their crop portfolio and food security: it should also contribute to their empowerment and to their livelihoods. Farmer Field Schools (FFSs) offer a highly suitable instrument for farmers to undertake PPB that is socially and gender inclusive.

This chapter discusses the socio-political position of smallholder farmers, the social challenges and ambitions in organizing an appropriate FFS, the tools available for farmers' understanding of their current farming systems, the setting of gender-differentiated breeding and selection objectives, and finally the various types of PPB that can be distinguished and applied. All findings reported are rooted in experience from the Sowing Diversity = Harvesting Security programme.[1]

Poverty, and the empowerment of the poor

Human beings and societies all struggle to achieve greater prosperity, peace, justice and fairness. The sustainable use of resources is instrumental in such efforts. Progress towards these goals builds and depends on the strengths of people and their institutions. Where institutions are weak, these goals get out of reach.

In *Why Nations Fail*, Acemoglu and Robinson (2013) argue that poverty and the lack of peace are not caused by the absence of resources or a population's propensity for violence, but by the ways a society is organized. They hold that rural poverty and low productivity in agriculture in the Sahel are not caused by lack of water and other natural stresses, but by the weakness of social institutions in the countries in question. Similarly, in his analysis of the roots of poverty and war in Sierra Leone, Richards (1998) concluded it is the lack of protection and security of land ownership, the lack of ensured access to these lands by smallholder farmers, and the lack of protection from labour exploitation, that have resulted in extremely low productivity in agriculture. Nobel Prize Laureate Douglass North (1990) wrote that the failure of institutions, especially of the nation state, to protect such basic human rights as the right to justice, a fair share in resources, and property rights results in poverty and societal instability.

Institutions become effective when anchored in and supported by a strong, critical and active citizenry. Therefore, the poor and weaker sectors of society need to build their own power by strengthening their organizations in order to advance their rights and interests. The 'rights–based approach' in development (Cornwall and Nyamu-Musembi, 2004; Uvin, 2007) has recognized that fairness and justice are not received as gifts by the poor but are results of exercising their power and asserting their rights. The entitlement of citizens to resources and fair rules and laws as an outcome of actions to assert rights for the poor is indeed the foundation of democracy and economic progress (Sen, 2009).

Community-level work on PPB should be understood in this wider context. Plant genetic resources and the farmers who manage them do not exist in isolation but are part of societies, so weaknesses of these societies will negatively impact on sustainable results in PGR management. In particular, women's access to and use of PGR tend to get marginalized by the same factors that discriminate against women by gender, class and ethnicity (Oxfam Novib, ANDES, CTDT, SEARICE, CGN-WUR, 2015). Women play major roles in food production and in the management of PGR, but these roles are often invisible, not recognized in agricultural institutions and policies, or rewarded by resource allocation. It is imperative to link efforts in managing PGR to the empowerment of farmers, including women, and to the reform and strengthening of the institutions of their societies. Conservation and utilization of PGR, while a major goal in its own right, becomes far more meaningful when improved management and use can contribute to social change.

In aiming to broaden the genetic base of crops and strengthening farmers' seed systems, efforts must be related to farmers' empowerment, gender equality and poverty alleviation. And here it is essential to understand the roots of poverty.

Basically, poverty is a structural phenomenon. People are poor because of how societies are organized. The triple burdens of women – productive, reproductive and social – are rooted in unequal gender relations. Access to and control over productive assets are unequal, power is concentrated in the hands of the few, and opportunities to achieve prosperity are not equally distributed. At the same time, poverty is behavioural, it is reproduced and justified. The poor tend to be dependent, to lack confidence and cohesion, and to focus on short-term goals. They are often blind to the abuse of power and selfishness of the powerful and rich. Indeed, persons who abuse their positions in unjust societies show behaviour and lead lives that may become, in the eyes of the poor, models of a successful life. And if the poor copy the behaviours of the rich and powerful, that may obstruct them from working closely together to solve their problems: 'the internalization of the oppressor image' (Freire, 1968). Many poor people fail to recognize the forces in society that make them poor, or how they themselves contribute to their poverty. They fail to understand 'context' and 'self'. This is the 'culture of silence' (Freire, 1972).

Therefore, the pedagogical aspects of an FFS on PPB are of major importance. Potentially, the greatest benefit of FFSs to PPB is their contribution to the building of critical thinking, helping farmers to understand the structures that make them and their communities poor, and to recognize how their own behavioural weaknesses ensure that they remain poor. Protecting and asserting the interest of farmers requires the development and exercise of power, because interactions between human beings and between sectors of societies are permeated by power (Bond, 2006). Contributing to the strengthening of farmers and their communities and building critical thinking are the ultimate goals of participatory action research, as in FFSs.

Thus, to be truly meaningful, PPB should serve as a platform for social reform. Smallholder farmers need to be empowered by government and other societal institutions, to intervene and participate in decision-making. FFSs in PPB can provide not just a technological methodology but a socio-political one as well. It is therefore important, when preparing and implementing PPB, to make clear the main two foci: base broadening of crop diversity, and the empowerment of farmers. These foci should inform and guide the development of gender-sensitive methodologies and tools in PPB. The importance of plant genetic diversity for resilience in agricultural production and the central role played by smallholder farmers are widely accepted, and have been included in the International Treaty on Plant Genetic Resources for Food and Agriculture (the Plant Treaty), in particular in its article on Farmers' Rights (see Chapters 16 and 17 in this book on the relationships between the Plant Treaty and PPB). A main task of institutions undertaking PPB remains: to develop effective, gender-sensitive frameworks for methodologies, activities and tools that combine the empowerment agenda with activities with technical rigour needed in PPB. These foci constitute the central elements in the following discourse.

The implications of changing production systems for PPB

Farming and the associated production systems are changing worldwide, also regarding most farming in the developing world. These changes in farming have redefined the use of PGR, and influenced the relative importance of crops, and of varieties and traits within these crops. They have also had major impacts on the livelihoods and challenges of smallholder farmers.

First, market forces are increasingly defining which crops are needed in national and global markets – and, importantly, which varieties are preferred. Markets are changing the use of PGR. For example, in the Philippines, three million hectares are planted to maize each year. After maize was introduced in the Philippines more than 300 years ago, mainly white, open-pollinated varieties (OPVs) were produced for human consumption. Today, almost all maize planted in the country consists of yellow varieties produced for livestock feed (Gerpacio et al., 2004). Similar developments have taken place in Vietnam. Only 30 years ago, maize was not a major field crop there. A few traditional varieties, mostly waxy types, were grown in small plots or in home gardens, and the ears were sold as snacks. Today, maize is planted to more than one million hectares, solely for livestock feed (Than Ha et al., 2004). Modern hybrid varieties are planted for industrial livestock farming. As a result, every year, farmers have to purchase seeds from commercial providers, requiring a substantial part of the family income.

Second, wider socio-economic and cultural changes are changing family livelihoods. Other sources of income and/or new economic opportunities outside agriculture have become increasingly important across the globe. Family livelihood assets, especially labour, are re-allocated in order to create more diverse sources of income, thereby reducing the amount of family labour available for farming. As a result, many traditional rice varieties that needed to be harvested by hand and per panicle have been replaced by more modern varieties that can be easily harvested mechanically. With this type of farming, varieties that fulfil the DUS requirements (distinctness, uniformity, stability) are in high demand. And, with reduced availability of family labour, farmers increasingly buy seeds instead of saving their own for the next season. As in the case of maize, farmers have become more dependent on commercial seed providers, and must spend more of their income on seed purchases.

Third, with improved availability of communication technology, farmers may now react quickly to market opportunities, and cultivate crops and varieties based on much better available market information. Good roads and other infrastructure allow more efficient movement of the seed and other produce. Where produce is sold in urban markets, this has often also increased the dependence of smallholder farmers on middlemen between farmer producers and urban outlets.

Fourth, cropping patterns have changed globally. For example, production of small cereals has decreased sharply as a result of underinvestment in breeding efforts in these crops. Whereas for centuries pearl millet had been the most important crop of Zimbabwe, anchored in the Shona civilization, its cultivation has decreased sharply – from 600,000–700,000 hectares per year as recently as only 30 years ago to barely 100,000 hectares today. Pearl millet has been replaced by maize (Rohrbach and Mutiro, 1998). Whereas originally traditional OPVs of maize were planted, today 90 per cent of maize fields is planted to hybrids, not only on big commercial farms but on smallholder farms as well. Climate change is also leading to changes in cropping patterns. Erratic onset of rains, the increasingly variable duration of the growing season and higher day temperatures are causing new and more serious stresses in smallholder farming systems across the globe. As a result, many farmers opt for short-duration varieties of their main crops. Long-duration varieties, mainly traditional ones, are lost and most often replaced by modern cultivars.

Another example of changing cropping patterns related to climate change is the effect of late blight in potato occurring at increasingly higher altitudes as temperatures rise, infecting cultivars in the Andes at altitudes not observed before. That explains why, after a devastating late blight outbreak, more than half of the cultivated potato acreage in the Andes involves modern varieties bred for late blight tolerance (Egúsquiza and Apaza, 2003).

In many countries, socio-political changes related to globalization have influenced cropping patterns as well. The case of Vietnam offers clear examples of these socio-political changes. Vietnam reformed its land-use policies from 1985 onwards (Pingali and Xuan, 1992). The reforms basically moved away from socialist models of collective and state farming to family farms, supported by massive investments in agricultural infrastructure, and coinciding with market liberalization. After 30 years, Vietnam has become the world's second-largest exporter of coffee and rice, and now cultivates more than 1 million hectares of maize. Rice-farm irrigation has soared, from less than 50 per cent to 92 per cent today, and 85 per cent of farmers have shifted to mechanized land preparation, harvesting and post-harvesting activities (see Ut and Kajisa, 2006; Minh and Long, 2009). In the flood-prone areas of the Mekong Delta, the 500,000 hectares planted to traditional deep-water rice and floating rice varieties only 30 years ago have decreased to less than 5,000 hectares, as a result of irrigation and floodwater control. Similarly, whereas the rice farms in the coastal areas of the Red River Delta in the north were formerly dominated by salt-tolerant traditional rice varieties, today these saline zones and the accompanying varieties are largely gone. Farmers in Vietnam now prefer rice varieties that are not photo-sensitive and can be planted and harvested in any season. Shorter-duration varieties are preferred, because irrigation allows two to three crops a year. Similar changes have occurred in many other regions of the world.

In summary, changes in farming systems, whether resulting from climate change or human-induced, also change the crops and varieties in these systems, as these need to fit new markets and new growing conditions. In turn, the new crop and variety portfolios have rendered farmers more dependent on commercial seed providers and exposed them to higher seed prices, influencing their budget allocation and livelihoods. The replacement of diverse local cultivars by modern varieties also means that farmers become mere end-users of the intellectual efforts of breeders in the formal sector – and that in turn erodes confidence in their own capacities, and weakens their position vis-à-vis the powers of other sectors in society.

The developments outlined above mean that PGR utilization is changing. Genetic erosion and a narrowing of the genetic base of crops has ensued, in particular in farming systems that are market-oriented and are characterized by intensified production. These are the farmlands and crops where farmers' varieties and their own associated knowledge have been replaced by externally produced seeds and knowledge, and where FFS on PPB are most needed.

This is not meant to imply that farmers should revert to traditional varieties that no longer fit current socio-economic realities and the changes that have occurred in agro-ecological systems. Rather, farmers should participate in the creation of new varieties that are best adapted and most useful under the new and changing conditions. In such a scenario, farmers can become a source of new cultivars, complementing and further adjusting the new varieties developed by research institutions and commercial seed companies. Strengthening the role of farmers as developers of modern cultivars that fit today's changing agricultural needs can reduce dependency, build confidence and prevent a production system in which farmers find themselves forced to cultivate only those crops and varieties for which seeds are available on the commercial market.

Conservation of PGR diversity should not aim at the conservation of traditional phenotypes and distinct local cultivars. Rather, it should aim at the conservation of a farmer-based system of management and innovation of genetic diversity as such. In this perspective, traditional cultivars may be used as sources of genetic diversity breeding, whether by specialized breeders or by farmers in PPB activities.

In dialogues with breeders of research institutions and private seed companies, often the advice is heard that work on PGR diversity should focus on marginal lands and minor crops. But to heed such advice would entail becoming marginalized, and would not address the major effects of genetic erosion in farming systems that have experienced increased market exposure and major infrastructural changes in their agro-ecosystems.

Genes or innovating capacities defining cultural identities

Traditional cultivars are by many seen as more than expressions of genes: they are recognized as cultural expressions as well (see Altieri and Merrick, 1987; McGuire, 2008; Westengen et al., 2014; Arce et al., 2015). It may also be argued that conserving and protecting these cultivars helps preserve the cultural and even spiritual identities of local people. These aspects need further analysis.

We hold that it is more important to conserve and develop the traditional role of farmers as the actors in innovating PGR – in other words, as actors involved in a system of innovation, rather than sticking or reverting to the cultivation and maintenance of specific traditional phenotypes that may no longer fit the changed circumstances. In this view, in order to preserve local cultures and identities, PPB should focus on enhancing farmers' knowledge and capacity for innovation of their crop portfolio. The following examples should illustrate and underpin that perspective.

Vietnam: At the height of the 'Patriotic War' that ended in 1975, the adoption of Green Revolution rice varieties in Vietnam was on a par with the rest of Asia, although these varieties came from the International Rice Research Institute, an institution largely funded by the USA. The Vietnamese protected their cultural identity in the war, but did not reject modern rice cultivars. After the war, the country strengthened its own institutions for plant breeding, resulting in less dependence on external sources of germplasm. At the same time, it supported and encouraged its farmers to participate in PPB and farmer seed production, as partners in innovation. For a ten-year period, 75 per cent of the Farmer Field Schools on PPB in the Mekong Delta were funded by the government. Soft loans and grants for seed processing facilities to seed producers' cooperatives and larger seed clubs were based on national budget allocation (worth more than USD 220,000 in monetary terms) (Searice, personal communication; Berg, 2016).

Mindanao, the Philippines: The peoples of Mindanao in the southern Philippines were able to maintain a culture distinct from the rest of the country for more than 300 years. They resisted the Spanish and the American colonizers, and today form an autonomous region with its own parliament. The rice farmers of Mindanao replaced their traditional rice cultivars decades ago, as the advantages of modern rice varieties became appreciated and as irrigation facilities expanded. However, this development did not influence their commitment to fight for an autonomous region in order to preserve their own cultural identity, keeping some aspects but losing others.

Such strengthening of the cultural identities of peoples lies in the creative use of different sources of knowledge, steeped in tradition. To this concept belongs appreciation of the strengths of innovation offered by other sources, complementing internal capacities. This can keep cultural identities alive, especially by the capacity to change while maintaining the core identity. Therefore, PPB should build on a partnership and fruitful interaction between local and institutional science, between farmer breeders and 'formal', specialized plant breeders.

The FFS framework applied in PPB

This section introduces how the principles of the Farmer Field School approach and its underlying perspectives on poverty and farmer empowerment are translated into methodologies, activities and tools and used in PPB. Activities undertaken to manage the technical challenges in plant breeding are designed to build critical thinking and to enhance self-confidence.

First, the approach is based on the concept of experiential learning. Behind the FFS stands the recognition that experience is the best way to learn. The

FFS avoid 'teaching' farmers; instead, they facilitate exercises in which farmers make their own observations and draw their own conclusions from their experiments. Second, the FFS respects and builds on existing local knowledge and skills, as the strongest platform for developing new knowledge (Feder *et al.*, 2003; Braun *et al.*, 2006). Recognition given to farmers' knowledge respects and affirms their capacity for intellectual innovation, in contrast to top–down teaching approaches that deliver the implicit message that what farmers know is inconsequential and weak, thereby eroding self-confidence. And third, very importantly, new knowledge is developed and partly co-developed in the FFS – and the farmers feel ownership over this self-acquired knowledge.

These principles help to foster critical thinking and confidence, and greater capacity for experimentation and decision-making. Such increased capacity assists farmers in recognizing the societal structures that keep them in poverty. Farmers build their own knowledge by creating their own breeding populations, by learning effective selection techniques for faster genetic progress, by learning how to design experimental plots to manage biases, and to differentiate between selection pressures exerted on self-pollinating and cross-pollinating species. But they also learn to analyse their position in society. When farmers develop locally adapted varieties through the FFSs on PPB, they often face interference in the form of seed policies and regulations – like the need for varietal registration and seed certification, requirements that are designed for the commercial sector, but are inappropriate for smallholder farming systems and are biased against farmers' varieties and seeds. What is initially a critique of power structures in the realm of seed production may develop into an agenda for examining wider structural forces maintaining poverty and injustice.

A major challenge for Farmer Field Schools is to ensure that the issues discussed and studied respond to farmers' needs and interests as closely as possible. Only when this is properly realized can the commitment and perseverance of farmers to contribute to FFS activities for the full season to be achieved. The FFSs designed for PPB has the intention of being 'holistic', in understanding that farmers are best motivated if the FFS deals with the issues that are closest to their livelihoods (gut feelings), that are intellectually stimulating (the mind), and that are emotionally relevant (the heart), like memories of their hungry children because of bad seeds. All participatory approaches need to address these motivational requirements.

The FFS on PPB: insights, principles and tools

The main challenge in implementing Farmer Field Schools on PPB is to develop robust methods, clear activities and effective tools that reflect the different perspectives and agendas of men and of women farmers. Important here are the identification and inclusion of female-headed households, a gender balance in focus group discussions, the selection of female rapporteurs, and the collection of gender-disaggregated data. This becomes increasingly important in the light of the ambition to promote external use by third parties of all developed methods and tools, in an attempt at upscaling farmer plant-breeding activities (see also chapter by Visser *et al.*, this volume). These paragraphs describe the main components and issues of attention of the FFS against its timeline defined by the cropping season. The first step is to diagnose the state of farmers' plant genetic resources.

The following insights and principles guide this diagnostic stage.

Communities are stratified, not homogeneous. The 'social pyramid' of inequality in society as a whole rests on smaller 'social pyramids' at the micro- or village level. There are richer, more powerful farmers and communities, and there are poorer and weaker ones (Chambers, 1983). Activities and data of the diagnostic stage must be segregated to ensure that the richer, more powerful farmers in the village will not identify and define the problems on behalf of everyone. Without segregation into subgroups in the diagnostic stage, the FFS will identify problems and possible solutions that are of interest of those who have more resources, to the detriment of the poor and powerless. The roles and interests of male and female farmers in the community also differ, and so it is essential that women's voices be heard and appreciated in the FFS. The major implication of community stratification is that the FFS needs to be organized in smaller subgroups, including in all-female and/or all-male subgroups, and that the organizational heart of the FFS, the unit that will gather data and make analyses, is the smaller subgroup. Organization into small and segregated groups can address the local stratification of the community – an important step towards self-examination and understanding society as a whole.

Subjective perceptions and objective conditions are both important. Initially, data collection and analysis may not be fully correct or 'exact' if the FFS exercise is 'coloured' by the intellectual and emotional contributions of the farmer-participants. However, in the FFS, it is a question of *their* data and *their* analysis. The major objective is for the data and the analysis to maintain high interest and motivation amongst the farmer participants. Which outcomes are influenced by perceptions, and to which extent the data collected and the data analyses are reliable, will normally become clear to all farmer participants during the course of the FFS. The initial data and analysis in the diagnostic stage will be questioned, verified, corrected and improved if needed, based on experiences in the course of the season. In reality, the data and analyses at the end of the FFS will be more reliable and more important for evaluation of the FFS and for future work than the initially reported findings (see Chevalier and Buckles, 2013).

Diagnostic tools play an essential role in the FFS. Whereas farmers are accustomed to memorizing observations they make as a basis for decision-making, the FFS involves deeper and more precise observations that warrant documentation of the quantitative collected data. In many cases, the lessons learned during FFS studies and experimentation will prompt the farmer participants to improve the quality of their analysis and subsequent decisions and actions.

The following key baseline tools have been developed under the Sowing Diversity = Harvesting Security (SD = HS) programme:

The timeline analysis tool, by which farmers visualize the agro-environmental and socio-economic changes occurring in their communities over the last 30 years (one generation). The tool focuses on changes in (1) the socio-economic sphere, especially market developments; (2) changes in agricultural infrastructure, e.g. the introduction of irrigation; (3) changes in government policies and programmes, e.g. the promotion of high-yielding modern varieties; (4) changes in livelihoods and lifestyles of farming families, e.g. increased alternative sources of income and increased education; and (5) perceived changes in climatic conditions and seasonal patterns. After analysing the changes in their production systems over the 30-year period, farmers analyse the related changes in the use of PGR, especially the changes in cultivated crops and in priority traits. This tool ensures that the crops, varieties and traits prioritized for FFS studies truly reflect the demands and preferences of the farmer participants.

The Diversity Wheel for crops and varieties[2] that visualizes the percentage of the family farm devoted to a particular crop (first stage), or to a particular crop variety (second stage) as an indicator of the importance of that crop or variety in the farming system. The underlying rationale is that farmland is a limited family asset, and its allocation to specific crops and varieties signifies their importance to the community. The Diversity Wheel also lists crops and varieties that are no longer cultivated locally but that farmers still remember. The use of this tool allows farmers to select which crops to take up in the FFS and the traits they wish to work on.

Scoring of traits in the order of importance facilitates the setting of breeding and selection goals. Each of the farmer participants (divided into small subgroups) identifies the traits that the farmer deems most important to him/her. The total number of points collectively allotted to each trait will rank its relative importance. 'Secret voting' is important, in order to avoid the negative influences of community stratification and gender imbalances outlined above. It is important to quantify requirements for improvement ('short' or 'long' duration should be detailed by number of days; plant height by actual inches), and that farmers understand that some traits are inter-related and cannot (or only with great difficulty) be changed independently (e.g. days to maturity and yield).

Deciding on the type of participatory plant breeding to apply in the FFS

Once the breeding objectives for a selected crop have been set, FFS participants must choose among three different types of PPB activities.

The best choice for a beginning FFS is to apply *Participatory Variety Selection* (PVS), in which stable lines (like formally released varieties, stable breeder lines, farmers' varieties) are compared. Participants regularly assess the performance of each of the 10 to 12 lines against a standard representing a popular variety widely grown in the community, and will evaluate this performance especially at harvest time. They will rank the lines according to their selection objectives. Experiences in the PVS trials may result in the adoption of new lines or varieties of a particular crop in local farmers' fields. This activity does not require major technical skills, and it can be performed in the course of a single growing season. Since the conditions in subsequent seasons may differ substantially (e.g. La Niña following El Niño), a second PVS season, in which the same set of lines is tested, can be warranted.

A second option is to develop *Participatory Variety Enhancement (PVE)*, an approach that is best chosen if the FFS participants favour a particular farmers' variety whose performance and seed-quality potentials have deteriorated, while the genetic potential of the same variety can exhibit most or all traits listed in the FFS breeding objectives. In PVE experiments, such a local variety is subjected to strong selection pressure over a period of two or three growing seasons. The final population is then used to produce seed lots that are planted and compared with the original population used to start the PVE studies. Here, FFS participants may not wish to fully regain the original phenotype of the variety, but to improve concomitantly one or two major traits in response to climate change and new ecological conditions. This option requires a longer commitment and basic selection skills of participants, who must identify plants that best fit their breeding objectives.

The third and most demanding option normally requires a *full plant breeding cycle*. Because of its intrinsic complexities (identification of good starting materials, well-established selection capacities, long duration) it is best implemented in close collaboration with plant breeding institutes. The approach may start with a crossing exercise aimed at establishing a diverse and segregating population that contains the traits identified in the breeding objectives. Farmers can make these crosses themselves. Alternatively, a plant breeding institute may conduct the crossing and create the segregating population, which is released to the FFS participants only in the F4 or F5 generation[3] for further selection. The advantage of this approach as opposed to a regular in-house breeding programme is that the distributed lines are selected under farmers' conditions and preferences, ensuring local adaptation and adoption of the end product. The approach requires several growing seasons to be completed, and the presence of good selection capacities amongst the FFS participants.

Organizing the FFS PPB plots and performing the FFS studies

An FFS on PPB should ideally involve approximately 25 participants, both women and men. In line with the FFS principle of working in small subgroups (in recognition of community stratification and to allow more inclusive collective learning), the participants are usually divided into five-member groups. These small subgroups each gather data and make initial analyses. Each group is also assigned a particular sub-plot in the FFS PPB study field which it manages and inspects throughout the entire growing season, to observe the crop stand and document plant development and agro-ecological conditions.

The season-long FFS curriculum runs from land preparation to harvesting stage, and normally involves one (morning) session per week. The weekly session should not last more than three hours to avoid becoming too demanding. Data related to growing conditions (including biotic and abiotic stresses) and to plant performance are collected by each group according to the Agro-Ecological Systems Analysis tool originally developed in Integrated Pest Management, and sometimes referred to as 'Gene by Environment Analysis'.

In the case of PVS, each subgroup would normally manage two of the PVS lines. In the case of PVE and selection from segregating populations the FFS field will also be subdivided, and subplots allotted to individual subgroups for the same reason, but genetically speaking these sub-plots cannot be distinguished. Replicates may be included or not, as the participants choose. Often the same lines will be tested in nearby villages running their own FFS, in which case the different PVS sites serve as replicates.

The selection pressure to be applied in PVE will differ between crops. With a self-pollinating crop, positive selection may be conducted just before harvest time, in which the best 10–20 per cent of plants will be chosen. In case of a cross-pollinating crop, both negative and positive selection should be applied. Negative selection is applied until flowering time, to prevent inferior plants from pollinating other plants in the stand. At the harvest time, positive selection is advised, as for the PVE of a self-pollinating species.

In full-cycle plant breeding involving self-pollinating plants, bulk selection is usually applied until the F4 or even F5 generation, after which pedigree selection is used in the ensuing generations. The more demanding pedigree selection is manageable by farmers only with the later generations when segregation levels are reduced. If seed of the F4 or F5 generation is provided by a plant breeding institute, bulk selection by farmers can be skipped entirely. In the case of FFS cross-breeding in a cross-pollinating crop, one option is to start with the creation of a composite population from at least six diverse parents, followed by bulk selection to create a new OPV; this may take up to ten generations to complete. Or, individual superior plants or ears may be selected and planted in 'ear-to-row' plots, in combination with controlled pollination.

The FFS study, whether PVS, PVE or selection from a heterogeneous population, will require a plot of approx. 500–600 m^2 for maize, sorghum or pearl millet; for rice, a plot of 250 m^2 will suffice.[4]

Examples of initial FFS PPB successes

The FFS PPB approach has resulted in significant outcomes. In Vietnam, farmers in the Mekong Delta, where the FFS PPB has been practised for a decade and longer, have established over 400 'seed clubs'. These are a major expression of farmer empowerment; they have contributed to improved live-lihoods for the members/seed producers and to higher yields for the users. The seed clubs provide some 25 per cent of certified rice seeds of the Mekong Delta. Two modern rice varieties bred by farmers in the Delta have been officially registered and released, and some others resulting from PPB experiments have been given 'provincial recognition' and can be marketed. A local glutinous rice variety called Nep Lech especially popular with women has been improved through PVE, and has spread all over the province of Yen Bai in North Vietnam.

In Zimbabwe, new lines have begun to diffuse into local communities from FFS PPB fields, as farmers adopt cultivars from their own FFS plots. In many wards of the UMP District, 100 km north of the capital Harare, 5 new sorghum lines have spread, and now occupy 30 per cent of all lands planted to sorghum on family farms. Farmers explain that being able to obtain new crop varieties helps them to cope better with climate change and to reduce the length and severity of the hunger period. The development of farmers' seed enterprise to respond to the needs of smallholder farmers has boosted self-confidence among the farmer-producers. Moreover, farmers involved in PVS in maize started to question the subsidized distribution of inappropriate maize varieties that served as political base-building for the previous Mugabe government.

In Laos, three PPB varieties are soon to be awarded 'provincial certificates as varieties', as these lines have become widely recognized in the provinces to which they are well adapted. This provincial certification follows the similar spread of a few varieties (like 'Pakchang 1', 'released' by the district of Pakchang of Vientiane province six years ago). In the absence of a strong breeding capacity in the country, the FFS efforts have contributed substan-tially to an improved portfolio of rice varieties and higher yields. Recognition of farmers' role in developing this portfolio has further added to their self-confidence.

The SD = HS programme has developed gender-sensitive FFS curricula that address gender roles and the underlying social structures that maintain these roles, thereby creating awareness of the triple burden of women. PPB is also being used as an entry point to approach women in their position of strength: as managers of biodiversity. The selection of female FFS participants and the inclusion of women in the Training of Trainers have proven essential for realizing more gender-sensitive FFSs, for reaching out to more women, and for breaking the traditional bias against women's participation in training sessions. The fact that FFSs are conducted in the local community avoids the traditional mobility constraints often faced by women. FFS are best conducted in the early morning due to crop conditions, but that is also the busiest time for women, with their additional household tasks. For that reason, household negotiations were undertaken, aimed at relieving women of household chores one morning per week, to participate in the FFS session. Further, varieties preferred especially by women, like the Nep Lech rice variety in Vietnam, were selected for improvement in the FFS. Better-empowered women can define, access, co-develop, select, store and exchange crop varieties that are important for household food and nutrition security, at the same time leading to greater recognition and respect for women within households and communities. Globally, there is increasing recognition of women's labour contribution to food production; PPB through FFS is generating awareness and appreciation of women's knowledge and contribution of skills. Better self-confidence enables women to engage in policy discussions, and to demand appropriate resources and services from local institutions.

Conclusions

This chapter has argued and justified why Participatory Plant Breeding cannot focus solely on the technical challenges of managing and improving crop diversity by farmers, but must also take into consideration the socio-economic, gender and political context in order to be relevant and have lasting effects. Properly conceived and implemented, a PPB approach and curriculum will necessarily lead to farmers' empowerment and contribute to the strengthening of political as well as socio-political capacities. Improvement of food security and farmers' livelihoods and farmers' empowerment go hand-in-hand. We have presented several examples showing why policies and legislation need to be addressed to support the improved functioning of local seed systems.

Participants in Farmer Field Schools on PPB set their breeding objectives on the basis of their own needs and conditions. They know how to identify traits for improvement and can engage in collaboration with plant breeding institutes aimed at enhancing farmers' systems of creating and maintaining diversity.

The supportive role of state plant breeding institutes in such initiatives and efforts may call for certain institutional reforms. PPB challenges traditional breeding programmes and breeders' attitudes that fail to include farmers. By fully involving farmers, PPB aims to provide a better adapted crop portfolio, which in turn can improve livelihoods. Through this approach, PPB contributes to stronger public institutions and addresses structural inequalities The critical thinking and empowerment established through the FFSs helps farmers to challenge seed policies, laws and regulations that are biased against farmers' seed systems, and to confront the institutions that create these rules. The assertion of their rights as breeders and creators of new varieties may then open their eyes to other structural issues that make and keep them poor. In particular, women farmers participating in the FFS will better understand how gender inequality is rooted in unequal rights biased against women's access to and control over productive assets like land. Landless tenants who are required to deliver to their landlords a portion of the products of their labour may begin to question the landownership system that had been taken for granted.

Finally, the FFS PPB experience, the associated generation and ownership of new knowledge through carrying out research together, builds confidence and unity. The farmers involved are better able to articulate their needs and press for support of their efforts to be recognized as actors in the creation of new crop varieties. Through FFS activities they begin to analyse the power structures and decision-making processes that affect their lives and their options to participate as active and critical citizens.

The most challenging and important element of a Farmer Field School is the process of reflection. The collective learning process in the FFS helps participants to analyse their own positions as citizens and as human beings, and examine their own behaviours, family relationships and community structures, as well as the functioning of society at large. Efforts to broaden the genetic base of crops, to strengthen the role of farmers in breeding new varieties, and to relate all such efforts towards social reforms must be based on the recognition of the poor, and how *all* members of the community have the right to realize their full potential as human beings.

Notes

1 The first phase of the SD = HS programme (2014–2018) was conducted by ANDES (Peru), Agricultural Research Centre (LAO P.D.R., Metta Foundation (Myanmar), Can Tho University (Vietnam), Community Technology Development Trust (Zimbabwe), Searice (the Philippines), and coordinated by Oxfam Novib (the Netherlands), with assistance of the South Centre (Geneva), Third World Network (Malaysia), ETCgroup (Canada) and GRAIN (Spain). See www.sdhsprogram.org/ for more information.
2 The Diversity Wheel has been derived from the 'four square' analysis of LiBird and Bioversity International, as developed by Bhuwon Sthapit.
3 F4 or F5 generation means the fourth or fifth generation of (selected) progeny after a crossing.
4 All these approaches are described in manuals for training the trainers and in FFS Facilitator Field Guides for the crops rice, maize, sorghum, pearl millet, groundnuts and potatoes, and have been published on the SD = HS website.

References

Acemoglu D, Robinson JA, 2013. *Why Nations Fail*. New York: Random House.
Altieri MA, Merrick L, 1987. In situ conservation of crop genetic resources through maintenance of traditional farming systems. *Economic Botany* 41, 86–96.
Arce A, Sherwood S, Paredes M, 2015. Repositioning food sovereignty: Between Ecuadorian nationalist and cosmopolitan politics. In Trauger A, (ed) *Food Sovereignty in International Context*, 125–143. London: Routledge.

Berg, T, 2016. *External Evaluation IFAD-Oxfam Novib Programme: Putting Lessons into Practice: Scaling up People's Biodiversity Management for Food Security.* The Hague: Oxfam Novib.

Bond P, 2006. Civil society on global governance: Facing up to divergent analysis, strategy, and tactics. *Voluntas* 17, 357–371.

Braun A, Jiggins J, Röling N, van den Berg H, Snijders P, 2006. *A Global Survey and Review of Farmer Field School Experiences.* Endelea, Wageningen, The Netherlands.

Chambers R, 1983. *Rural Development: Putting the Last First.* London: Routledge.

Chevalier JM, Buckles DJ, 2013. *Participatory Action Research: Theory and Methods for Engaged Inquiry.* London: Routledge.

Cornwall A, Nyamu-Musembi C, 2004. Putting the 'rights-based' approach to development into perspective. *Third World Quarterly* 25, 1415–1437.

Egúsquiza RB, Apaza WT, 2003. Peru late blight profile. In Forbes G, *Global Blight Initiative* (ed) https://research.cip.cgiar.org/confluence/display/GILBWEB/Peru.

Feder G, Murgai R, Quizon J, 2003. *Sending Farmers Back to School: The Impact of Farmer Field Schools in Indonesia.* World Bank Group E-library. https://doi.org/10.1596/1813-9450-3022.

Freire P, 1968 (various later editions) *Pedagogy of the Oppressed.* https://pdfs.semantic scholar.org/64a7/9f89a714dc4d2d845597bb342a72443ace71.pdf.

Freire P, 1972. *Cultural Action for Freedom.* Harmondsworth: Penguin.

Gerpacio RV, Labios JD, Labios RV, Diangkinay EI, 2004. *Maize in the Philippines: Production Systems, Constraints, and Research Priorities.* Mexico DF: CIMMYT.

McGuire S, 2008. Path-dependency in plant breeding: Challenges facing participatory reforms in the Ethiopian Sorghum Improvement Program. *Agriocultural Systems* 96, 139–149.

Minh NK, Long GT, 2009. Efficiency estimates for the agricultural production in Vietnam: A comparison of parametric and non-parametric approaches. *Agricultural Economics Review* 10, 62–78.

North D, 1990. Institutions, Institutional Change, and Economic Performance. New York: Cambridge University Press.

Oxfam Novib, ANDES, CTDT, SEARICE, CGN-WUR, 2015. From lessons to practice and impact: Scaling up pathways in peoples' biodiversity management. Submission to the Sixth Governing Body Meeting of the International Treaty for Plant Genetic Resources for Food and Agriculture. www.planttreaty.org/content/farmers-rights-submissions.

Pingali P, Xuan V-T, 1992. Viet Nam: Decollectivization and rice productivity growth. *Economic Development and Cultural Change* 40, 697–718.

Richards P, 1998. *Fighting for the Rain Forest: War, Youth and Resources in Sierra Leone.* Oxford: James Currey.

Rohrbach DD, Mutiro K, 1998. Sorghum and Pearl Millet Production, Trade, and Consumption in Southern Africa. ICRISAT, Bulawayo, Zimbabwe. http://oar.icrisat.org/1674/1/ISMN39_33-41_1998.pdf.

Sen A, *The Idea of Justice.* 2009. Cambridge, MA: Harvard University Press.

Thanh Ha D, Thao T Dinh, Khiem N Tri, Trieu M Xuan, Gerpacio RV, Pingali PL, 2004. *Maize in Vietnam: Production Systems, Constraints, and Research Priorities.* Mexico DF: CIMMYT.

Ut TT, Kajisa K, 2006. The impact of Green Revolution on rice production in Vietnam. *The Developing Economies*, 44 (2), 167–189.

Uvin P, 2007. From the right to development to the rights-based approach: How 'human rights' entered development. *Development in Practice* 17, 597–606.

Westengen OT, Okongo MA, Onek L, Berg T, Upadhyaya H, Birkeland S, Khalsa SDK, Ring KH, Stenseth NC, Brysting AK, 2014. Ethnolinguistic structuring of sorghum diversity. *Proceedings of the National Academy of Sciences of the USA* 111, 14100–14105.

12 Building collaborative advantages through long-term farmer–breeder collaboration

Practical experiences from West Africa

Anja Christinck, Fred Rattunde and Eva Weltzien

Collaborative advantages through farmer-researcher collaboration

We reflect in this chapter on long-term collaboration between farmers and plant breeders based on theoretical considerations and practical experience gained through two decades of farmer–plant breeder collaboration[1] in West Africa. Here we go beyond the technical breeding aspects discussed in Weltzien *et al.* in this volume, and focus on the collaboration itself. We examine the theoretical concepts in light of practical experience to help explain what made the collaboration work, and which underlying aspects may be helpful for similar initiatives in other contexts. To this end, we extend the focus beyond the immediate field of participatory plant breeding, and include concepts developed in non-agricultural businesses or other contexts.

The basic assumption behind participatory approaches in agricultural research and technology development is that, through collaboration, farmers and researchers can achieve more than either group alone. This is what in business settings is referred to as 'collaborative advantages': by building alliances, different types of 'organizations' can overcome limitations that may be due, for example, to lack of resources, expertise or skills (Huxham and Vangen, 2005). Such approaches aim to create synergies based on identified complementarities regarding the potential contributions of each partner, and to manage these towards a shared outcome or vision. While collaboration is commonly established in order to pursue some joint activity, it will often involve mutual learning – as an explicit aim, or something that 'happens' along the way (Huxham and Vangen, 2005).

Hoffmann *et al.* (2007) have adapted the concept of collaborative advantages for farmer–researcher collaboration. They suggest complementary roles for farmers and researchers in setting research priorities: whereas only farmers can identify the problems that they see as relevant, researchers can assess if and how their knowledge and methodological approaches can contribute to problem solving and translate objectives and methodologies into research. Making use of farmers' own capacities for decentralized experimentation and dissemination can also help to bridge the gap between 'technology develop-ment' and 'adoption', particularly in cases where it is difficult for professional researchers to know farmers' preferences and to understand the complexity of their situation. Researchers can learn from farmers' observational skills and diagnostic capacities, including the indicators they use to observe phenomena relevant to a specific context (Hoffmann *et al.*, 2007).

Restrepo *et al.* (2014) have highlighted the importance of collaboration and social learning between diverse stakeholders for developing solutions that can help to transform food and farming systems towards desired and sustain-able outcomes. Joint situation and problem analysis and design of collabora-tive learning processes, where the learning capacities of individuals are strengthened through interactions with others, increase the likelihood of achieving results that are relevant and can be implemented in practice.

Based on these considerations, this chapter presents how we experienced and conceptualized complementarities among and within the groups of 'farmers' and 'plant breeders'. Subsequently, ways of managing the collaborative process towards yielding 'advantages' are described. We then summarize important developments, achievements, and challenges for sustainability. In conclusion, we reflect on the importance of trust and tangible outcomes for maintaining longer-term collaboration and for the creation of collaborative advantages.

Understanding complementarities – or who cooperates, and in whose system?

The diversity of sorghum and pearl-millet varieties managed and maintained by West African farmers today result from selection activities performed by countless generations of farmers over hundreds or thousands of years. Their practices are based on contextual knowledge and are deeply rooted in cultural concepts and norms.

The 'landrace' varieties that farmers maintain have co-evolved with the local people and their agricultural production systems over long periods. They often show adaptation to certain key constraints and particular use factors, such as traits relevant for local storage conditions or specific dishes. Farmers tend to observe varieties over long periods in one specific local 'setting', including the full complexity and variability of factors that may influence their performance in different fields or years. They make observations frequently – while working in the fields, harvesting, cleaning and processing grain – but not necessarily intentionally or systematically. And they evaluate their observations immedi-ately against key 'requirements' for the given context.

Farmers are also keenly aware of the social context, including specific seed-related values and norms, or local seed-quality requirements. Farmer-managed seed cooperatives have become recognized actors in the evolving seed sector in Mali due not only to the quality and quantity of seed they provide, but also because of how such cooperatives can distribute seed and varietal information in ways that build on local traditions and that respect cultural norms.

In comparison, science-based plant breeding operates within shorter periods. The aim is usually to develop varieties that perform well over a target population of environments, which requires making observations at multiple locations, in a clearly structured, systematic manner. To be able to understand specific relations between environments and trait expression, plant breeders tend to reduce the complexity of influencing factors in their experimental designs. A major challenge is to ensure that the measurements and observations that plant breeders make are pertinent and predictive of varietal performance in farmers' fields and post-harvest use. These complementarities between farmers' and plant breeders' experiences in a research context are similar to those described by Hoffmann *et al.* (2007).

Further, the differences *within* the two groups, 'farmers' and 'plant breeders', offer potentials for creating collaborative advantages. An important complementarity within the group of breeders has resulted from cooperation between research staff of the international research centre ICRISAT (International Crops Research Institute for the Semi-Arid Tropics) with colleagues from the national breeding programmes of several West African countries. International scientists contribute experience from work at other locations, connections and experiences with the international research community and donors. In turn, breeders at national research institutes are familiar with national policies for agricultural development, procedures for variety release and seed certification, administrative structures; they have important networks for linking with extension services and media outreach, and can also contribute their long-term institutional memory, including perceived successes and failures of projects and programmes, for the specific context of each country.

Differences and complementarities also exist within the large group of 'farmers'. Methodologies developed in the past decades for characterizing variations in agro-ecological conditions and genotype by environment interactions are increasingly used to address the variability of conditions under which farmers work. However, most breeding programmes cannot rely on such established methodologies to address sociocultural variation. Hence, a specific challenge is to not only understand the diversity of farmers, both women and men, but to turn existing socio-cultural diversity into collaborative advantages.

Throughout West Africa, the post-harvest processing of sorghum and pearl-millet grain is usually performed manually by female household members, often younger women. Integrating these women in variety-evaluation activities has helped to shift the focus from 'yield harvested' towards 'food yield', a concept that had not been considered in breeding programmes. 'Food yield' includes consideration of losses occurring during storage and processing, along with quality aspects like the swelling capacity of flour and the acceptability of using leftover food for consumption the next day. Understanding 'food yield' proved critical for assessing the potential of new varieties for large-scale adoption.

Women often grow small plots of sorghum, generally considered a 'men's crop', in order to have supplementary food for young children outside of the main meals, or for the whole family when the men's granary is emptied. This insight led to appreciation of the roles of women in family food security, and facilitated reaching children with micronutrient biofortified varieties to help alleviate the high prevalence of anaemia.

Further, working with farmers over time made it possible for researchers to learn who had specific skills and expertise. Certain farmers were particularly skilled at single-plant selection, with progenies derived from their selections often showing superiority in performance trials. Others could effectively assess the acceptability of panicles for their form and ease of threshing. Some women with extensive experience in processing grain could rapidly score grain quality, with the repeatability of their observations being comparable to much slower and costly laboratory procedures, even over thousands of plots in early-generation progeny trials. Cultivating relationships, seeking recommendations and recognizing individuals with particular expertise thus allowed more effective collaboration over time and very probably increased the genetic progress achieved by the breeding program.

To summarize, individuals and groups of farmers have their own systems for enhancing and assessing varietal performance on the basis of experience embedded in broad agro-ecological and socio-cultural contexts. In contrast, science-based plant breeding methods tend to under-emphasize the influence of such context-related factors, while increasing compatibility with experiences made at other locations. Collaboration can create opportunities to overcome important gaps in many international and national breeding programmes that often hinder adoption of new varieties. These gaps can be described as both failures to link with the knowledge and requirements of people acting in a specific local context, as well as the lack of awareness of the relevant expertise and skills of key actors in those local systems.

Seven steps towards fruitful long-term collaboration between farmers and plant breeders

To turn complementary expertise and skills into collaborative advantages, a methodology is needed that can facilitate engagement of the partners and can steer their contributions towards the identified goal(s). Such a methodology that sustains fruitful collaboration and joint learning by farmers and researchers can be described as involving seven steps: (1) identify partners; (2) institutionalize the collaboration and build professional relationships; (3) engage in joint situation and problem analysis; (4) agree on goals and priorities; (5) engage in a learning and action process to develop possible solutions; (6) test 'prototypes' of possible solutions to gain practical experience and further adapt/refine them; and (7) reflect jointly on the state of the learning process and the relevance of the options explored (Christinck and Kaufmann, 2018).

Reflection on the work in West Africa revealed that the seven steps described by Christinck and Kaufmann (2018) were all touched on, but more 'by doing' and continued dialogue with a variety of partners rather than explicitly planned as distinct steps per se. Also, the seven 'steps' were not necessarily taken in sequence. For example, new partners were identified – e.g. new farmers' organizations or partners in other countries – not only in the beginning, but also as the collaboration evolved. Since breeding usually requires several cycles, steps 3–7 were implemented more on a continual basis, year after year, and often several steps were taken on one occasion, as with workshops where participants agreed on goals and priorities after reflecting on experiences and progress to date. Hence, the 'steps' described below should be understood as methodological elements that require attention, bearing in mind that their implementation may proceed in various ways.

Identify partners

Working towards collaborative advantages with new partners should be based primarily on common interests. Curiosity about the performance and qualities of new varieties is an interest readily shared by plant breeders and farmers alike. Sorghum and pearl-millet farmers in Mali, Burkina Faso and Niger were highly motivated to compare new varieties with their own local ones, especially if there was hope that the new varieties could provide benefits or new options. On-farm variety trials were found to offer excellent initial entry points for joint discussion and action.

It became clear, however, that an international breeding programme, operating in several countries, could not be based on the engagement of individual farmers. Formal organizations with legal status and transparent rules regarding financial management and decision-making processes were required as partners. Farmers' cooperatives and unions of cooperatives proved to be ideal allies. These organizations generally facilitate grain marketing, access to basic inputs and information on behalf of their members. Such farmer-managed enterprises, whose members – individual farmers – are keenly interested in grain production and marketing, were highly willing to work together with breeders to develop new varieties. The subsequent need for seed production and marketing, and the business opportunities emerging from these activities, made this choice of partners advantageous for both sides.

Institutionalize the collaboration and build professional relationships

Elaborating and signing an agreement for collaboration between the research team and a farmers' organization is necessary once the two decide to work together, especially in the context of a research grant. The agreement clarifies basic roles and responsibilities, schedules and rules for reporting and funds disbursement, as well as options for later revisions of the agreement. The more specific details of the technical collaboration would evolve through a regular process of joint planning. All partner organizations shared a need to learn about the specificities of managing the complexities of production and marketing quality seed of new varieties. However, each farmers' organization had its own history, strengths and weaknesses, especially with respect to managing communications with large numbers of members, assets management or general functioning.

We facilitated such learning by organizing training programmes on specific topics such as hybrid seed production or cooperative- and seed-enterprise management. Planning meetings often included training. Also occasional exchange visits were organized to bring farmers' organizations from different areas or countries together for mutual learning and inspiration. Such exchange visits revealed that certain problems might need different solutions, depending for example on the legal and administrative framework in each country.

The leaders of the collaborating farmers' organizations were mostly active farmers who had to balance their organizational responsibilities with their own farm activities, often including their own trials and seed production plots. Many found it difficult to represent their cooperatives effectively at national meetings due to language barriers. Thus, funding for farmers' organizations to hire technically trained staff was included in research proposals. These technical facilitators were paid and supervised by the farmers' organization, whereas training was organized by the participating research organizations. The facilitators helped to conduct trials and outreach to diverse agencies and development actors, and supported seed production and other activities as these evolved.

The ability of farmers and technical field personnel to pose questions, raise concerns, and obtain answers helped to build the rapport essential for effective collaborative relationships with individual farmers and their organizations. Joint visits by farmers and breeders to individual trials, and meeting with all those engaged in collaborative activities, were crucial to this process. Remaining within reach when questions or problems arise was very important and now is feasible since many farmers, including women, own mobile phones.

Engage in joint situation and problem analysis

Our collaborative breeding experience involved various activities that fostered shared understandings of problems relating to varieties and seed. These situation and problem analyses were closely linked to the following step of setting goals and priorities.

Open-ended variety evaluations that encourage the farmer to express her or his observations, experiences and assessments about a range of varieties are an ideal tool for researchers to gain insights into variety preferences, needs and farmers' visions of useful improvements. Such conversations also help to build a common understanding of terms and concepts, as well as confidence between farmers and researchers. Inclusion of local varieties was important to enable comparison of the advantages and constraints of experimental varieties relative to those of varieties the farmers are familiar with. Discussions about the diversity of locally grown varieties, their histories and roles in the production system also provided useful information and helped to develop a shared understanding of the situation as well as the constraints and opportunities for cropping system evolution.

Another effective activity for joint situation and problem analysis were the annual village-level meetings for evaluating the cooking characteristics of the varieties being tested in the on-farm trials. Such meetings provided a relaxed and informal setting for discussing issues, experiences, observations and ideas with a wide range of farmers in the villages.

Discussions about specific varieties, traits or plans for new trials can reveal 'underlying' concepts, values and rationales to be further explored. However, deeper understanding of the agro-ecological and socio-cultural 'setting' was necessary for developing appropriate research questions and activities in certain cases, like for issues relating to seed system development and nutrition, or complex combinations of agro-ecological and socio-economic constraints. The research teams engaged with researchers from other disciplines to explore certain aspects of the context with methods and expertise that could complement those of the breeders. For example, doctoral students worked on the societal aspects of seed systems and seed markets (Jones, 2014; Siart, 2008) or the importance of genetic variation in relation to phosphorus-limited environments (Leiser *et al.*, 2015). Other researchers contributed economic assessments and market analyses, as well as analyses of food consumption patterns and post-harvest processing (Bauchspies *et al.*, 2017; Isaacs *et al.*, 2018; Smale *et al.*, 2018). Most of this work built on the ongoing collaboration with farmers and their organizations. Through these combined efforts, much clearer understandings of the context were developed than is usual in breeding programmes that have not engaged in long-term collaboration with farmers.

Agree on goals and priorities

Some participatory plant-breeding projects have clear objectives from the beginning, but that was not the case for our work in West Africa. Defining objectives and exploring ways to do so formed an essential part of the collaborative work.

Insights into farmers' decision-making about which type of varieties would be most useful came first from discussions in individual variety trials and village-level post-harvest culinary tests, in which the results of that season's trials were presented and discussed. Annual feedback and planning workshops with partner organizations across villages within a given region were the most important way of obtaining farmers' inputs for setting priorities for the upcoming year. Participating farmers were chosen to represent the groups conducting varietal trials and seed production activities in each village, with attention being given to gender balance. Representatives of cooperatives and unions, including technical and farmer facilitators, also participated. The presentation of grain-yield performance and farmers' appreciation scores of the experimental varieties, differentiated by gender, spurred in-depth discussions about the next steps for the research. The conditions under which the new varieties would be grown and tested, as well as reasons for choosing specific varieties for seed multiplication and sale were intensely discussed. The trade-offs that farmers are prepared to make between certain traits, and the relative level of interest in varieties with specific advantages (such as resistance to *Striga*, specific nutritional qualities or enhanced fodder quality of the stover) were often revealed through such discussions.

Trial modifications or new proposed activities were typically discussed in separate small groups, where each person could make suggestions and voice his or her concerns. A research team member accompanied each small discussion group to help provide input (such as information about certain varieties or the breeding processes) and perhaps serve as note-taker or discussion facilitator. These meetings were vital for joint understanding of the overall goals and specific objectives, as well as for clarifying, revising, and adapting trials and activities to address the needs and realities of the farmers and their organizations.

Engage in a learning and action process to develop possible solutions

Large-scale collaborative on-farm variety testing was a key activity which facilitated exchange between farmers and breeders. On-farm trials allowed farmers to observe varieties under their own management conditions, while also enabling the evaluation of grain-yield performance using advanced statistical methods. Such networks of trials are the only way to assess performance of new varieties over a diversity of soil and management conditions that reflect the target environments where the 'improved varieties' are supposed to exhibit their superiority. Therefore these trial networks are vital for making genetic gains for traits like grain yield that are sensitive to environment and display large genotype by environment interactions.

The design and management of these trials were crucial for obtaining results relevant to farmers and breeders alike. The concept of repeatability and ways of conducting trials that could reduce environmental heterogeneity while allowing farmers to test varieties under their usual management practices were extensively discussed at annual meetings. Among the trial-design modifications made as a result of these discussions were: to reduce the number of entries per trial to include replications, conduct separate trials with taller and shorter varieties, and test the same set of entries over two consecutive years. An alternative, smaller, trial option with fewer entries was also initiated, to allow farmers with smaller fields (mostly women) to participate.

Farmer facilitators, one or two per village or cooperative, assisted with trial installation, data collection and organizing village-level discussions. These facilitators, selected for their literacy, ability to engage with others, and their interest and commitment to varietal and seed issues, received training on trial installation and choice of fields, conducting village-level trial evaluations, and the requirements for certified seed production. They were provided with basic supporting materials, such as measuring tapes, ropes, waterproof bags and raincoats. As their responsibilities grew, the farmers' union provided motorcycles and money for their running costs and phone communications. These facilitators could effectively relay messages, collect feedback from members and propose priority actions to the union leadership, thereby facilitating networking not only in 'their' set of villages, but also for the farmer organization as a whole.

Examining the yield performance of a variety across different farmers' trials in a given region was important, but proved challenging for the participating farmers. Variety-yield means from each trial in the region were therefore presented jointly in a table in the local language with (a) the varieties (rows) ranked in order of overall mean yield performance, and (b) the individual farmers' trials (columns) ordered from highest to lowest repeatability, with colour-coded cells (white, yellow or red) to indicate whether a particular variety yielded more than, within, or less than one standard deviation from the trial mean, respectively. Farmers could readily identify the varieties with the most 'white cells' and observe the stability of varietal performance across different farmers' fields.

Realistically managing expectations proved critical in this stage. Farmers needed to be aware that the new varieties might not necessarily provide any advantages, that conducting trials requires extra work, and that many visitors might view these trials. Such issues can best be addressed in the context of clarifying the goals and objectives of the trials and the collaboration – which type of data will be collected by whom, who will do what with the results, etc. Also needing prior clarification and agreement were whether farmers were willing to have their names made public along with their trial results, and what benefits farmers could expect from conducting trials: for instance, which inputs would be provided, and what would be the opportunities for participating in training programmes or workshops.

Each farmer-participatory breeding activity was generally conducted as a group activity. The researchers set a minimum of four farmers per village for a particular activity, thus covering a range of growing conditions and enabling the farmers to discuss their findings as a group. The groups presented their results to the assembly at the annual meeting and discussed options for new activities. When women began conducting trials, the same approach was followed. The women, often not members of the farmers' organization, thereby gained roles in reporting and planning activities. The farmers collaborating in breeding activities could engage in group exchanges and networking in their own villages, and increasingly in neighbouring villages and districts with other participating farmers. The women's varietal testing groups provided a valuable basis for subsequent work on nutrition issues (Bauchspies *et al.*, 2017).

Various monitoring activities indicated that the adoption and spread of new varieties from these trials was mostly limited to the villages in which the trials were conducted, and that information about the new varieties spread only slowly. Concurrently, farmers' interest in growing variety trials grew exponentially, to the point where the scientists and the farmers' organizations could not manage the rising demands. To address this dilemma, a Malian farmers' organization (Union Locale des Producteurs Cereales) decided to test selling seed packets via the local extension office. This positive experience prompted the farmers' union to discuss whether they should sell seed, and their subsequent decision to undertake this activity even though it was not a traditional practise.

The hurdles relating to commercial seed distribution were high, since selling and buying seed of staple grain crops like sorghum and pearl millet were not culturally acceptable. A solution to the seed dissemination problem was found: to approach it as a group activity serving the common good, rather than as an individual activity. However, the changing political and regulatory framework for seed dissemination at this time led to insecurity about rules and procedures. In 2008, the Economic Community of West African States (ECOWAS), of which Mali, Niger and Burkina Faso are members, adopted a 'harmonized' seed legislation whereby variety registration and seed certification became mandatory for marketing seed legally (see de Jonge *et al.*, this volume). The emerging famer-managed seed businesses had to be acquainted with the new seed–certification rules, standards and procedures to sustain this initiative (Christinck *et al.*, 2014).

In the first year, the focus was on the technical aspects of seed production and compliance with certification standards. Later, the emphasis shifted towards improving the organizational capacities of farmers' cooperatives and expanding the marketing and dissemination activities. Practical steps included registering producer groups at certification agencies, field inspection and seed harvest according to norms, processing and packaging in mini–bags with labels, and establishing 'distribution channels', for instance by organizing field days, seed fairs, offering seed at local shops or markets as well as via 'mobile seed shops' (motorcycles travelling among villages). Several cooperatives also explored partnerships with NGOs and the few existing commercial seed shops.

Many of these activities, like seed fairs, included exchange and dissemination of information about the varieties among members of cooperatives and beyond. Information flow was further supported by radio broadcasts and technical information sheets produced in local languages. The cooperative members' knowledge about the varieties, gained through their participation in developing and evaluating those varieties, formed the basis on which these seed–marketing activities could develop.

Test 'prototypes' of possible solutions

Possible solutions to varietal constraints faced by farmers were, as indicated above, (1) to develop new varieties offering additional options to farmers; and (2) to enhance availability and access to these varieties in rural areas.

The purpose of the collaborative on-farm variety trials described above was to find out whether the newly developed varieties could be considered 'solutions', and in which contexts. Hence, conducting these trials at many locations and involving various farmers was important for testing varieties as 'prototypes'. The development of trial designs, protocols and methodologies (Weltzien *et al.*, this volume) with our first partners in Mali made the process of 'upscaling' these trials much easier. Building on these experiences, similar collaborative variety testing could be established in other regions in Mali and other countries as additional funding and partners entered the picture. A second stage of on-farm varietal evaluations of experimental varieties chosen by participating farmers further tested these 'prototypes' at a still larger scale. The much smaller unreplicated 'Adaptation Trials' (Weltzien *et al.*, this volume), conducted by farmers with support from local facilitators, were intended to enable many more farmers, both men and women, to see these new varieties in their fields under their own management. These trials, numbering in the hundreds, were prepared by the breeders, but with minimal involvement thereafter. However, it has not yet been possible to establish a satisfactory system for gathering varietal yield and appreciation data from these trials, to make this information readily accessible and searchable by farmers, seed producers and researchers.

Seed production and dissemination activities by farmers' seed enterprises are complex endeavours. Continued collaboration with breeders and exchanges among cooperatives and countries was important for facilitating iterative learning. For example, with research support, farmers' seed cooperatives established committees and procedures for setting seed prices so that everyone involved would benefit. The costs of seed production, conditioning, storage and marketing all need to be covered, while balancing the interests of seed producers and customers and ensuring sustainability. Additional funds were raised for enhancing physical storage and processing facilities for some of the larger, fast-growing cooperatives. Reserve funds were built up by some cooperatives to help overcome payment delays and to bridge the interval between harvesting, seed certification and marketing. Only by actually implementing farmer-managed seed production and marketing in the 'real world' could the experience and capacities, organizational as well as physical, be gained to bring this activity up to scale and help solve the challenges.

Reflect jointly on the state of the learning process and the relevance of options explored

The results from variety trials, seed production and marketing activities were presented and jointly evaluated at the annual feedback and planning meetings.

These discussions – often facilitated first in small groups and then presented in plenary to help the view of all participants to be heard – were vital for planning and revising the upcoming season's activities. They also served to acknowledge the collaborators and their efforts in conducting and facilitating activities, further inspiring the participants as achievements were shared.

The manner in which the trials, seed production and marketing activities were implemented further facilitated reflection on a continual basis. For example, village-based groups conducting an activity would discuss the results among themselves prior to presenting them at the annual meetings. Also the frequent farmer–researcher interactions and the numerous visitors to village-based trials enabled extensive informal and formal reflection and learning.

Important developments and achievements

Establishing large-scale, decentralized variety-testing networks

Large-scale collaborative networks for variety testing have been successfully established for sorghum in Mali and Burkina Faso, and for pearl millet in Niger. The approach has further inspired breeding programmes in neighbouring countries to experiment with some of the elements that have been developed, as in Senegal and Ghana.

Farmers' cooperatives, breeders and development organization staff have jointly conducted 20 to 30 'first stage' variety trials annually in Mali since 2003, covering the three major sorghum production zones of the country. Also approximately 200 farmers, including about 40 women, have grown the smaller 'second stage' variety trials annually. The large and well-established network of trials proved vital for developing innovative varietal options such as landrace-based hybrids, and reliably assessing their yield performance and stability across widely diverging field conditions and farmers' management practices (Kanté *et al.*, 2017; Rattunde *et al.*, 2013). Farmers simultaneously gained experience with new types of variety and contributed to their development.

A recent impact study of participatory sorghum breeding in Burkina Faso over a 15-year period (Trouche *et al.*, 2016) reports that eight varieties developed by the collaborative programme have been registered in the national variety catalogue. These varieties yield 7–30 per cent more than traditional ones and possess a range of farmer-preferred traits. Further, the participating farmers acquired technical skills in varietal selection as well as knowledge essential for choosing the varieties best-suited to their production constraints and uses.

Decentralized trial networks enable farmers to assess the performance of new varieties with their input and under their own on-farm conditions. Such experiential learning is more relevant for farmers than information written on seed packets or leaflets or orally communicated by extension agents or shop owners. Malian farmers explain that '*seeing varieties*' is important for deciding to grow a new variety in their own fields (Christinck *et al.*, 2018). Collaborative trial networks enable many farmers to '*see varieties*' in their vicinity, and the corresponding meetings and workshops facilitate information sharing with plant breeders and among themselves. These trial networks also provide a springboard for establishing and sustaining farmers' seed enterprises.

Enhancing farmers' access to seed of improved varieties through farmer-managed seed cooperatives

Farmers' cooperatives have become relevant actors in the seed sectors of Mali, Burkina Faso and Niger. A project evaluation conducted in 2014 showed that seven farmers' cooperatives (two in Mali and Burkina Faso, respectively, and three in Niger) had each produced and distributed between 12 to more than 100 tons of seed in 2013. The seed produced by these seven cooperatives alone was sufficient for sowing 16,000 ha of sorghum and 11,500 ha of pearl millet (Christinck *et al.*, 2014). As of 2018, at least 30 cooperatives were producing and selling seed in Mali alone; and, in all three countries, the number of persons engaged in seed production and marketing is growing steadily as the cooperatives increase their capacity to sell seed in local markets.

In Burkina Faso, the number of farmers purchasing certified seed increased 25-fold over 5 years in areas targeted by a participatory breeding programme (Trouche *et al.*, 2016). Several very large farmers' unions in Burkina Faso and Niger now sell sorghum and pearl-millet seed; major government programmes further contribute to widespread distribution of certified seed from farmers' seed enterprises. The potential for these actors to achieve broad geographic coverage and serve farmers on a regional and country basis is growing with the increasing adoption of new open-pollinated and hybrid varieties (Smale *et al.*, 2016).

Over time, many cooperatives have improved their capacity to manage finances and increase the volume of grain they market, with higher yields of new varieties contributing to this development. Based on experiences with pre-financing and recovering costs for fertilizer and foundation seeds for hybrid seed producers, some cooperatives have started providing seed on credit to other cooperative members, adapting the traditional practice of providing seed in exchange for grain at harvest. The recipients of these 'seed credits' are not expected to return money to the cooperative, but an amount of grain of equivalent value which the cooperative can market. This practice helps farmers to access seed of new varieties without the constraints of cash availability.

Recipients of such 'seed credits' are obliged to sow at least some of the seed in a plot that can easily be visited by other farmers on organized visits. Through such measures, the cooperatives address the challenge of providing benefits to their members while increasing their marketing options as well as respecting traditional social norms – for instance, that access to seed should not be denied. Some cooperatives have also managed to increase their 'revolving fund' or 'social fund', making their business model more resilient and accepted within in the community.

The concrete outcomes of collaborative breeding networks have included opportunities for generating income from seed production, conditioning and sales, in addition to benefits from growing the new varieties (Smale *et al.*, 2016, 2018). In Burkina Faso, for example, the production and marketing of seed has become an important source of income for sorghum farmers associated with such networks (Trouche *et al.*, 2016).

Several farmer-managed cooperatives have recently begun engaging with researchers in order to enhance varieties of other crops in a similar way and to produce seed of other crops, such as groundnuts, cowpeas, soybeans, Bambara groundnut, maize or rice. They thereby increase their business opportunities while contributing to the diversification of the farming and food systems in their communities (Smale *et al.*, 2018).

'Upscaling' participatory breeding: planning for replicable approaches from the outset

Scale and sustainability have been identified as potential problems of participatory plant-breeding programmes, with institutionalization being a major challenge (Almekinders and Hadorn, 2006). In the West African example presented here, the partnering of an international research institute, ICRISAT, with national breeding programmes and farmers' organizations of several countries was a major step towards large-scale implementation. The overall coordination by ICRISAT guaranteed the continuity and coherence of research, while the cooperation between national research institutions and farmers' organizations was crucial to implementation (Trouche *et al.*, 2016).

Working with farmers' organizations, particularly cooperatives engaged in grain marketing, proved central to the sustainability and dynamic evolution of the collaboration. Relationships with farmers' organizations can be broader and longer-lasting than those involving individual farmers or NGOs whose programme or personnel may shift depending on external funding. Emphasising a model of collaborative engagement that allows both replicability and flexibility to adapt to local needs and conditions – such as choice of trial entries or the application of locally relevant management practices – enabled upscaling and networking across locations and even across countries.

Expanding networks and strengthening relationships can be an important outcome of collaborative research projects. These relationships can facilitate creation of advantages beyond the lifetime of a specific project (Restrepo *et al.*, 2014). Both horizontal networking (e.g. among several farmers' cooperatives) and vertical networking (e.g. between farmers' cooperatives and international and national research programmes) also played an important role in the work presented here.

Challenges for sustainability and needs for future study

Farmer capacity-building and collaborative advantages

Strengthened capacities among local farmers have been a major outcome of long-term researcher–farmer collaboration in West Africa and a key element in building and sustaining collaborative advantages for breeding and seed initiatives. Continuous attention has been paid to the capacities of farmers to conduct variety-evaluation and seed activities, with training programmes and discussions regularly organized at feedback and planning meetings. After several years of collaboration, farmers not only understood experimental error and could conduct trials effectively: they were increasingly experimenting on their own. For example, they sowed new varieties in alternative intercropping patterns or adjusted the sowing dates to optimize food and feed objectives or target niche markets.

Well-functioning farmers' organizations have proven vital to building and sustaining collaborative advantages for variety development and seed dissemination in West Africa. It was often necessary to strengthen members' understandings of cooperative governance and of their responsibilities, to sustain their cooperatives. Farmers' cooperatives engaged in seed production and dissemination need additional capacity building (communication, marketing, business planning and technical aspects of seed quality) to continue scaling up their operations.

Strengthening farmers' organizations is a major unrecognized development opportunity. The need is great, the benefits accrue to communities and 'pay' continued dividends – but such support has been rare, despite numerous seed- and aid-investments. Educating and training members of cooperatives and elected officials to be able to contribute effectively to the development of their cooperatives constitutes the fifth of seven globally recognized cooperative principles.[2] Studies that document the benefits of strengthening farmers' cooperatives and how the global cooperative movement and others can best contribute to building the capacities of West African farmers' cooperatives engaged in variety- and seed-activities would enable and sustain the upscaling of these collaborative efforts.

Building tools that ensure sustainable funding for decentralized breeding and variety evaluation

The number of farmers participating in experimental variety evaluations was and is largely driven by the funding available. In order to reduce dependency on external funds, one farmers' cooperative in Mali reduced the costs of conducting trials by hiring and training local secondary school graduates to collect data and conduct village-level meetings. However, for the foreseeable future, the longer-term sustainability of collaborative breeding networks will still depend largely on external funding. Simply earmarking a tiny percentage of food- and seed-aid for these countries to support in-country breeding research could fulfil that goal in the mid-term and help to end the need for such aid in the long run.

Developed countries like Germany have a long history of decentralized on-farm variety evaluation and communication with farmers undertaken by publicly funded regional agencies. A similarly sustainable financial basis has not yet been created for such activities in the countries of West Africa – but it could be developed, perhaps based on national funds to which donors contribute. Such approaches seem worth exploring. Further study of the benefits and costs of alternative efforts to 'develop' seed systems in West African countries would help guide future investments.

Final remarks

Trust is recognized as a necessary precondition or enabling factor for the creation of collaborative advantages (Huxham and Vangen, 2005). Building and maintaining strong relationships with farmers' cooperatives and their members in West Africa was critical for building trust, honest dialogue and joint learning. However, trust is not something that is simply present (or not) in collaborative projects: Restrepo *et al.* (2014) found that trust among participants emerges as a result of well-structured and well-facilitated processes. Furthermore, clarification of roles and potential benefits from participation in a collaborative project can help to avoid the dangers of opportunistic behaviour and unrealistic expectations.

A general rule followed in the West African work presented here was that farmer collaboration was voluntary, based on the farmers' own interests. Monetary benefits were limited to reimbursing farmers' costs, such as for travel to meetings. This helped to maintain and strengthen collaborative partnerships over time. It also meant that farmers made a 'private investment' in the collaborative process; to sustain the collaboration, rewards in terms of tangible outcomes were needed within a foreseeable period.

The chances to see new varieties and access seed of the most interesting ones were initially the primary 'incentives' for farmers' participation in collaborative trials. Later, the opportunities for farmers to participate in the production, processing and marketing of seed strengthened their motivation, along with the opportunities of cultivating new varieties, including landrace-based hybrids, with clear yield advantages.

Farmer–researcher collaboration should be recognized as involving 'alliances' between different types of 'organizations' engaged in a long-term partnership to create collaborative advantages. This perspective helps to counteract the widespread view of farmers as 'beneficiaries' or passive adopters of new varieties. Especially in countries like Mali, Niger and Burkina Faso, where breeding is publicly funded with little or no private sector investment, such alliances between farmer-managed cooperatives and breeding programmes can be highly advantageous for both sides.

Notes

1 See Weltzien *et al.* (this volume) for names and description of participating organizations.
2 See www.ica.coop/en/cooperatives/cooperative-identity.

References

Almekinders C, Hadorn J, 2006. Bringing farmers back into breeding. Experiences with Participatory Plant Breeding and challenges for institutionalisation. *Agromisa Special 5*, Wageningen: Agromisa.

Bauchspies WK, Diarra F, Rattunde F, Weltzien E, 2017. 'An Be Jigi' Collective cooking, whole grains, and technology transfer in Mali. *FACETS* 2, 955–968.

Christinck A, Kaufmann BA, 2018. Facilitating change: Methodologies for collaborative learning with stakeholders. In Padmanabhan M, (ed) *Transdisciplinary Research and Sustainability: Collaboration, Innovation and Transformation*, 171–190. Abingdon: Routledge.

Christinck A, Diarra M, Horneber G, 2014. Innovations in seed systems: Lessons from the CCRP-funded project 'Sustaining Farmer-Managed Seed Initiatives in Mali, Niger, and Burkina Faso'. Minneapolis, MN: The McKnight Foundation.

Christinck A, Rattunde F, Kergna A, Mulinge W, Weltzien E, 2018. 'You can't grow alone': Prioritized sustainable seed system development options for staple food crops in sub-Saharan Africa: Cases of Kenya and Mali. Final Project Report. Program for Accompanying Research for Agricultural Innovation (PARI). Bonn: Center for Development Research (ZEF).

Hoffmann V, Probst K, Christinck A, 2007. Farmers and researchers: How can collaborative advantages be created in participatory research and technology development? *Agriculture and Human Values* 24, 355–368.

Huxham C, Vangen S, 2005. *Managing to Collaborate. The Theory and Practice of Collaborative Advantage.* Abingdon: Routledge.

Isaacs K, Weltzien E, Diallo C, Sidibe M, Diallo B, Rattunde F, 2018. Farmer engagement in culinary testing and grain-quality evaluations provides crucial information for sorghum breeding strategies in Mali. In Tufan HA, Grando S, Meola C, (eds) *State of the Knowledge for Gender in Breeding: Case Studies for Practitioners.* Working Paper No. 3, 74–85. Lima: CGIAR Gender and Breeding Initiative, International Potato Center (CIP).

Jones K, 2014. *Emerging Seed Markets, Substantive Seed Economies and Integrated Seed Systems in West Africa: A Mixed Method Analysis.* Dissertation, Department of Agricultural Economics, Sociology, and Education. University Park: Pennsylvania State University.

Kante M, Rattunde HFW, Leiser WL, Nebié B, Diallo B, Diallo A, Touré AO, Weltzien E, Haussmann BIG, 2017. Can tall guinea-race sorghum hybrids deliver yield advantage to smallholder farmers in West and Central Africa? *Crop Science* 57 (2), 1–10.

Leiser WL, Rattunde HFW, Piepho H-P, Weltzien E, Diallo A, Toure A, Haussmann BIG, 2015. Phosphorous efficiency and tolerance traits for selection of sorghum for performance in phosphorous-limited environments. *Crop Science* 55 (3), 1152–1162.

Rattunde HFW, Weltzien E, Diallo B, Diallo AG, Sidibe M, Touré AO, Rathore A, Das RR, Leiser WL, Touré A, 2013. Yield of photoperiod-sensitive sorghum hybrids based on guinea-race germplasm under farmers' field conditions in Mali. *Crop Science* 53, 2454.

Restrepo MJ, Lelea MA, Christinck A, Hülsebusch C, Kaufmann BA, 2014. Collaborative learning for fostering change in complex social-ecological systems: A transdisciplinary perspective on food and farming systems. *Knowledge Management for Development Journal* 10 (3), 38–59.

Siart S, 2008. Strengthening local seed systems: Options for enhancing diffusion of varietal diversity of sorghum in Southern Mali. *Communication and Extension Series* No. 85. Weikersheim: Margraf.

Smale M, Kergna A, Assima A, Keita N, Traoré A, Haggblade S, Témé B, 2016. Use and adoption of sorghum improved varieties and hybrids in Mali: Economic impact. *Policy Research Brief 21, Feed the Future Innovation Lab for Food Security Policy.* East Lansing: Michigan State University.

Smale M, Assima A, Kergna A, Thériault V, Weltzien E, 2018. Farm family effects of adopting improved and hybrid sorghum seed in the Sudan Savanna of West Africa. *Food Policy* 74, 162–171.

Trouche G, vom Brocke K, Temple L, Guillet M, 2016. Analyse de l'impact des programmes de sélection participative du sorgho conduits au Burkina Faso de 1995 à 2015. Rapport final validé par le chantier Impress. Montpellier: Centre de coopération internationale en recherche agronomique pour le développement (CIRAD).

13 Sourcing and deploying new crop varieties in mountain production systems

Bhuwon Sthapit, Devendra Gauchan,
Sajal Sthapit, Krishna Hari Ghimire,
Bal Krishna Joshi, Paola De Santis and
Devra I. Jarvis

Introduction

Around the world, considerable crop genetic diversity continues to be maintained on-farm in the form of traditional crop varieties (Fenzi *et al.*, 2017; Jarvis *et al.*, 2008; Mulumba *et al.*, 2012; Thomas *et al.*, 2015). However, these smallholder farming systems are increasingly threatened by national and international pressures to produce genetically homogeneous crops. Government extension services focus on subsidies for modern varieties and associated agricultural inputs; and lack of access to sufficient quantities of high-quality diverse crop biodiversity planting materials limits the potentials for optimal use of this important asset to prevent pre- and post-harvest crop loss (Frison and IPES-Food, 2016). In developing countries, modern varieties released by the formal seed sector cover only a small part of the total crop acreage (Coomes *et al.*, 2015; McGuire and Sperling, 2016; Tripp, 2001), but increasingly centralized variety release and stringent seed regulatory frameworks make it difficult for farmers to access crop genetic resources (Halewood, 2016; Tripp, 1997). The result is a shrinking of genetic diversity of traditional crops, with farmers losing options to meet their needs and preferences (FAO, 2010; Frison *et al.*, 2011; Frison and IPES-Food, 2016). Precisely at a time when more new diversity is needed to cope with climate and market change, farmers find themselves with fewer alternatives available (Atlin *et al.*, 2017).

Deployment of new crop varieties occurs through diverse, innovative and evolving methods. Conventional processes of deploying new crop varietal diversity are time-consuming, offering limited choices and often targeting high-production potential environments (Witcombe *et al.*, 1996, 1998). Moreover, conventional plant-breeding approaches make varieties available to farmers late in the development process, whereas participatory breeding approaches, using new sourcing and deployment methods, can provide farmers with access at a much earlier stage of development. These new methods are intended to remove barriers to farmers' access to greater varietal diversity; to fast-track plant breeding, variety release and registration; align research to the needs of smallholder farmers and promote local-level seed innovation (Gyawali *et al.*, 2010; Johnson, 1972; Joshi and Sthapit, 1990; Joshi *et al.*, 1997; McGuire, 2008).

Field practitioners in research and extension agencies are often not fully aware of the pros and cons of these new methods. Moreover, practitioners may need to decide quickly which strategies to adopt, without enough time for thorough analysis of the strengths and weaknesses. Selecting the appropriate method for sourcing new crop diversity involves four major aspects (Jarvis *et al.*, 2016): whether there is sufficient diversity of traditional crop varieties within the production system; whether farmers can access this diversity; whether key performance information is available; and finally, the ability of farmers and communities to realize the true value of the materials they manage and use. Examining the pros and cons of strategies for sourcing and deploying varieties from conventional and non-conventional breeding programmes, in this chapter we offer multiple options to support practitioners in identifying the best methods for sourcing and deploying new crop diversity in various situations.

We test the potential of an heuristic framework as a decision-making tool, as proposed by Jarvis *et al.* (2011) and presented in Figure 13.1, to provide

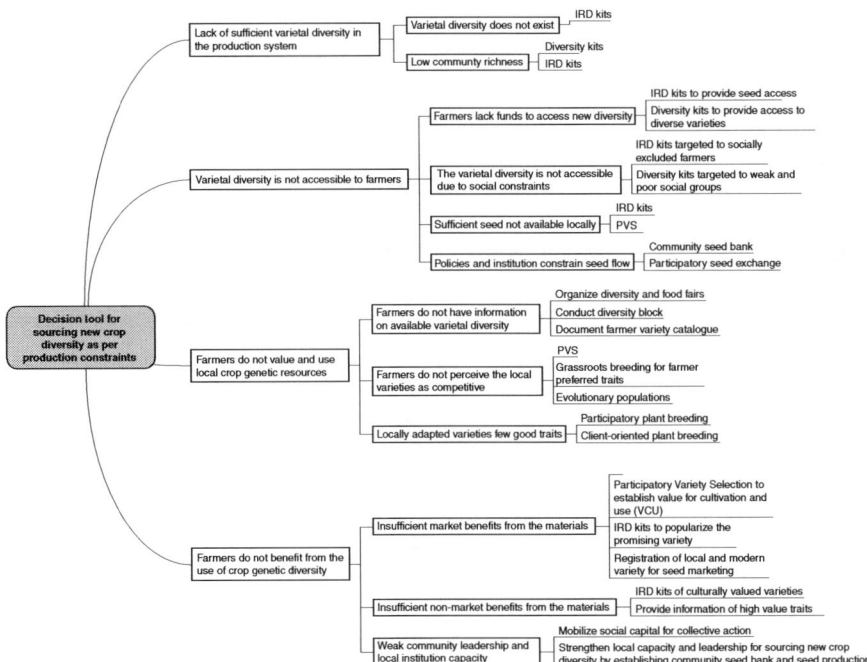

Figure 13.1 Decision tool for sourcing and deploying new crop diversity as per production constraints.

Source: Figure modified from Jarvis *et al.*, 2011.

farmers with diversity-rich solutions for managing environmental variability, biotic and abiotic production constraints, and to broaden the genetic base of traditional crop diversity for enhanced system resilience. This tool provides a set of options based on comparative analysis of the methods available for sourcing and deploying new crop varieties, which can be applied in various situations and to face a range of constraints related to on-farm genetic diversity.

In Nepal, a country characterized by extreme and diverse systems, releasing new varieties has often been challenging. The adoption rates of newly released technologies are often rather low, due to the wide range of different environments where new varieties do not always perform as well as expected. The adoption of participatory approaches and the options presented in the framework presented here, aimed at improving farmers' adoption rates and better meeting farmers' needs, have resulted in somewhat higher acceptance rates.

Sourcing and deploying new varieties

Methods for sourcing and deploying new varieties are grouped under two broad categories, conventional and participatory approaches, to highlight the differences. The conventional system takes a linear view of technology development and transfer. Varieties are developed in research centres, where breeders focus on a major functional trait (e.g. yield, resistance to a given pathogen) without considering farmers' needs and constraints; it is only when the variety and technology are fully developed and tested that they are made available to farmers. By contrast, in participatory approaches, farmers and end-users are involved at an earlier stage, as key actors in the technology development and testing process even before dissemination. This results in active farmer participation in the development of the variety itself, as researchers will take farmers' knowledge and needs into consideration. This is why, when using participatory approaches, the terms 'sourcing' and 'deployment' are relevant: they reflect the active role of famers in developing new varieties and their contributions in modelling them.

In practice, the distinction between the two systems today is less pronounced than earlier. Agricultural researchers in Nepal often, although not always, incorporate lessons from both systems into their work.

Conventional methods for sourcing and deploying new varieties

The conventional system of variety deployment follows the centralized model developed during the Green Revolution. Newly developed and released varieties from research stations are passed on to extension agencies for popularization and dissemination along with modern agronomic practices. Variety development is seen as the responsibility of plant breeders; extension workers play a major role in disseminating seed and information on new varieties. Only finished products are used for popularization and dissemination, one argument being that it is unfair to expose farmers to under-researched materials that might cause them avoidable losses (Witcombe *et al.*, 1998). Three methods are used for popularization and dissemination of final products of conventional breeding: farmer field trials, frontline demonstration and mini-kits (Table 13.1).

In conventional plant breeding, multi-location varietal testing is conducted, to see how a given variety performs in different target environments and how broadly or narrowly new cultivars can be recommended due to genotype by environment interaction (G × E) (Ceccarelli, 1989). Much of the funding allocated to crop improvement is devoted to establishing value for cultivation and use (VCU), which is done through multi-location varietal testing under the close supervision of researchers. While well-intentioned, establishing VCU delays farmers' access to new pipeline materials for decentralized and participatory evaluation (Ceccarelli, 2009; Sah *et al.*, 2016; Witcombe *et al.*, 2005). Of the many genotypes evaluated in the multi-location varietal testing system, only a few will eventually become successful varieties (Witcombe *et al.*, 1998).

Farmer field trials (FFTs)

Farmer field trials (FFTs) are used to evaluate on-farm performance of advanced lines with respect to local check varieties. Under the conventional plant breeding system, an FFT is the earliest stage at which farmers can access new varieties. In Nepal, FFTs are a key step in conventional plant breeding that generates the data required for variety release or registration. An FFT includes four to six best-performing advanced lines, a few popular released varieties and a local check variety in a randomized complete block design, using a farmer's field. Data on flowering and maturity time, plant height, diseases/pests tolerance, grain yield and farmer preferences are recorded, and an analysis of variance is computed to compare means.

First introduced in 1973 in Nepal, FFTs were rapidly mainstreamed under the national crop commodity programmes (Farrington and Mathema, 1991). Experimental lines that have been fixed after hybridization typically undergo an additional five to seven years of evaluation (one to two years in observation nurseries, two in initial evaluation trial, and two to three in coordinated variety trials) before being included as advanced lines in FFTs (Sah *et al.*, 2016). Since most of the genetic variation generated through hybridization has been de-selected by the time advanced lines enter FFTs, the scope of farmer choice at this stage is very limited. Farmers who host the trials on their land get access to information and seeds of the new lines – but only three to six replicates of the trial are set in any given village, so at most six farming households and their neighbours or friends get access to seeds of five to ten new varieties, including the released checks.

Table 13.1 Methods of sourcing diversity: pros and cons

Methods of sourcing and deployment	Pros	Cons
Conventional methods		
Farmer field trials (FFT)	• generates on-farm data including farmer feedback that is necessary for release of crop varieties in Nepal • greater researcher control, resulting in arguably more reliable scientific data • coordinated trials provide high-quality data from breeders around the country • farmers get access to seeds and information on new varieties • use of standard checks allows comparison of varietal performance over long durations	• little flexibility for farmers to manage and make decisions suitable for their conditions • unlikely to represent the diverse and often marginal production environments • check varieties are selected by researchers and may not include local varieties of relevance to farmers, although this can be remedied easily
Frontline demonstration	• simple to design and manage • effective way of demonstrating and communicating the superiority of new varieties	• only a few elite farmers have access to participation, seeds and associated technology • the variety is demonstrated under ideal management conditions and may not reflect the realities of farmers' fields
Mini-kits	• can be a powerful tool for disseminating new varieties if all components of the kit are included	• may be limited to elite farmers • fertilizers and pesticides need to be included as part of the kit, but in Nepal supply of these inputs has been inconsistent
Participatory methods		
Participatory varietal selection (PVS)	• makes use of existing varieties (local, farmer variety, escapees and released) to give farmers more choice	• relatively expensive to set up mother-baby trials

Methods of sourcing and deployment	Pros	Cons
	• is both an extension and research method; saves time to reach farmers • helps in setting breeding goals in PPB and COB • uses farmers' knowledge to identify relevant varietal traits • identifies non-acceptable varieties very quickly (effective screening) • creates awareness of and market demand for acceptable varieties • farmer involvement helps to identify problems in variety and to set breeding goals for PPB • allows evaluation of multiple traits • increased genetic diversity by PVS in high potential production system in India and Nepal	
Informal research and development (IRD kits)	• increased uptake of new varieties • higher farmer-to-farmer seed dissemination • mechanism for testing seed marketing of a new variety • provides information on overall acceptability of variety • cost-effective	• does not determine the eventual acceptance of varieties • human resources required to package and distribute large numbers of packets in remote locations can be challenging
TRICOT (Triadic comparisons of technologies)	• generation of large dataset and use of Bradley-Terry model for ranking data allows for evaluating varieties without setting up on-farm variety trials • combining weather data with the analysis can help measure genotype/ environment interaction to determine appropriate recommendation domain for varieties • use of ClimMob software and digital data collection can minimize error	• farmers use a common (shared) check instead of their own variety as local check • challenges in receiving feedback voluntarily from crowds/farmers using mobile phones (e.g. Ethiopia, India and Nepal) • use of ClimMob software requires skill development

continued

Methods of sourcing and deployment	Pros	Cons
Diversity kits	• revalorize rare and unique local varieties by improving access to them • low-cost activity that can strengthen community institutions	• concept of diversity kits has been misunderstood and used to provide seeds of commercially available improved and hybrid varieties as well
Participatory seed exchange (PSE)	• low-cost, quick way to consolidate community-level seed exchange by organizing it seasonally • especially effective for vegetable crops, as not so many seeds are required • many crops and varieties can be exchanged in the course of a one-day event	• challenging to organize if more than 50 farmers participate in one day; data recording may also be challenging • possibly less effective for cereal crops where farmer will need greater quantities of seeds, heavier to carry to the exchange event

Frontline demonstration

Frontline demonstration is a method used to demonstrate a new technology (e.g. a newly developed or notified variety) or an improved management practice to farmers, in order to increase awareness and thereby adoption. Frontline demonstrations are carried out by the extension services with the participation of researchers and selected collaborating farmers. Plots chosen for demonstration are larger than those used in farmer's field trials. For instance, a typical plot in an FFT in the hills measures around 12–$20\,m^2$, while a demonstration plot can be around 250–$500\,m^2$ or more, depending upon the crop. There will normally be one to six new notified varieties to be promoted. Large plot sizes and highly trafficked locations are chosen to demonstrate the impressive performance gains of the new varieties. Specific technical data recording is not needed, but farmer preference observations may be recorded through anecdotes or surveys to use in reporting. With demonstrations, farmers gain access to new varieties quite late in the variety development process, as only notified (released or registered) or improved varieties are used. With only a few demonstrations in a village, fewer than five farmers will normally gain access to seed in a given year. However, a successful demonstration can generate seed locally, which can be distributed to more farmers in the following year.

Mini-kit

A mini-kit is a packet of improved seed of one to three recently notified varieties plus the necessary chemical fertilizers and pesticides for planting on a parcel of land (roughly $500\,m^2$), to determine the suitability and popularity of the variety. In theory, seed, inputs and feedback–card together constitute a mini–kit; in practice in Nepal, however, the provisioning of fertilizer and pesticides has been inconsistent. Feedback cards allow crop coordinators to identify location–specific and widely adapted crop varieties for future recommendation and seed marketing.

Mini–kits have traditionally been a main source of modern varieties developed by public–sector breeding programmes, although there have been few published peer–reviewed articles on implementing mini–kits and their impact. They have been successfully implemented in remote areas of Nepal where farmers had access to new varieties (APROSC, 1983). The concept was successfully introduced by the USAID Integrated Cereal Project in Nepal[1] (Chaurasia and Perdon, 1981). As each mini–kit provides 25 to 50 farmers with seeds, more farmer can reap the benefit of access to seeds. However, such access may be limited to progressive farmers who have good connections with the local extension workers. Moreover, the seeds included in the mini–kit are recently notified varieties, which limits the scope of farmer choice. Farmer feedback data are not required for variety release or registration, but are helpful for forecasting demand for the variety and for progress reporting.

Participatory methods for sourcing and deployment of new varieties

Participatory methods incorporate the perspectives of farmers, usually by inviting them to participate in varietal evaluation of activities and make decisions about varietal choice. Success here depends on the researchers' ability to incorporate the knowledge and preferences of the farmers themselves (Burman *et al.*, 2018). The most popular participatory methods of testing new varieties with farmers' check are Participatory Variety Selection (PVS) and other forms of Participatory Plant Breeding (PPB), including client–oriented breeding, grassroots breeding and evolutionary plant breeding (Ceccarelli, 2009; Döring *et al.*, 2011; Murphy *et al.*, 2016; Sthapit and Rao, 2009; Sthapit *et al.*, 1996; Witcombe *et al.*, 1996; Witcombe *et al.*, 2005). We focus here on the methods used for sourcing and deploying diversity and disseminating the products of PPB as well as conventional breeding, such as PVS, IRD, diversity kits and triadic comparisons of technologies (TRICOT). Although participatory seed exchange (Shrestha *et al.*, 2013) is not part of the breeding process as such, it has been included here because it provides mechanisms for sourcing of new diversity and allows farmers to select the varieties they want to evaluate.

Participatory varietal selection (PVS)

PVS involves selection of fixed genotypes by farmers in their target environments using their own selection criteria (Joshi and Witcombe, 1996). There may be several notified (released or registered) varieties still not accessible to farmers: PVS is intended to remedy this. When fixed experimental lines are used in PVS, it also helps generate data to support the case for variety registration, by demonstrating that participating farmers are interested in adopting them.

A successful PVS consists of four steps: (i) identification of farmers' requirements for a new variety, (ii) search for genetic materials (landraces, escape varieties,[2] notified and pipeline varieties), (iii) testing varieties in farmers' fields and (iv) scaling farmer-preferred varieties (Sthapit and Jarvis, 1999). PVS typically employs intensive systems of on-farm participatory evaluation which involve frequent field visits by researchers, to interact with farmers and assess their preferred traits in new varieties (Joshi and Witcombe, 1996). Recently, 'mother and baby' trials (MBT) have replaced the third step in PVS, testing varieties in a target environment in order to generate data on value for cultivation and use for variety release. A 'mother' trial is equivalent to an FFT in the conventional system: they share the same experimental design, and the data generated are used in the same way for variety release or registration. In a 'baby' trial, a small quantity of seed of the varieties being evaluated in the mother trial is distributed to individual farmers to compare with their own local checks under their own management. Typically, a PVS will have two to three mother trials with five to six or more varieties being evaluated per village, and 25 to 50 baby trials per variety (Witcombe *et al.*, 2001). Informal research and development (IRD) kits (described below) have been deployed as step (iv) to scale up farmer-preferred varieties in large geographic areas.

Baby trials in a PVS provide access to seeds of all the pipeline entries to a larger number of farmers at the same time as the mother trial is being conducted, thereby greatly improving farmer access to diversity. When the mother and baby trials are conducted simultaneously, yield and agronomic data as well as farmer perception data for variety registration and release can be generated in the same year. Importantly, farmers have the opportunity to select and spread sister-populations of the varieties through informal channels, even before the varieties are notified – thus fast-tracking the potential benefits of new varieties.

Informal research and development (IRD) kits

IRD is an informal research approach to popularizing recently notified or pipeline varieties at low cost. It is similar to a baby trial in the mother/baby trial set-up described above, but the number of farmers is higher and feedback collection requirements lower. The main intention is to provide better farmer access to new crop genetic diversity than under conventional plant breeding (Joshi *et al.*, 1997). IRD evolved out of necessity in Nepal at the Lumle Agriculture Centre (Joshi and Sthapit, 1990). Lack of transport in the mid-hills of Nepal in the late 1980s and early 1990s had limited farmers' access to promising new varieties being tested or developed at the Centre. Hence, researchers distributed large numbers of IRD kits, consisting of a small packet of seeds (0.25 to 1 kg for cereals) and an information leaflet, to introduce and deploy promising varieties to farmers in remote areas. Monitoring visits were conducted two or three years later, to assess the spread of the varieties.

IRD is an efficient, appropriate and cost-effective means of enabling large numbers of farmers in otherwise inaccessible areas to obtain new genetic diversity (Joshi and Sthapit, 1990; Joshi and Witcombe, 2002; Witcombe et al., 2017). If the amount of seed provided is kept small, many farmers can test the varieties under their own management. As small quantities help keep possible losses to a minimum, this approach allows researchers to test experimental lines as well. Joshi and Witcombe (2002) found that IRD kits resulted in greater farmer-to-farmer dissemination than baby trials conducted within a PVS programme. Since conducting a PVS entails more field staff, farmers see the researchers regularly and therefore expect to get seeds in the following year too. Such a luxury is not available to farmers receiving IRD kits, so they share seeds with each other as a backup for their own supply.

Diversity kit

A diversity kit is a set of seeds of three or more unique, rare or culturally useful landraces in small quantities made available to farmers (Sthapit et al., 2012). The objective is to deploy threatened diversity in farmers' fields, in the hope that these landraces will come into use again. Four-cell analysis, community biodiversity register or other participatory rural appraisal tools are used to identify varieties and landraces that have become rare but might become popular again if seed access could be restored. The diversity kit approach is a methodological descendent of IRD in Nepal, first tested by Bioversity International's global project 'Strengthening the scientific basis of *in situ* conservation of agricultural biodiversity on farm'. The method has also been applied in Brazil as a means of restoring the traditional farmer practices of seed saving for home consumption. Hybrid varieties are not used (Canci et al., 2013).

Similar to the IRD method, feedback on the acceptance of each new variety and reasons for acceptance or rejection are not always collected, as diversity kits are often not part of a variety release or registration process. However, sample surveys similar to IRD feedback can be used to assess the adoption of varieties. The method promotes farmer experimentation by deploying a portfolio of varieties; it encourages farmers to select, exchange and disseminate the varieties most preferred for different situations. This informal research task is shared by many farmers (50–500 sets) who choose location-specific best varieties. For vegetable crops, a diversity kit will include many varieties or even multiple species, for greater dietary diversity from home gardens; with cereals and pseudo-cereals, each household receives three varieties to compare with the local check.

Triadic comparisons of technologies (TRICOT)

On-farm triadic comparisons of technologies – TRICOT – is a modified version of IRD that uses the principle of crowd-sourcing to gather data from a much larger number of farmers than typically done in other methods. In the TRICOT method, each participating farmer is supplied with a set of three varieties and a simple feedback card. The set includes two varieties to be tested, which can be farmer or improved varieties, and a local check variety. The names of the varieties are withheld, for blind testing. Feedback is collected from farmers' self-reported ratings of the worst and best of the three varieties as regards performance traits of pest and disease resistance, abiotic stress tolerance, yield, market value and taste. Since the combinations of the three varieties overlap, statistical methods are used to construct an overall performance ranking of the complete pool of new varieties. This provides results that can be statistically analysed and interpreted. In addition, weather data are logged, using automatic loggers like iButton in the working sites. Voluntary feedback of large numbers of farmers utilizing mobile technology adds a crowd-sourcing dimension to this method, which has been implemented in pilot studies in India, East Africa and Central America (Steinke and van Etten, 2016; van Etten *et al.*, 2016; van Etten *et al.*, 2019).

The TRICOT method provides seeds of many varieties to large numbers of farmers. It also employs local distribution networks and is open to include pipeline, notified and landrace varieties, making it valuable in providing access to materials. If successful in identifying location-specific recommended varieties, TRICOT could significantly reduce the costs of varietal testing, release and distribution, and the farmer seed system would become highly efficient and resilient. However, in some countries, the regulatory frameworks for variety release would have to be changed to accept such a dataset for variety release and registration.

Although this approach is attractive as regards wider testing and scaling up, it requires a structured system to collect information and feedback from farmers. Farming communities in remote mountainous areas often lack such organization; even telephone contact can be difficult. Researchers are still testing the effectiveness and appropriateness of the method in countries where farmers have little formal education, and access to information technologies and networks is not reliable. In developing countries it can be difficult to engage people in voluntary research. The authors tested this approach with a small pilot study conducted in Nepal's Humla and Jumla areas in 2014 and 2015. There were many challenges: not all participating smallholder farmers had mobile phones; mobile-phone networks in mountainous areas are not reliable and coverage is limited; very few farmers responded voluntarily, as they did not see the value of providing feedback; and farmers often got confused with the terminology used (e.g. numbers instead of variety names). This significantly reduced the number of reliable data points. While the initial concept paper on TRICOT emphasized data collection using mobile-phone technologies (van Etten, 2011), later publications include the option of collecting pictorial feedback cards from participants, for later digitalization by research staff (Steinke and van Etten, 2016).

This method relies on externally supported systematic seed-multiplication programmes by research stations for pre-released and released varieties, and requires local institutions, like community seed banks or cooperatives, to supply the large quantities of local seed needed for multi-locational testing by participants. Because significant quantities of seeds are required, that tends to exclude farmers' varieties, where seed availability is often very limited.

Participatory seed exchange (PSE)

Most smallholder farmers still use informal seed systems, self-saved seed in particular. Farmers rely on personal networks and kinships to obtain seed in cases of seed loss or new varieties. Participatory seed exchange (PSE) consolidates such one-on-one exchanges by organizing a village-wide, preferably seasonal, seed exchange where many farmers can share germplasm among themselves (Shrestha *et al.*, 2013). Participating farmers bring seeds and planting materials that they have and are willing to share, and an inventory of the varieties and the amount of seed is prepared. Then farmers examine the seeds on display, ask questions about the varieties to the donor farmers, and register their name in the request sheet if they are interested in a variety. Hearing about the experiences of fellow farmers can help farmers to decide whether a variety will fit their needs or not. When all farmers have had the opportunity to observe, interact and register their requests, the available seeds are divided equally among those who requested them.

In Nepal, PSE was first piloted by the Western Terai Landscape Complex Project (WTLCP) in 2008. In 4 exchanges in 2 years, 126 farmers participated and 96 received seeds of 1 to 10 varieties of their choice from 74 donating farmers. The total number of varieties and accessions of these four exchanges amounted to 110 and 600 respectively (Shrestha *et al.*, 2013). PSE was also used successfully in Nepal's 2015 post-earthquake seed-relief efforts in six earthquake-affected districts. Altogether 503 farmers received seeds from 366 fellow farmers, with each farmer receiving between 1 to 38 varieties (Sthapit and Gautam, 2016).

Of the methods discussed in this chapter, PSE is the most disconnected from plant breeding. It does not generate data for variety release or registration proposals, or for adoption and diffusion studies. However, the process of variety registration does help create an inventory of different seasonal diversity in an area.

Heuristic framework for selecting appropriate methods

Strategic interventions are needed to source and deploy new diversity using an available suite of tools and approaches to meet the diverse seed-source needs using a range of actors and supply channels. Which tool to use will depend on the production constraints faced by farmers as regards sourcing and deploying new materials, as described in Figure 13.1. Often more than one tool, or a suite of tools, can be used to resolve a given sourcing constraint. This will depend on the local agro–ecology, farming systems and seed systems, and the facilitation capacity of the research and extension system (Table 13.2).

Farmers need a genetically diverse portfolio of crop varieties, suited to a range of agro–ecosystems and farming practices and resilient to climate change. Farm size is also a factor; fewer nutrient-dense and climate–resilient crops, vegetables and fruits are found on larger farms, whereas species (e.g. cereals, sugar and oil crops) that are readily cultivated with mechanized techniques are likely to increase (Herrero *et al.*, 2017). Lack of sufficient diversity in the production system (Constraint 1) can be addressed by using FFTs, mini-kits, PVS, IRD, diversity kits and TRICOT to identify and introduce genetic diversity from elsewhere into the system. IRD and diversity kits are cost-effective and easy to implement. Crowd-sourcing principles can be applied to collect feedback from IRD and diversity kits as well, if the methodological rigour of double-blind variety testing is not required.

Genetic diversity in Nepal's traditional highland crops (buckwheat, foxtail millet, proso millet, naked barley, and amaranth) are limited to a few varieties at the community level, so farmers have limited options for selection (Gurung *et al.*, 2017; Palikhey *et al.*, 2018; Parajuli *et al.*, 2018; Pudasaini *et al.*, 2017). At the national level there is greater diversity, but, given the weak national and international crop improvement programmes for these crops, options for testing new crop varieties through conventional methods are not available. Participatory methods like IRD and diversity kits can be used to offer new crop diversity from national and international genebanks and from other communities for decentralized testing and selection. If funding is available, PVS can be used to generate data for variety registration or release, to allow these varieties to be disseminated through the formal system. These participatory methods offer excellent opportunities to disseminate traditional and under-researched crop varieties, and to improve their access to farmers. Moreover, in the Nepalese system for variety release, less data are required in the case of known historical landraces.

With traditional minor and neglected local crops, where diversity exists the crops are cultivated by few farmers in small areas and the supply of seed is insufficient to meet the needs of all interested farmers, both participatory methods such as IRD (one variety) or diversity kits (three varieties) can be employed for rebuilding the local seed system. Local organizations can organize participatory seed exchanges to promote seasonal exchange for several farmer-maintained varieties. Also community seedbanks can regularly deploy diversity kits of rare but unique varieties, and can organize PSEs.

If diversity is available in the production system, but access is the issue (Constraint 2), then diversity kits, IRD and PSE can be used, depending on the number of varieties available. Adoption also depends on identifying

Table 13.2 Conventional and participatory variety sourcing methods: comparison

	Conventional methods			Participatory methods				
	FFT	Frontline demonstration	Minikits	PVS	IRD	TRICOT	Diversity kits	PSE
Purpose	test pipeline varieties for on-farm performance and farmer acceptance	showcase new varieties or technology to promote adoption	popularize new varieties and package of practices	test farmer variety preference on experimental lines and new varieties	increase access to new varieties and generate adoption data	generate variety preference data on experimental lines on a large scale	provide access to seeds of several farmer varieties that are rare, unique or special	multiplex farmer to farmer seed exchanges
Tool used for	Research	extension	extension	research and extension	extension	research and extension	extension	extension
Type of varieties used	Pipeline	registered	registered	pipeline, registered	pipeline, registered	pipeline, registered, landraces	landraces	any farm saved seed
Types of on farm trials	yield trial only	demo plot only	farmer trials only	both yield (mother) trial and farmer (baby) trials	farmer trials only	farmer trials only	farmer trials only	farmer trials only
Design	RCBD	non replicated	–	RCBD in mother trials	–	–	–	–
Trials per village	3 to 6	2 to 5 demos	5 to 25	3 to 6 mother trials and 25 to 50 baby trials	500 to 1,000	1,500 to 2,000	50 to 500	60 to 150
Entries per trial	4 to 6	1 to 2	1 to 3	5 to 6 plus checks in mother trials and 1 to 2 per farmer in baby trials	1 to 3 per village, 1 per farmer	8 to 12 per study, 3 per farmer	6 to 12 per village, 3 or more per farmer	100 to 500 per exchange, a farmer typically chooses 2 to 5
Farmer feedback collection	preference ranking after trial visit	typically, not collected	feedback collected by JTA	preference ranking in mother trial, household survey a 2–3 months after harvest in baby trials	informal anecdotes during field visit, household surveys optional	collected using mobile devices or collection of feedback card.	anecdotes and household surveys optional	anecdotes and household surveys optional
Use of feedback	Variety registration or release proposal	–	Monitor variety adoption	Variety registration or release proposal	Monitor adoption and spread of variety	variety registration or release proposal	Monitor adoption and spread of variety	Monitor adoption and spread of variety

seed-supply sources readily available to farmers, extension workers, agro-vets, and seed companies (Gauchan, 2017; Pautasso *et al.*, 2013). If access is limited because of resource problems, weak social ties, formal systems that do not offer options, and lack of a seed-retailer network, then periodic free seed distribution under IRD (for varietal diversity) and diversity kits (for species diversity) can be cost-effective ways of introducing new diversity (Joshi *et al.*, 1997; Witcombe *et al.*, 2017).

If the problem is that farmers do not value or use local crop diversity (Constraint 3), then conventional methods can be used for modern varieties. Conventional approaches such as front-line demonstrations and mini-kits can offer notified varieties as a source of new crop diversity; however, that automatically excludes non-registered/farmers' best landraces and thus excludes farmers' varieties. In Ethiopia, crowd-sourcing experiments together with field trials resulted in the identification of a large portfolio of barley and durum wheat landraces preferred by farmers. Nationally bred varieties were compared with commercial improved varieties. Some 60 per cent of the traditional durum wheat populations grown in Ethiopia outperformed improved varieties: the latter were based on Mediterranean parents that lacked local adaptation to Ethiopian conditions (Mengistu *et al.*, 2016).

If farmers do not benefit from the existing diversity (Constraint 4) then sourcing new crop diversity alone will not be sufficient. Strong local community-based organizations that support commercial seed production, adding value to local crop diversity, and collective action for public awareness and marketing, will be necessary (see Jarvis *et al.*, 2011 for a review).

Discussion

This presentation of how various sourcing and deployment methods are typically employed has been descriptive, not prescriptive. Significant innovation can be achieved by modifying these methods to fit local needs and contexts. Farmers are competent innovators, often willing to try new seeds and practices if diversity is available (Johnson, 1972; Richards, 1986; Sperling and Loevinsohn, 1993; Sthapit *et al.*, 2015). Selection studies have demonstrated that decentralization (direct selection by farmers in the target environment) is almost always more successful in terms of response to selection (Almekinders and Elings, 2001; Ceccarelli *et al.*, 1996; Packwood *et al.*, 1998; Simmonds, 1991; Witcombe *et al.*, 1996; Witcombe *et al.*, 2005). Participatory methods such as PVS, IRD and diversity kits have greater relevance in the context of a change in the model of research and extension. In the conventional approach, breeders do the research (develop new varieties), and extension agencies conduct the extension work (transfer of the new varieties). In this two-stage research-plus-extension model, the breeder is responsible until the stage of breeder-seed production and variety release. By contrast, in participatory approaches, research and extension go hand in hand: having the breeder responsible for rapid adoption of varieties reduces the time-lag before new technologies can reach large numbers of farmers (Witcombe *et al.*, 2017). The matching between the selection environment and the target production environment may be the main reason why participatory methods have achieved rapid success in Nepal and India.

The power of participatory methods comes from their capacity to source and deploy new diversity to many farmers, spreading it through an informal farmer–to–farmer seed exchange system. Over 117,000 IRD kits, including over 70,000 of three new rice varieties (Barkhe 3004; Sunaulo Sugandha, and Barkhe 2014) developed through client-oriented breeding (COB), were distributed in 18 districts of Nepal from 2008 to 2011. A follow–up survey in 2011 showed that the three COB varieties had been adopted by almost twice as many households than had the three varieties (Loktantra, Ram Dhan and Mithila) of the contemporaneous National Rice Research Programme (NRRP) (Witcombe *et al.*, 2017).

Proponents of the TRICOT method argue for blind–testing in varietal trials, to reduce possible biases about local varieties and motivate farmers to complete the experimental cycle in order to get the names of the tested varieties (van Etten *et al.*, 2016). In principle, a blind test (using a code instead of the variety name) aids methodological rigour, but in practice it can entail new errors: farmers found it confusing to use codes (such as A, B, C or 1, 2, 3) and not the names of the variety. (In stark contrast, IRD distribution programmes actively encourage rural FM radio and testing networks to share the names of their varieties.) As using codes instead of variety names removes the opportunity for farmer–to–farmer transmission of knowledge and social connection, blind testing in the TRICOT process may delay the adoption process further down the chain.

The diversity kit has a function similar to that of baby trials in PVS and IRD, but basically relies on farmer innovation and informal seed systems to spread the preferred varieties. This method offers opportunities to select varieties from options provided at the household level. Unlike the PVS and IRD methods, the effectiveness and efficiency of the diversity kits and their importance in view of climate change, disease, and pests are yet to be established. This method can be powerful in areas where access to diversity is limited due to lack of information, resources to acquire the materials, and weak social ties and connections (Jarvis *et al.*, 2011). Diversity kits require few resources for formal evaluation compared to the PVS approach but can rapidly identify farmer-preferred varieties. They can also determine more precisely where new varieties should be marketed, as the tests are greater in number and involve a wider area (Witcombe *et al.*, 2017). Both diversity kits and IRD methods can serve the purpose of crowd–sourcing as they are simple, low–cost and easy to manage – but many farmers must be reached, and be willing to provide voluntary feedback by mobile phone.

Conclusions

Using participatory methods such as PVS, diversity kits and IRD can accelerate the adoption of new varieties and boost crop genetic diversity (Witcombe *et al.*, 2001), while providing information on overall acceptability and the areas for which such varieties are best suited. PVS has the added advantage of generating the variety-trial data required for release or registration, while offering farmers early access to best lines and fast-tracking adoption and associated benefits. However, PVS involves much higher costs than with IRD and diversity kits; these methods can be used if yield-trial data are not needed. Crowd-sourcing principles employed in TRICOT can also be used for feedback collection in IRD and diversity kits, to improve data-generation capacity without greatly increasing complexity and researcher costs. Although the TRICOT method is theoretically attractive, farmers and practitioners in Nepal often prefer to use diversity kits and IRD for sourcing and deploying new crop diversity. These methods have been shown to be cost-effective and simple to adopt in risk-prone diverse and complex mountain environments. Their value increases if the scale of testing is similar to that of the TRICOT method and if there is a voluntary mechanism to receive ample feedback from farmers, removing the need to rely on questionnaire surveys of households.

Acknowledgements

The development and writing of this chapter was initiated by Dr Bhuwon R. Sthapit, who passed away before its completion. A true pioneer, leading expert, and mentor on participatory methods to assess and use agricultural biodiversity, participatory crop improvement, and community-based biodiversity management, Dr Sthapit always put farmers at the centre of his work. This chapter represents only a small portion of the extensive experience and observations of Dr Sthapit in his collaborative work.

Notes

1 The Integrated Cereals Project (ICP) (367–0114) was a five-year grant-funded project to assist in strengthening the capacity of the Government of Nepal to generate improved production technology for the major food grain crops, and to transfer that technology to Nepali farmers in such a way that it would be readily adopted. The project supported a major rice and maize crop improvement programme, 1976–1981. http://pdf.usaid.gov/pdf_docs/PDAAQ726.pdf (accessed 15 June 2017).
2 Escapees are usually non-released or rejected varieties because of some perceived weaknesses. Masuli, a high-yielding Malaysian rice variety, is arguably the best-known escapee in India.

References

Almekinders CJM, Elings A, 2001. Collaboration of farmers and breeders: Participatory crop improvement in perspective. *Euphytica* 122, 425–438.
APROSC, 1983. Evaluation of minikit programme (rice, maize, wheat and soybean). Volume 1, main text. Kathmandu: Agricultural Projects Service Centre (APROSC).

Atlin GN, Cairns JE, Das B, 2017. Rapid breeding and varietal replacement are critical to adaptation of cropping systems in the developing world to climate change. *Global Food Security* 12, 31–37.

Burman D, Maji B, Singh S, Mandal S, Sarangi SK, Bandyopadhyay BK, Bal AR, Sharma DK, Krishnamurthy SL, Singh HN, delosReyes AS, Villanueva D, Paris T, Singh US, Haefele SM, Ismail AM, 2018. Participatory evaluation guides the development and selection of farmers' preferred rice varieties for salt-and flood-affected coastal deltas of South and Southeast Asia. *Field Crops Research* 220, 67–77.

Canci A, Guadagnin CA, Henke JP, Lazzari L, 2013. The diversity kit: Restoring farmers' sovereignty over food, seed and genetic resources in Guaraciaba, Brazil. In de Boef WS, Subedi A, Peroni N, Thijssen M, O'Keeffe E, (eds) *Community Biodiversity Management: Promoting Resilience and the Conservation of Plant Genetic Resources*, 32–36. New York: Routledge.

Ceccarelli S, 2009. Evolution, plant breeding and biodiversity. *Journal of Agriculture and Environment for International Development (JAEID)* 103, 131–145.

Ceccarelli S. 1989. Wide adaptation: how wide? *Euphytica* 40, 197–205.

Ceccarelli S, Grando S, Booth RH, 1996. International breeding programmes and resource-poor farmers: Crop improvement in difficult environments. In Eyzaguirre P, Iwanaga M, (eds) *Proceedings of Workshop on Participatory Plant Breeding*, 99–116. 26–29 July 1995, Wageningen, The Netherlands. Rome: IPGR.

Chaurasia PC, Perdon ER, 1981. The integrated cereals project experiences with technology transfer. Minikit and cropping systems pilot production program. In *Proceedings of a Seminar on Appropriate Technology for Hill Farming Systems*, 219–257, 22–26 June 1981, Kathmandu, Nepal. Kathmandu: DoA, HMG/Nepal, ADC/Nepal and ICP/Nepal.

Coomes OT, McGuire S, Garine E, Caillon S, McKey D, Demeulenaere E, Jarvis D, Aistara G, Barnaud A, Clouvel P, Emperaire L, Louafi S, Martin P, Massol F, Pautasso M, Violon C, Wencélius J, 2015. Farmer seed networks make a limited contribution to agriculture? Four common misconceptions. *Food Policy*, 56, 41–50.

Döring TF, Knapp S, Kovacs G, Murphy K, Wolfe MS, 2011. Evolutionary plant breeding in cereals – into a new era. *Sustainability* 3, 1944–1971.

FAO, 2010. *Second Report on the State of the World's Plant Genetic Resources for Food and Agriculture*. Rome: FAO.

Farrington J, Mathema SB, 1991. *Managing Agricultural Research for Fragile Environments: Amazon and Himalayan Case Studies*. London: Overseas Development Institute (ODI).

Fenzi M, Jarvis DI, Arias Reyes LM, Latournerie Moreno L, Tuxill J, 2017. Longitudinal analysis of maize diversity in Yucatan, Mexico: Influence of agro-ecological factors on landraces conservation and modern variety introduction. *Plant Genetic Resources Characterization and Utilization* 15, 51–63.

Frison EA, IPES-Food, 2016. *From Uniformity to Diversity: A Paradigm Shift from Industrial Agriculture to Diversified Agroecological Systems*. Louvain-la-Neuve (Belgium): IPES. www.ipes-food.org/images/Reports/UniformityToDiversity_FullReport.pdf.

Frison EA, Cherfas J, Hodgkin T, 2011. Agricultural biodiversity is essential for a sustainable improvement in food and nutrition security. *Sustainability* 3, 238–253.

Gauchan D, 2017. Research and support services in seed production and supply in Nepal. In Khanal M, Adhikari R, (eds) *Seed Industry Development in Nepal*, 2–27. Kathmandu: Ministry of Agricultural Development.

Gurung R, Sthapit SR, Gauchan D, Joshi BK, Sthapit BR, 2017. Baseline Survey Report: Ghanpokhara, Lamjung. Integrating traditional crop genetic diversity into technology: Using a biodiversity portfolio approach to buffer against unpredictable environmental change in the Nepal Himalayas. Pokhara, Nepal: LI-BIRD, NARC and Bioversity International.

Gyawali S, Sthapit BR, Bhandari B, Bajracharya J, Shrestha PK, Upadhyay MP, Jarvis DI, 2010. Participatory crop improvement and formal release of Jethobudho rice landrace in Nepal. *Euphytica* 176, 59–78.

Halewood M, (ed) 2016. *Farmers' Crop Varieties and Farmers' Rights: Challenges in Taxonomy and Law*. New York: Routledge.

Herrero M, Thornton PK, Power B, Bogard JR, Remans R, Fritz S, Gerber JS, Nelson G, See L, Waha K, Watson RA, West PC, Samberg LH, van de Steeg J, Stephenson E, van Wijk M, Havlík P, 2017. Farming and the geography of nutrient production for human use: A transdisciplinary analysis. *The Lancet Planetary Health* 1, e33–e42.

Jarvis DI, Hodgkin T, Sthapit BR, Fadda C, Lopez-Noriega I, 2011. An heuristic framework for identifying multiple ways of supporting the conservation and use of traditional crop varieties within the agricultural production system. *Critical Reviews in Plant Sciences* 30, 125–176.

Jarvis DI, Brown AHD, Cuong PH, Collado-Panduro L, Latournerie-Moreno L, Gyawali S, Tanto T, Sawadogo M, Mar I, Sadiki M, Thi-Ngoc Hue N, Arias-Reyes L, Balma D, Bajracharya J, Castillo F, Rijal D, Belqadi L, Rana R, Seddik S, Ouedraogo J, Zangre R, Rhrib K, Chavez JL, Schoen D, Sthapit B, De Santis P, Fadda C, Hodgkin T, 2008. A global perspective of the richness and evenness of traditional crop-variety diversity maintained by farming communities. *Proceedings of the National Academy of Sciences* 105, 5326–5331.

Jarvis DI, Hodgkin T, Brown AHD, Tuxill J, Lopez Noriega I, Smale M, Sthapit B, 2016. *Crop Genetic Diversity in the Field and on the Farm: Principles and Applications in Research Practices*. New Haven, CT: Yale University Press.

Johnson AW, 1972. Individuality and experimentation in traditional agriculture. *Human Ecology* 1, 149–159.

Joshi A, Witcombe JR, 1996. Farmer participatory crop improvement. II: Participatory varietal selection, a case study in India. *Experimental Agriculture* 32, 461–477.

Joshi KD, Sthapit BR, 1990. Informal Research and Development (IRD): A new approach to research and extension. Discussion Paper. Pokhara: Lumle Agricultural Centre.

Joshi KD, Witcombe JR, 2002. Participatory varietal selection in rice in Nepal in favourable agricultural environments – a comparison of two methods assessed by varietal adoption. *Euphytica* 127, 445–458.

Joshi KD, Subedi M, Rana RB, Kadayat KB, Sthapit BR. 1997. Enhancing on-farm varietal diversity through participatory varietal selection: A case study for Chaite rice in Nepal. *Experimental Agriculture* 33, 335–344.

McGuire S, 2008. Securing access to seed: Social relations and sorghum seed exchange in eastern Ethiopia. *Human Ecology* 36, 217–229.

McGuire S, Sperling L, 2016. Seed systems smallholder farmers use. *Food Security* 8, 179–195.

Mengistu DK, Kidane YG, Catellani M, Frascaroli E, Fadda C, Pe ME, Dell'Acqua M, 2016. High-density molecular characterization and association mapping in

Ethiopian durum wheat landraces reveals high diversity and potential for wheat breeding. *Plant Biotechnology Journal* 14, 1800–1812.

Mulumba JW, Nankya R, Adokorach J, Kiwuka C, Fadda C, De Santis P, Jarvis DI, 2012. A risk-minimizing argument for traditional crop varietal diversity use to reduce pest and disease damage in agricultural ecosystem of Uganda. *Agriculture, Ecosystem and the Environment*, 157, 70–86.

Murphy KM, Bazile D, Kellogg J, Rahmanian M, 2016. Development of a world-wide consortium on evolutionary participatory breeding in quinoa. *Frontiers in Plant Science* 7, 608.

Packwood AJ, Virk DS, Witcombe JR, 1998. Trial testing sites in all India co-ordinated projects. How well do they represent agro-ecological zones and farmers' fields. In Witcombe JR, Virk DS, Farrington J, (eds) *Seeds of Choice: Making the Most of New Varieties for Small Farmers*, 7–26. New Delhi and London: Oxford and IBH Publ.

Palikhey E, Sthapit SR, Gautam S, Gauchan D, Bhandari B, Joshi BK, Sthapit BR, 2018. Baseline Survey Report: Hanku, Jumla. Integrating traditional crop genetic diversity into technology: Using a biodiversity portfolio approach to buffer against unpredictable environmental change in the Nepal Himalayas. Pokhara, Nepal: LI-BIRD, NARC and Bioversity International.

Parajuli A, Subedi A, Adhikari AR, Sthapit SR, Joshi BK, Gauchan D, Bhandari B, Sthapit BR. 2018. Baseline Survey Report: Chhipra, Humla. Integrating traditional crop genetic diversity into technology: Using a biodiversity portfolio approach to buffer against unpredictable environmental change in the Nepal Himalayas. Pokhara, Nepal: LI-BIRD, NARC and Bioversity International.

Pautasso M, Aistara G, Barnaud A, Caillon S, Clouvel P, Coomes OT, Delêtre M, *et al.* 2013. Seed exchange networks in agrobiodiversity conservation: Concepts, methods and challenges. *Agronomy for Sustainable Development* 33 (1), 151–175.

Pudasaini N, Sthapit SR, Gauchan D, Bhandari B, Joshi BK, Sthapit BR, 2017. Baseline Survey Report: Jungu, Dolakha. Integrating traditional crop genetic diversity into technology: Using a biodiversity portfolio approach to buffer against unpredictable environmental change in the Nepal Himalayas. Pokhara, Nepal: LI-BIRD, NARC and Bioversity International.

Richards P, 1986. *Coping with Hunger: Hazard and Experiment in a West African Farming System.* London: Allen & Unwin.

Sah RP, Dass S, Kochhar S, Bhooshan N, Kumar A, Joshi PK, 2016. *Fast-track breeding, variety release, registration, and intellectual property rights mechanisms for the seed sector in Nepal.* PRIP-Nepal Report 2, New Delhi: IFPRI.

Shrestha P, Sthapit SR, Paudel IP, 2013. Participatory seed exchange for enhancing access to seeds of local varieties. Pokhara. www.libird.org/app/publication/view.aspx?record_id=109. Accessed 26 January 2018.

Simmonds NW, 1991. Selection for local adaptation in a plant breeding programme. *Theoretical and Applied Genetics* 82, 363–367.

Sperling L, Loevinsohn ME, 1993. The dynamics of adoption: Distribution and mortality of bean varieties among small farmers in Rwanda. *Agricultural Systems* 41, 441–453.

Steinke J, van Etten J, 2016. *Farmer experimentation for climate adaptation with triadic comparisons of technologies (TRICOT): A methodological guide.* Bioversity International, USAID, CCAFS.

Sthapit B, Jarvis D, 1999. Participatory plant breeding for on-farm conservation. *LEISA* 15, 40–41.

Sthapit BR, Rao VR, 2009. Consolidating community's role in local crop development by promoting farmer innovation to maximise the use of local crop diversity for the well-being of people. *Acta Horticulturae* 806, 669–676, DOI: 10.17660/ActaHortic.2009.806.83, https://doi.org/10.17660/ActaHortic.2009.806.83.

Sthapit BR, Joshi KD, Witcombe JR, 1996. Farmer participatory crop improvement. III: Participatory plant breeding, a case study for rice in Nepal. *Experimental Agriculture* 32, 479–496.

Sthapit B, Gyawali S, Gautam R, Joshi BK, 2012. Diversity kits: Deploying new diversity to farmers. In Sthapit B, Shrestha P, Upadhyay M, (eds) *On-farm Management of Agricultural Biodiversity in Nepal: Good Practices*, 33–36. Pokhara: NARC, LI-BIRD, Bioversity.

Sthapit B, Vasudeva R, Rajan S, Sripinta P, Reddy BMC, Arsanti IW, Idris S, Lamers HAH, Rao RV, 2015. On-farm conservation of tropical fruit tree diversity: Roles and motivations of custodian farmers and emerging threats and challenges. *Acta Horticulturae* 1101, 69–74, DOI: 10.17660/ActaHortic.2015.1101.11, https://doi.org/10.17660/ActaHortic.2015.1101.11.

Sthapit S, Gautam S, 2016. Rebuilding family farms: Seed Rescue Project. Final Progress Report, Pokhara: LI-BIRD.

Thomas M, Verzelen N, Barbillon P, Coomes OT, Caillon S, McKey D, Elias M, *et al.* 2015. A network-based method to detect patterns of local crop biodiversity: Validation at the species and infra-species levels. *Advances in Ecological Research* 53, 259–320.

Tripp R, 1997. *New Seed and Old Laws: Regulatory Reform and the Diversification of National Seed Systems*. London: Intermediate Technology Publ.

Tripp R, 2001. *Seed Provision and Agricultural Development: The Institutions of Rural Change*. London: ODI/James Currey/Heinemann.

van Etten J, 2011. Crowdsourcing crop improvement in sub-Saharan Africa: A proposal for a scalable and inclusive approach to food security. *IDS Bulletin* 42, 102–110.

van Etten J, Beza E, Calderer L, van Duijvendijk K, Fadda C, Fantahun B, Kidane YG, *et al.* 2016. First experiences with a novel farmer citizen science approach: Crowdsourcing participatory variety selection through on-farm triadic comparisons of technologies (TRICOT). *Experimental Agriculture* 1–22.

van Etten J, Sousaa K, Aguilar A, Barriose M, Cotoa A, Acqua M, Fadda C, *et al.* 2019. Crop variety management for climate adaptation supported by citizen science. *Proceedings of the National Academy of Sciences (PNAS)* 2019, USA. www.pnas.org/cgi/doi/10.1073/pnas.1813720116.

Witcombe JR, Virk DS, Farrington J, 1998. *Seeds of Choice: Making the Most of New Varieties for Small Farmers*. New Delhi and London: Oxford and IBH Publ.

Witcombe JR, Joshi A, Joshi KD, Sthapit BR, 1996. Farmer participatory crop improvement. I. Varietal selection and breeding methods and their impact on biodiversity. *Experimental Agriculture* 32, 445–460.

Witcombe JR, Joshi KD, Rana RB, Virk DS, 2001. Increasing genetic diversity by participatory varietal selection in high potential production systems in Nepal and India. *Euphytica* 122, 575–588.

Witcombe JR, Joshi KD, Gyawali S, Musa AM, Johansen C, Virk DS, Sthapit BR. 2005. Participatory plant breeding is better described as highly client–oriented plant breeding. I. Four indicators of client-orientation in plant breeding. *Experimental Agriculture* 41, 299–319.

Witcombe JR, Khadka K, Puri RR, Khanal NP, Sapkota A, Joshi KD. 2017. Adoption of rice varieties – 2. Accelerating uptake. *Experimental Agriculture* 53, 627–643.

14 Expanding community support in genetic diversity management

The FFS approach

Bert Visser, Hilton Mbozi, Patrick Kasasa,
Anita Dohar, Rene Salazar, Andrew Mushita
and Gigi Manicad

This chapter focuses on experiences with the Farmer Field School (FFS) approach in facilitating participatory plant breeding at the community level. Participatory plant breeding (PPB) brings together local farmers and staff of extension services, non-governmental organizations and public-sector breeders, to select and create plant varieties best adapted to local farmers' needs and preferences. This requires capacity building on the part of community members, to provide suitable learning conditions for small-scale farmers. It also requires interest and relevant skills on the part of other partners who are expected to contribute to the FFSs in various ways. Several other participatory approaches have been promoted and implemented, but, unlike top–down teaching approaches, the focus of FFSs is on direct learning in the field by and between farmers on the basis of their own experimentation, supported by other stakeholders. Through this approach, farmers develop new knowledge and skills that contribute to their empowerment. This empowerment is a crucial outcome: the farmers themselves can search for crops with new traits and adapt available varieties to new needs and preferences.

Small-scale farmers' seed systems[1]

Small-scale farmers provide a major share of our global food supply, and their agriculture forms the backbone of global food security in the developing world (Tscharntke et al., 2012). Against the backdrop of population growth and climate change, strengthening small-scale farmer food production can make a major contribution to current and future food security. Seed security – access to appropriate seed in sufficient quantities at affordable prices and at the right time – is a major precondition for food security in rural areas of the developing world. In many developing countries, seed systems are predominantly farmer-managed (Richards et al., 2009), and may encompass activities ranging from seed collection to selection, crossing, testing, multiplication and storage of seeds, normally without formal oversight or support.

Many small-scale farmers produce their own seed, setting aside part of their harvest for the next growing season for that specific purpose. Informal markets are further important sources of seed for most food crops, except perhaps maize and vegetables (Sperling and McGuire, 2010). In their study of six seed-system security assessments of sites where government, UN or NGO interventions were being implemented, McGuire and Sperling (2016) documented the degree to which seed acquisition depended on the informal sector. They found that, in Africa, farmers access 90 per cent of their seed from informal systems, with half of that coming from local markets. Further, they found that 55 per cent of seed is paid for in cash, indicating that smallholders are already making important investments in this area. These informal markets offer a diversity of seeds, ranging from traditional farmers' varieties to modern formal-sector varieties, as well as selections from the latter or from crossings between various materials. Seed from formal sources may offer new and important traits relating to yield and resistance not available in traditional varieties. However, formal-sector seed is often not readily available to small-scale farmers. They may face logistical and financial barriers, like having to travel to a town where there is an agrochemical outlet. Moreover, many farmers lack information on where seed can be purchased, and are not well-informed about cultivars, quality or price (Visser, 2017).

Access to diverse seeds

Small-scale farmers can produce and sell high-quality and well-adapted seed lots, but a major problem of small-scale seed-production systems is access to genetic diversity within and across crops in order to broaden the variety portfolio, to enable adaptation to changing climate conditions, and to improve nutrition security (Oxfam *et al.*, 2015). Female and male farmers generally have different needs and preferences here (Oxfam *et al.*, forthcoming). Further, small-scale farmers often lack access to high-quality breeder and foundation seed for further multiplication, as observed in project sites in Zimbabwe and Vietnam (personal communication). Despite investments in seed-sector development programmes and technology adoption, access to good-quality seed remains a challenge in many developing countries.

In recent decades, various strategies have been developed for increasing small-scale agricultural production, and improving the functioning of seed systems in particular. Here we may mention the Alliance for a Green Revolution in Africa (AGRA) initiatives aimed at promoting the commercial seed sector, the adoption of the Integrated Seed System Development by the African Union (Louwaars and de Boef, 2012), and the improvement of extension, training and education offered to small-scale farmers (Davis, 2008).

This chapter focuses on the development of FFSs aimed at widening the diversity and improving the quality of crop seeds available to small-scale farmers within the framework of the global multi-year programme Sowing Diversity = Harvesting Security (SD = HS), conducted by a consortium of partners coordinated by Oxfam Novib. In particular, we examine whether the FFS can be developed so as to become sustainable and cost-effective, and what conditions need to be fulfilled to that end – a topic heavily debated and researched over the past two decades (see for example Tripp *et al.*, 2005; Braun *et al.*, 2006; Davis *et al.*, 2012; Larsen and Lilleør, 2014). Major and divergent experiences in two countries, Vietnam and Zimbabwe, are described and analysed.

Farmer Field Schools as an approach to education

Many agricultural development strategies have shown that uniform technologies coupled with a linear process of technology transfer do not result in the production increases and behavioural changes that were foreseen (Tripp *et al.*, 2005). Challenges come from the considerable variations in agro-ecosystems, farm household assets and strategies, and not least the importance of recognizing farmers' roles and responsibilities in technology generation.

As an alternative approach, the concept of FFSs was developed in the 1990s within the UN Food and Agriculture Organization. It was initially applied to introduce the principles of Integrated Pest Management to rice farmers in Asia, but later also for dealing with a range of other topics within and outside agriculture, such as soil fertility, community forestry, livestock production, gender issues, and even the containment of HIV/AIDS in rural areas. FFSs have been operating in more than 75 countries (Braun *et al.*, 2006). The FFS approach is based primarily on participation and empowerment, with rights-based approaches such as those of Oxfam and its programme partners. Impressive numbers of FFSs have been organized – over 150,000 between 1992 and 2004 in only 13 countries (Braun *et al.*, 2006).

A typical FFS organizes a group of 20 to 30 farmers from a given community over at least one growing season. They meet regularly, often once a week, in the field to make observations and discuss findings and responses, in order to improve crop yields. The FFS is supported by a trainer, who may be an employee of a development organization, an extension agent or a farmer-trainer. An important element of the FFS is the season-long field curriculum that supports farmers in their experimentation. The underlying principle is mutual learning, in which farmers set the learning agenda and learn from each other in small groups, guided by the trainer (see also Salazar *et al.*, this volume). The FFS is not a universal solution for development challenges, nor a substitute for extension or market-driven approaches, but it is especially well-suited for dealing with complex problems that require local solutions and behavioural change based on location-dependent knowledge (Braun *et al.*, 2006). For instance, issues like strategies for coping with climate change requiring a shift in crops appear more amenable to a FFS approach than to standard extension campaigns. The FFS is generally seen as a time-bound effort, not necessarily to be sustained, although many farmer groups have continued their activities after the end of the FFS, based on the human and social capital developed in its context.

The wider participatory plant breeding approach

In recent decades, the FFS approach has also been applied to seed-system development, especially for promoting and facilitating participatory plant breeding (PPB), a somewhat loose concept that has co-developed with the FFS approach to a considerable extent. Sperling *et al.* (2001) provide a framework for differentiating among PPB approaches. Key variables explored in their analysis include the institutional context, the bio-social environment (both agro-ecological and socio-economic), the goals set, and the kind of 'participation' achieved, (including the stage and degree of participation and the roles taken by various actors). In some cases, institutional actors take the lead and farmers contribute to the process of variety development and selection; in other cases, the farmers may organize themselves and request incidental inputs from external experts.

The SD=HS programme places farmers in the centre. Farmers set the goals of their FFS, and all FFS activities take place in farmer fields in the communities, with breeders and staff from extension services assisting the farmers and playing an essential role in achieving the outcomes.

Participatory varietal selection

Various goals have been set regarding the output of an FFS season in terms of new plant materials in participatory plant breeding initiatives. Specific terminology has been developed to distinguish between goals stemming from differences in farmers' interests, each requiring different activities and capacities. For instance, Participatory Varietal Selection (PVS) is an activity that can be undertaken with modest means (Witcombe *et al.*, 1999). This entails the scoring of stable lines and varieties of a particular crop by growing these in close proximity, with reference to criteria or properties – e.g. drought tolerance or plant height – identified and prioritized by participating farmers at the start of the FFS. The scoring may also involve aspects like storage properties, processing or taste, to be ascertained after harvesting. FFS participants normally compare ten lines or varieties against a local variety that is widely grown in the community. Such lines and varieties might have been accessed from farmers' own fields or from local markets, or perhaps supplied by breeders who have associated themselves with the FFS programme. A single growing season is normally sufficient for this basic experiment.

An alternative activity, which is more demanding as it sets higher capacity requirements to farmers and requires the FFS to continue over several seasons, involves the selection of best plants by farmers from an acquired segregating population (normally F3 or F4)[2] or even from a self-performed crossing, e.g.

between a modern line with a major preferred trait or set of traits and a local variety preferred for other reasons. At the end of this activity the remaining population is stable and relatively uniform.

A third activity in the SD=HS programme has been termed Participatory Varietal Enhancement (PVE). It focuses on the restoration of a popular local variety that may have lost some of its preferred traits or is no longer well adapted to changing climate conditions. The resultant line may be quite similar to the original but now-deteriorated variety, or it may have acquired essential new properties, like earlier maturation or increased drought toler-ance. The source of a PVE activity is by definition a local variety that is or has been popular in the community.[3]

Further, the SD=HS programme's FFSs often feature 'diversity plots' intended mainly to familiarize farmers with crops or crop varieties that they do not know and might be interested in, whether through direct adoption or for inclusion in later FFS experiments. An example is the demonstration of the hybrid *nerica* ('New Rice for Africa') rice in sub-Saharan Africa.

Requirements for a successful FFS

Farmers choose the crops to work with, identify traits of importance to them, and the type of activity (PVS, PVE or PPB). Breeders are approached to assist the FFS by providing breeding materials or finished new lines and varieties that answer farmers' needs. Genebanks can assist in the re-introduction of older varieties that farmers remember as valuable but which have since been lost in the community.

Although female smallholder farmers play a major role in maintaining and reproducing agricultural biodiversity, a recent study found that women were minor participants in the PPB programmes reviewed (ACB, 2018). The reasons cited included gendered decision-making norms, unreflective exclu-sion from projects, and lack of expressed interest. Seeking to address this issue, the SD=HS programme has ensured that the majority of its FFS parti-cipants are female, and that documentation for the FFSs takes a gender perspective.

PPB is only one part of a bigger picture, however. Plant breeding on its own, no matter how democratically and inclusively it is done, cannot resolve all the problems and challenges facing smallholder farming communities. PPB needs to be situated within a wider agenda of agro-ecological programming and support (ACB, 2018).

In order to provide support to large numbers of FFS running simultan-eously in several communities, trainers attached to the FFS are themselves trained in a Training of the Trainers (ToT) course. The ToT is facilitated by a core group of master trainers. This 'staggered' approach is needed to enable upscaling to a large number of parallel FFSs. The FFS curriculum is docu-mented in the form of a farmer field curriculum used during FFS implemen-tation, further supported by a specific and dedicated curriculum for the training in the ToT. In the case of the SD = HS programme, the ToT curriculum consists of several core chapters that are discussed in all ToTs. These may include the setting of objectives, crop-specific instructions for field experimentation and gender issues, and additional modules, for instance on climate-change resilience and policy change requirements that can be integrated into the ToT as appropriate.[4]

Impacts of FFS

Considerable attention has been given to the impacts of FFS. Demonstrating extension impacts as such is very difficult, especially regarding attribution issues and linking cause and effect quantitatively. Various attempts have been made at measuring the impact of FFSs more specifically. However, there is no agreement as to exactly what to measure, how to measure, and how to evaluate the results of such measurements: some researchers see FFSs primarily as an extension activity, whereas others consider them as an educational investment (Braun et al., 2006). Some studies have attempted to measure the effects of FFSs on crop yields and pesticide use as well as livelihoods, whereas others have focused on the level of dissemination of behaviour and practices, and a third group has been particularly interested in empowerment and its effects on knowledge, skills and well-being (Davis et al., 2012). In line with this last group, the SD = HS programme has developed indicators measuring farmers' capacity to articulate demands for resources and services, to continue FFS activities independently, and to engage in policy discussions. Farmers have participated actively in national and global policy events like the meetings of the Governing Body of the International Treaty on Plant Genetic Resources for Food and Agriculture (the Plant Treaty), and at the Svalbard Global Seed Vault, presenting their case themselves. Surely that reflects the preliminary results of these empowering efforts.

A particular concern regarding impact relates to the diffusion effects, or alternatively, the options for increasing impact by expanding the numbers of FFSs. There is some evidence to indicate that, whereas information and changes in practice may diffuse to non-participating farmers, the acquired underlying knowledge and analytical skills in problem solving do not spread, or only to a very limited extent. Many conditions influence the long-term results of the FFS approach, including weather and climatic conditions, the development of produce markets, as well as conducive policy in relation to FFS topics. These issues are also reflected in the two cases discussed below. Nevertheless, ample evidence exists that varieties tested and preferred in the context of FFS have spread rapidly across large distances (see, for example, Joshi et al., 2001).

Several studies have reported positive short- and medium-term impacts of FFSs. Farmers have managed to improve agricultural productivity, food security and/or income, and to increase their knowledge and leadership involvement in community-based activities (Braun et al., 2006; Bunyatta et al.

2006; Davis *et al.*, 2012; Larsen and Lilleør, 2014). Moreover, FFSs offer an approach that targets women and poorly literate farmers (Davis *et al.*, 2012) as well as focusing on the production and distribution of quality seed (Alme-kinders *et al.*, 2007). However, some studies report that poverty has remained unaffected (Larsen and Lilleør, 2014) and that there is little evidence that the skills learned are passed on to non-participants (Tripp *et al.*, 2005).

Costs of FFS

Some researchers argue that a major drawback of the FFS approach is its cost, which is associated with its financial sustainability. FFS trainings are expensive in terms of each farmer trained (Andersen and Feder, 2004) as opposed to tra-ditional extension efforts. Anadajayasekaram *et al.* (2007) report costs for running an FFS in Kenya from USD 5,000 in the first year to less than USD 1,000 in subsequent years, which translates into USD 200 and 40 respectively per participant. Simpson and Owens (2002) reported that in Ghana a decen-tralized FFS approach allowed the reduction of costs per farmer to USD 8–10 per farmer, whereas estimates of costs per farmer for FFS training in several East African countries varied between USD 9 and 35. An estimation of costs to run a season-long FFS in 10 countries across continents showed averages ranging from USD 150 (FFSs in Nigeria with farmer-trainers) to USD 650 (food security in Mozambique) (Braun *et al.*, 2006). Overall, costs per farmer-participant in a season-long FFS appear to vary between USD 8 and 40, although the costs for an initial year in a new context may be higher.

The SD=HS programme experienced that, although the initial invest-ments are substantial, the costs of subsequent FFS cycles can be reduced considerably through the formation of a core group of master trainers and the creation of farmer trainers, so that farmer trainers and local extension staff can take over from trainers who must be paid from the programme budget. Financial sustainability comes within reach as soon as required inputs for established FFSs in the next growing seasons can be reduced to policy support and the occasional provision of new breeding and selection materials. In addi-tion, financial support from other sources based on proven concepts, in par-ticular from local government, can take over from programme funding and even exceed the financial contribution from programme sources. Once estab-lished, FFSs in the SD=HS programme can be continued at an average cost of 10–30 per farmer per season, although the initial costs are several times higher.

Evidence from the Philippines and Indonesia has shown that fiscal unsus-tainability of the FFS if applied on a large scale is a risk, as the amount of national funding for extension services may not provide for sufficient expan-sion (Quizon *et al.*, 2001). Farmer-led field schools may be viewed as one way out of this dilemma, since part of the cost is shifted to the beneficiaries. However, that requires proper training of lead-farmers in one or more ToT sessions, to ensure FFS quality.

Conditions for expansion: experience from two case countries

As noted, one limitation of the FFS approach has been the intensive capacity support these season-long efforts require, allowing direct support of only a limited number of FFSs in a single season. Whereas the spread and adoption of FFS on Integrated Pest Management has been well documented, such expansion has appeared more cumbersome in FFSs on genetic diversity and plant breeding. The SD = HS programme alliance and its predecessors[5] have gained experience in facing this challenge in two countries with very different agro-ecological, economic and political conditions: Vietnam and Zimbabwe.

Vietnam

Since 1995, activities have been undertaken by SD = HS predecessor programmes (CBDC, BUCAP) to improve crop genetic diversity to better meet the needs and preferences of small-scale farmers in the Mekong Delta in Vietnam, in collaboration with the Mekong Delta Development Research Institute of Can Tho University (see also Salazar *et al.*, this volume). The Mekong Delta is a major global producer and exporter of rice. Infrastructure in the Mekong Delta, including for agricultural production, has been developed considerably in recent decades, and rice cultivation has changed from a system in which many farmers had to deal with annual inundation of their fields towards a well-irrigated system with up to three growing seasons per year. Although the Mekong Delta can be regarded as highly productive, salt intrusion as an effect of rising sea levels is an increasing problem in coastal areas.

Over the years, many thousands of farmers have been trained in the evaluation of and selection in locally adapted and new public sector developed rice varieties, as well as in the proper production of their own seeds. Such rice-seed production by small-scale farmers, organized in seed clubs, has now become a major economic activity. Such seed clubs normally consist of between 10 and 20 farmers. More than 400 seed clubs are now operational, functioning largely without substantial external support. They supply approximately 30 per cent of local seed demand, surpassing the contribution of the private sector, which covers less than 20 per cent of demand. The seed quality realized by the seed clubs equals or surpasses that of the private sector; certification of seed-club seed lots has not presented major problems, and prices have remained substantially below those of the private sector. The local and provincial authorities have long recognized the major contribution of the seed clubs to rice-seed supply in the Mekong Delta, and have offered various forms of economic and policy support to the functioning of the seed clubs, including the provision of seed processing equipment and storage facilities, as well as by accommodating the registration of farmer producers and the certification of seed lots of registered varieties. A remaining problem is the arduous variety registration process, which represents a major barrier to the registration of new varieties developed by farmers of the seed clubs (for an analysis, see de Jonge *et al.*, this volume).

As regards the expansion of crop diversity, as of the end of 2017 the breeding and selection efforts of farmers in the Mekong Delta involved the distribution of 35 rice varieties (both commercial and farmers' varieties for the purpose of PVS) as well as 30 segregating populations (generated by the Mekong Delta Institute for the purpose of PPB) used as breeding materials. This has resulted in the multiplication and further use of 32 entries, of which five have been planted over large areas of the Delta. Three have subsequently been released for variety testing. Released varieties have also been subjected to seed certification, and ten local rice varieties has been subjected to participatory varietal enhancement as well. In addition a substantial number of samples of the crops maize (seven), mung bean (six) and sesame (ten) have be made available for farmer experimentation (Searice and MDI, personal communication).

Our own analysis of the development of the seed clubs shows that they have been able to fill a vacuum. The private and public sector lacked the capacity to produce sufficient quantities of rice seed, and the local and provincial authorities were requested to deliver certain rice-seed quotas set by government policy. Growing market demands and improved infrastructure led farmers to look for quality seed in the market rather maintaining their own seed stocks. With the seed clubs, it became possible to provide the seed demanded, in increasing quantities. Extension services supported farmers in rehabilitating local varieties in addition to supporting the application of Integrated Pest Management, and seed-production protocols were developed that guaranteed seed of good quality.[6]

In conclusion, economic and political conditions, including the improvement of the infrastructure in the Mekong Delta, gave rise to the development of farmer seed clubs that, with relatively little external support, could develop out of the FFSs on Integrated Pest Management (IPM) and crop diversity. Thanks to substantial support from local government, in turn supported by development aid provided by DANIDA, seed clubs gained access to processing equipment and storage facilities. The regional authorities assisted seed clubs in the marketing and registration of their seed produce. We lack financial data that would allow further analysis of the costs and benefits relating to the development of these seed clubs. However, they can be seen as the expression of greater farmer capacity in seed production and variety development, and thus as a major empowerment of small-scale producers, building on the introduction of FFSs. This development took place gradually, over a period of two decades from the first FFSs.

Zimbabwe

In Zimbabwe, recent developments have been based on two decades of farmer support activities by the Community Technology Development Trust

(CTDT), since 2012 funded by the International Fund for Development and the Swedish International Development Agency and the Dutch Postcode Lottery. In the 2016/2017 growing season, 318 FFSs were simultaneously managed. In that season FFSs conducted within the framework of the SD = HS programme tested more than 70 stable lines of maize, sorghum, pearl millet, groundnut and cowpea, using the PVS approach described above. Farmers decided to work with sorghum and pearl millet in particular, in view of increasing drought. In the 2017/2018 growing season, FFS tested 160 lines and populations, including 16 segregating populations (F3) of sorghum and of pearl millet across 8 districts, 18 stable lines of sorghum and pearl millet and 10 stable lines of maize sourced from the Crop Breeding Institute (CBI).

Our own analysis identified four conditions necessary for a dramatic increase in the number of FFSs running in parallel in a country in a single season, while retaining FFS quality. These conditions are: (1) the availability of a baseline tool-kit and well-established, season-long, and flexible curricula for trainers as well as farmer-participants, (2) the successful establishment of a core group of master trainers, (3) the involvement of extension service staff (at national and local levels) in facilitating the FFS, and (4) the availability of new and adapted germplasm, both stable and segregating, from participating breeding institutions.

The development of a baseline tool-kit and the development of season-long trainer and field curricula required substantial initial investments by the SD = HS programme and its partners, in particular the CTDT, based in Zimbabwe, and Oxfam Novib. These investments can be regarded as a necessary but one-time condition, with the maintenance and further improvement and expansion of the tool-kit and curricula requiring only modest future funds. With the development and elaboration of the curricula, ample use was made of experience and documentation from earlier programmes operational in the country, like those of CBDC and IFAD.

A core group of 14 master trainers has been established, consisting of officers of CTDT and allied local NGOs, the extension service Agritex, and lead farmers. Agritex has also become an essential ally in the field implementation of the FFSs, offering important support in the organization and implementation of the season-long schools, facilitating the FFSs at the local level through the involvement of local extension staff, and by promoting the FFS approach at the national level, and adopting the approach and the concepts involved in its own plans and reports.

Finally, the breeding institutions collaborating with the programme were ready to provide mainly stable breeding lines but also segregating populations to the FFS for farmer experimentation. Breeders visited several FFSs to support the participants and monitor their germplasm. The CBI as well as the regional stations of the CGIAR centres ICRISAT (International Crops Research Institute for the Semi-Arid Tropics) and CIMMYT (International Maize and Wheat Improvement Center) saw the FFS participants as major evaluators of their breeding products; for Agritex adoption of the FFS approach added to the national importance of the agency. It was this intensive collaboration involving the SD = HS programme partners, extension services and agricultural research and breeding institutions that made possible the simultaneous organization of a large number of FFSs.

Zimbabwe offered this opportunity and the need for a FFS approach because of the prolonged economic difficulties facing the country, resulting in a severe weakening of government institutions and a greater role for civil society. Exposure to climate change, especially the severe drought following the 2015/2016 El Niño, led many farmers to realize the need to change their cropping systems by testing and adopting new crops and varieties better adopted to drier conditions and shorter rainy seasons. Breeders from national and international institutions based in Zimbabwe have increasingly been looking for ways of directly discussing in depth their breeding efforts with their clientele, the small-scale farming sector. As in Vietnam, these agro-ecological, economic and political conditions have made it possible for the FFS approach to fill a gap and play a major role in improving the supply of diverse seeds and in expanding crop diversity in farmers' fields.

Conclusions

From the findings from these two highly diverse countries, can we conclude that the FFS approach is feasible and needed in more developing countries, to promote participatory plant breeding and to assist farmers in staying ahead of the climate-change curve? Does it represent a better alternative to other, mainstream, approaches for promoting the adoption of new varieties? In our opinion, and granted that all countries and farming systems are unique, the use of the FFS approach in many more countries certainly seems warranted. It offers small-scale farmers in all developing countries the opportunity to gain better access to seeds of a larger diversity of crops and varieties; to review, change and widen their cropping patterns; and to adjust to the effects of climate change – thereby improving their capacities and helping to empower them to cope with rapidly changing agro-ecological and socio-economic conditions. This approach tackles a major weakness of farmers' seed systems: sufficient access to appropriate genetic diversity. The introduction of new rice varieties in Vietnam, of sorghum and pearl millet varieties in Zimbabwe, as well as the repatriation of native potatoes in Peru (not discussed here), all with the help of national breeding institutes and the international CGIAR Centres, serve as examples of such diversification of farming systems. In addition, the empowerment offered by the FFS approach has helped farmers to enhance some of their own highly valued farmers' varieties, such as sticky rice (Vietnam) and sorghum (Zimbabwe).

In many countries, the private breeding sector focuses on only a limited number of staple crops. Public-sector breeding institutions need direct exchange with small-scale farmers regarding their needs and preferences, and better opportunities for on-farm testing of their newly developed products. Currently underfunded extension services may benefit from involvement in a wider coalition supporting farmers in improving their production and making their crop portfolio more resilient to the effects of climate change. Although the costs of establishing a large number of FFSs may be substantial, coalitions can drastically reduce such costs by leveraging support from a multi-actor platform. Training lead farmers to become trainers and actively supporting them through a group of master trainers lowers costs and increases farmer-community ownership of the FFS approach. The benefits of the FFS approach are indisputable, in general as well as in the sphere of improving crop diversity. The need to improve the livelihoods of small-scale farmers, to guarantee food security (local and global), and to better manage crop diversity better on-farm – this constitutes the rationale for future FFS programmes in crop diversity. As in the SD = HS programme, the focus of such programmes may vary. Some FFSs may opt to concentrate on increasing crop diversity and access to diverse and good-quality seed; others, on seed production by farmers for crops and markets not covered by the private sector, on improving diets and nutrition security, on rescuing crop genetic resources, or on supportive national and international policy that recognizes and strengthens the role of the small-scale farming sector in all these areas.

Acknowledgements

The SD = HS programme has been generously funded by the Swedish International Development Cooperation Agency Sida (global) and the Dutch National Postcode Lottery (Zimbabwe).

Notes

1 In the literature also referred to as 'informal seed systems'.
2 The third or fourth generation of offspring acquired from the performance of a crossing.
3 A similar approach aimed at improving local maize varieties through stratified mass selection has been promoted in Meso-America by the farmers' cooperative ASOCUCH (see http://asocuch.com/index.html for more information).
4 See also SD = HS website at www.sdhsprogram.org/.
5 The predecessors include the CBDC programme, active in both countries from 1995 until 2005, the BUCAP programme active in Vietnam from 2000 until 2008, and the IFAD funded programme implemented in Zimbabwe from 2013 until 2016.
6 Here it should be noted that these seed clubs have not been formalized or registered as farmer cooperatives.

References

Almekinders CJM, Thiele G, Danial DL. 2007. Can cultivars from participatory plant breeding improve seed provision to small-scale farmers? *Euphytica* 153, 363–372.

Anadajayasekaram P, Davis KE, Workneh S. 2007. Farmer field schools: An alternative to existing extension systems? Experience from Eastern and Southern Africa. *Journal of International Agricultural and Extension Education* 14, 81–93.

Andersen JR, Feder G, 2004. Agricultural extension: Good intentions and hard realities. *World Bank Research Observer* 19: 41–60.

African Centre for Biodiversity (ACB), 2018. *A Review of Participatory Plant Breeding and Lessons for African Seed and Food Sovereignty Movements.* Johannesburg: African Centre for Biodiversity.

Braun A, Jiggins J, Röling N, van den Berg H, Snijders P, 2006. *A Global Survey and Review of Farmer Field School Experiences.* Final Report to ILRI. Wageningen: Endelea.

Bunyatta DK, Muteithi JG, Onyango CA, Ngesa FU, 2006. Farmer field school effectiveness for soil and crop management technologies in Kenya. *Journal of International Agricultural and Extension Education* 13, 47–63.

Davis KE, 2008. Extension in Sub-Saharan Africa: Overview and assessment of past and current models, and future prospects. *Journal of International Agricultural and Extension Education* 15, 15–28.

Davis K, Nkonya E, Kato E, Mekonnen DA, Odendo M, Miiro R, Nkuba J, 2012. Impact of farmer field schools on agricultural productivity and poverty in East Africa. *World Development* 40, 402–413.

Joshi KD, Sthapit BR, Witcombe JR. 2001. How narrowly adapted are the products of decentralized breeding? The spread of rice varieties from a participatory plant breeding programme in Nepal. *Euphytica* 122, 589–597.

Larsen AF, Lilleør HB. 2014. Beyond the field: The impact of farmer field schools on food security and poverty alleviation. *World Development* 64, 843–859.

Louwaars NP, de Boef WS. 2012. Integrated seed sector development in Africa: A conceptual framework for creating coherence between practices, programs, and policies. *Journal of Crop Improvement* 26, 39–59.

McGuire S, Sperling L. 2016. Seed systems smallholder farmers use. *Food Security* 8, 179–195.

Oxfam Novib, ANDES, CTDT, SEARICE and CGN-WUR. 2015. *From Lessons to Practice and Impact: Scaling up Pathways in People's Biodiversity Management.* Briefing Note. The Hague: Oxfam Novib.

Oxfam Novib, ANDES, CTDT, SEARICE. Forthcoming. Women's roles in biodiversity management. From lessons to practice and impact: Scaling up pathways in people's biodiversity management. Case study submitted for publication in the FAO *State of the World on Biodiversity.* The Hague: Oxfam Novib.

Quizon J, Feder G, Murgai R, 2001. *Fiscal Sustainability of Agricultural Extension: The Case of the Farmer Field School Approach.* Report 28213. Washington, DC: World Bank.

Richards P, de Bruin-Hoekzema M, Hughes SG, Kudadlie-Freeman C, Offei SW, Struik PC, 2009. Seed systems for African food security: Linking molecular genetic analysis and cultivator knowledge in West Africa. *International Journal of Technology Management* 45, 197–214.

Simpson BM and Owens M, 2002. Farmer field schools and the future of agricultural extension in Africa. *Journal of International Agricultural and Extension Education* 9, 29–36.

Sperling L, McGuire S, 2010. Understanding and strengthening informal seed markets. *Experimental Agriculture* 46, 119–136.

Sperling L, Ashby JA, Smith ME, Weltzien E, McGuire S. 2001. A framework for analyzing participatory plant breeding approaches and results. *Euphytica* 122, 439–450.

Tripp R, Wijeratne M, Piyadasa VH. 2005. What should we expect from farmer field schools? A Sri Lanka case study. *World Development* 33, 1705–1720.

Tscharntke T, Clough Y, Wanger TC, Jackson L, Motzke I, Perfecto I, Vandermeer J, Whitbread A. 2012. Global food security, biodiversity conservation and the future of agricultural intensification. *Biological Conservation* 151, A 53–59.

Visser B. 2017. *The Impact of National Seed Laws on the Functioning of Small-scale Seed Systems. A Country Case Study.* The Hague: Oxfam Novib.

Witcombe JR, Petre R, Jones S, Joshi A. 1999. Farmer participatory crop improvement IV: The spread and impact of a rice variety identified by participatory varietal selection. *Experimental Agriculture* 35, 471–487.

15 From participatory to evolutionary plant breeding

Salvatore Ceccarelli and Stefania Grando

Introduction: what is participatory plant breeding?

The literature on participatory plant breeding (PPB), including that on participatory varietal selection (PVS), shows that there are several ways of implementing a PPB programme. This is hardly surprising, as there are many different ways of implementing a breeding programme – depending, for example, on the mating system of the crop (i.e. self-pollinated, cross-pollinated or vegetatively propagated), on the clients, and on the final product sought (for example hybrids versus open–pollinated varieties).

On the other hand, as noted by Schnell (1982), all breeding programmes involve three main stages: (1) *generating genetic variability* (includes selection of parents, making crosses, crossing techniques, choice of type and number of crosses, induced mutation, introduction of germplasm from genebanks or other breeding programmes, or from farmers); (2) *selection of the best genetic material within the genetic variability created or acquired in stage 1* (can be implemented using various methodologies), and (3) *testing of breeding lines* (includes comparisons between existing cultivars and the breeding lines emerging from stage 2, and the experimental and analytical methodologies appropriate for conducting such comparisons). The most common way of generating variability in stage 1 is to make crosses among parental lines selected for specific traits. The number of crosses generated at the beginning of each cycle can vary from a few hundred to several thousand. During stages 2 and 3, genetic variability is gradually reduced, and breeding lines are identified. While the number of breeding lines decreases, the amount of seed per line increases, as does the number of locations in which the material can be tested (Ceccarelli, 2009). A breeding programme handles considerable amounts of material on a yearly basis, whether a new cycle starts every year, or twice a year.

Other important stages in a breeding programme are *social targeting and demand analysis* (Weltzien and Christinck, 2009), and *dissemination of cultivars* (Bishaw and van Gastel, 2009). A breeding programme can be schematically represented as in Figure 15.1.

A stage by itself is not the process: for example, farmer evaluation of a germplasm collection remains just that, and is not PPB, unless it is part of the

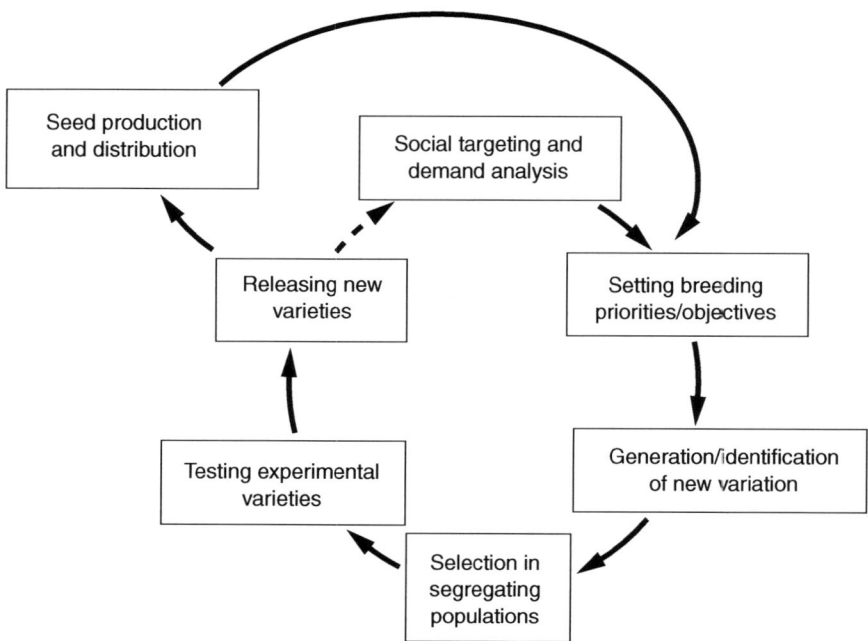

Figure 15.1 Main stages of a breeding programme.
Source: Modified from Tufan *et al.*, 2018: 9.

entire process involved in identifying useful diversity and/or parental material. Molecular tools such as marker-assisted selection and genomic selection do not change the picture, as they are designed to improve efficiency in identifying new genetic variation and in selection, and can be fully integrated in a PPB programme. However, as we will show, they do not necessarily translate into greater breeding efficiency.

A breeding programme becomes *participatory* when farmers (as the most common clients of plant breeding programmes) share with scientists all major decisions made throughout *all* the stages shown in Figure 15.1. A distinction has been made between PPB and PVS. The latter term is used when farmers' participation begins in stage 3 – testing of breeding lines. PVS is technically easier to organize because farmers are involved only in expressing their opinion on the limited number of lines that usually reach that stage (Ceccarelli *et al.*, 2000); however, they are left with a very limited number of possible choices. Further, with PVS there is the risk that breeding material that farmers might find desirable could be discarded before it is even seen by them.

However, because PVS is simple to organize, it can be a useful entry point to start experimenting with farmer participation, assuming that PVS is fully decentralized.

Successes and recognitions of PPB

Highly successful PPB programmes include rice in India (Witcombe and Yadavendra, 2014) and in Nepal (Sthapit *et al.*, 1966), pearl millet in West Asia (Omanya *et al.*, 2007), barley in several countries (Ceccarelli and Grando, 2009) and others; one of the papers on barley PPB (Ceccarelli *et al.*, 2000) received the 2000 CGIAR Chairman's Excellence in Science Award for Outstanding Scientific Article, and sorghum and pearl millet breeders were awarded the Justus von Liebig Prize for World Nutrition by the Fiat Panis Foundation. Some varieties developed through PPB have been accepted for formal release (Laurie and Magoro 2008; Gyawali *et al.*, 2010; Gibson *et al.*, 2011). PPB has received wide coverage at several conferences (see Eyzaguirre and Iwanaga, 1966; PRGA, 2000; Galié *et al.*, 2009). The establishment in 1996 of PRGA (the CGIAR Systemwide Programme on Participatory Research and Gender Analysis, concluded in 2011) was further recognition; likewise the Report prepared for the World Bank (Walker, 2006) which mentioned several success stories. The World Development Report 2008: Agriculture for Development (World Bank, 2007: 160) noted: 'it was found that participatory plant breeding and varietal selection speeds varietal development and dissemination to 5–7 years, half the 10–15 years in a conventional plant-breeding programme'; and 'participatory plant breeding is now paying off with strong early adoption of farmer-selected varieties that provide 40 percent higher yields in farmers' fields'. PPB and decentralized participatory research were among the key recommendations in the Report of the UN Special Rapporteur on the Right to Food (De Schutter, 2014).

Why then has PPB so rarely become institutionalized? As we discuss below, one reason could have been perceptions of weak scientific basis.

Why 'decentralized'? – participatory plant breeding was not the breeders' idea

This question is far from rhetorical, and the answer lies in the *breeder's equation*:

$$R = Sh^2 \tag{1}$$

where the response to selection (R), or genetic gain, depends on the selection differential (S=the difference between the mean of the selected individuals and the mean of the whole population), and on the heritability (h^2) of the target trait (Falconer, 1981). The selection differential depends on the intensity of selection (i), which in turn depends on the percentage of individuals selected, and is equal to S/σ_p, where σ_p is the square root of the phenotypic variance. Therefore, the breeder's equation may also be expressed as

$$R = ih^2\sigma_p \tag{2}$$

It should be borne in mind that

$$h^2 = \sigma_g^2/\sigma_p^2 = \sigma_g^2/\sigma_g^2 + \sigma_{ge}^2/e + \sigma_e^2/(re) \tag{3}$$

where σ_p^2, σ_e^2, σ_{ge}^2, and σ_g^2 are the phenotypic, environmental, geno-typic × environment interaction (GEI), and genotypic variances, respectively, and r and e are the number of replications and the number of environments (locations, years, or location–years combinations), respectively. GEI plays a key role, as we will see below.

The breeder's equation, expressed as (1) or (2), estimates the 'direct' response to selection or genetic gain. In practice, however, the selection environment (one or more research stations) is usually different from the target environment (farmers' fields in one or more countries). That makes the breeder's equation irrelevant, because a research station cannot possibly represent the multitude of often diverse target environments, except in highly specific situations, such as when no or only trivial GEI exists.

Far more relevant is the correlated response to selection (*CR*) in the target environment(s), which is equal to

$$CR_t = R_s h_t^2/h_s^2 r_g \text{ or } ih_t h_s r_g \sigma_{pt} \tag{4}$$

where R_s is the genetic gain in the selection environment (direct response), h_t^2 is the heritability in the target environment, h_s^2 is the heritability in the selection environment, r_g is the genetic correlation coefficient between the measures of the trait object of selection in the two environments, and σ_{pt} is the phenotypic standard deviation of the trait in the target environment (Ceccarelli, 2009).

The relationship between CR_t and R_t is given by

$$CR_t/R_t = r_g(h_s/h_t) \tag{5}$$

which tells us:

a   When $h_t = h_s$, the maximum value of CR_t/R_t is 1, when $r_g = 1$; when her-itabilities are the same, direct selection will always be more effective $(R_t > CR_t)$ because a genetic correlation coefficient equal to 1 has very low probability.

b   With low genetic correlations (0.1–0.2), which are often found between high-yielding breeding nurseries and low-yielding target environments (Atlin et al., 2001), h_s must be at least 5–10 times higher than h_t for CR_t to be greater than R_t. The ratios between h_s and h_t reported in the liter-ature (Ceccarelli, 1996) indicate that generally $R_t > CR_t$.

c   Heritability alone is not sufficient to determine the optimum selection environment, because when r_g, is negative, as in the case of geno-type × environment interactions (GEI) of crossover type, the ratio h_s/h_t becomes irrelevant.

As one distinctive feature of PPB is decentralized selection, the arguments above provide a theoretical framework for discussing the advantages and disadvantages of centralized versus decentralized selection. It has been argued that centralized selection (selection in the research station) maximizes heritability, whereas decentralized selection (selection in the target environment) maximizes correlation (Ceccarelli, 1996). However, that argument is based on the assumption that heritability is intrinsically higher in the favourable and controlled conditions of a research station – which is not always the case (Ceccarelli, 1994; Dawson *et al.*, 2008). Furthermore, low heritability poses a problem for genomic selection as well, because its accuracy depends on the size of the training population, the heritability and the effective number of loci (Bassi *et al.*, 2017).

When heritability is considered across the whole range of the target population of environments, it can be manipulated by conducting Multi-Environment Trials (MET) to divide GEI into Genotype $\times$ Locations (GL) and Genotype $\times$ Years within Locations (GY/L). This allows a measure of the repeatability of GL and hence a subdivision of the target population of environments into subgroups, such that within each subgroup GEI is minimized, h^2 is maximized and hence R is also maximized by selecting for specific adaptation between and for wide adaptation within each subgroup.

Long before the pioneering paper by Rhoades and Booth (1982), *selection theory* supplied breeders with the instruments for deciding when and for which traits it is more efficient to select in the precise and controlled (but often unrepresentative) conditions of the research station, as against in the less controllable but more representative target environments, which can be made more precise with suitable experimental design and statistical analysis. In addition, selection theory allows a precise definition of the target population of environments, thereby optimizing genetic gains in each target environment.

Genetic gains and breeding efficiency

One major issue in research for development programmes (and here plant breeding is no exception) has always been how to ensure adoption of the final products, among farmers' communities and other clients. See Rhoades and Booth (1982) and Rhoades *et al.* (1986), who introduced the 'farmers first' concept: 'successful agricultural research and development must *begin and end* with the *farmer*' (italics in one of the two original papers). Selection theory, as discussed thus far, cannot predict whether the product obtained by maximizing genetic gains, even in the target environment, is what farmers actually need – and this is an additional limitation of genomic selection which merely increases genetic gains as such (Bassi *et al.*, 2017).

Adoption has been negatively affected by the disconnect between breeding activities and seed production, and the inefficiencies of several formal seed-supply systems (Bishaw and Turner, 2008). Therefore, in many breeding programmes, 'breeding efficiency' has been measured as the number of varieties accepted for official release. That, however, means applying the breeder's equation to the variety release system, which is notorious for being based on severely flawed trials (Tripp *et al.*, 1997). Variety release, unrelated to adoption, still drives the professional careers of breeders in several countries.

The ample literature on adoption notes that it is difficult to predict adoption in a centralized, non-participatory breeding programme in the absence of what, in the private industry, is a marketing department (Ceccarelli, 2015). The deeper reasons behind adoption (or lack of it) are not always easy to ascertain: for instance, the adoption of hybrid corn in the USA between 1933 and 1945 has been attributed not only to the superiority of the hybrids but also to a successions of droughts that made obtaining corn seed difficult, and to the policy of corn acreage reduction, which tempted farmers to try the more productive hybrids (Fitzgerald, 1993). Whereas private breeding companies can drive adoption by both advertising and seed-market control, public plant breeding can increase adoption rates through fully non-discriminatory, gender-inclusive, participation of farming communities. Such participation boosts adoption because the potential adopters are involved in the selection work – that is, if the participants are representative of the wider client community.

Centralized participatory plant breeding, where famers take part in selection work at a research station, is not a valid alternative, because at a research station farmers select entries differently than they do in their own fields (Ceccarelli *et al.*, 2000; Ceccarelli *et al.*, 2001; Reguieg *et al.*, 2013).

Why farmers' participation has remained marginal

The origin of PPB, in the widest sense, can be traced back to the seminal paper by Rhoades and Booth (1982), which presented an 'alternative approach to solving farm-level technological problems'. The research was conducted at the Centro International de la Papa (CIP) based in Lima, one of the CGIAR centres, and supported by funds from CIP, the Rockefeller Foundation and the International Development Research Centre, Canada. Their study emphasized the advantages of an *interdisciplinary* approach (research teams working together) over a *multidisciplinary* one (teams filling independent specific disciplinary roles and passing on information).

That publication, together with the one which followed (Rhoades *et al.*, 1986), made clear the importance, when developing a new agricultural technology, of involving the farmers from the start – rather than ignoring them and then handing over a 'beautiful', ready-to-use technology. Applied to plant breeding, this represented a reversal of the model that had been termed 'delegative' (from the French *délégatif*) by Bonneuil and Demeulenaere (2007) and Thomas *et al.* (2011), in which agricultural production, seed production, varietal innovation and conservation of genetic resources shifted, from being part of farmers' normal activities to being functionally separated and delegated to specialized scientists, while farmers lost responsibility for innovation and conservation. By the time the Rhoades and Booth (1982) paper was published, this new model had already become well established; and the system, described by Kloppenburg (2010) where 'they (*the farmers*) decided what seeds to plant, what seeds to save and who else might receive or be allocated their seed as either food or planting material. Such decisions were made within the overarching norms established by the cultures and communities of which they were members' (2010: 370–371) had nearly been forgotten. Therefore, the idea of reversing such a well-established model, implied changes in matters such as power, authority and control (Fitzgerald, 1993). That made it very radical, perhaps even subversive (Crane, 2014).

Several almost-concomitant factors might help in explaining why professional breeders were not enthusiastic about the idea of participation (Belay, 2009). Firstly, as indicated above, the proposal for participatory research came from social scientists who also conducted early PPB experiments in which the institutions and practices of the biophysical sciences were often either left invisible or assumed to be purely technical (Crane, 2014). At about that time, a conventional, centralized-non participatory plant breeding had already become well established, geared for an industrialized model regarding the seed market and the food industry. In parallel, a new vocabulary had been developed: it described, for example, participation in plant breeding as being 'conventional or contractual, consultative, collaborative, collegial, farmer experimentation' (Sperling *et al.*, 2001). This gave the impression of PPB as a static phenomenon – whereas field experience has shown that, in a truly participatory programme, as the farmers become more empowered, the process itself evolves further (Desclaux *et al.*, 2012).

As a result, a confidence gap that still persists today was generated between the breeders, who felt they had been expropriated of their science, and scientists (social and gender scientists, anthropologists and socio-economists) whom the breeders saw as trespassers on 'their' territory. Several breeders felt that the proponents of PPB were actually driven by a social agenda, where the technical issues related to breeding were subordinate (Ashby, 2007; Scoones *et al.*, 2008).

On the part of professional breeders, this marked a considerable attitudinal change. In fact, as early as in 1908, Herbert J. Webber had explicitly advocated farmers' breeding, recognizing their ability to undertake selection (Webber, 1908). Also several early professional breeders at the US Department of Agriculture, such as H.A. Wallace, in the 1920s recognized the importance of farmers' skills and knowledge to the work of breeders; Wallace asserted that the only way for breeders to discover new corn strains was to rely on the expertise of the knowledgeable corn farmers themselves (Fitzgerald, 1993).

As a result of this emerging confidence gap, although the proposal for participatory breeding was originally formulated from within the CGIAR, very few 'CGIAR breeders', defined as the scientists formally in charge of the breeding programme of one or more CGIAR-mandated crops (i.e. having responsibility for such fundamental choices as target environments, product profiles, breeding methodologies and strategies) modified their breeding programmes entirely or partially into a participatory programme, or added a participatory component to their breeding programmes, to verify the advantages, disadvantages, costs and benefits of the new approach. Even at CIP, the birthplace of participatory research, its implementation was still limited, some 20 years later (Thiele *et al.*, 2001). On the other hand, within the CGIAR there were several 'non-breeders' (as per the definition above) using participatory approaches. Many more scientists within and outside the CGIAR started to use participatory approaches of various type, in several countries and with several crops. A recent inventory has identified 254 studies (experimental and conceptual) covering 69 countries and 47 crops (Ceccarelli and Grando, 2019).

However, while some form of farmers' participation is routinely used by private breeding companies (mostly in the form of market surveys), PPB as such has virtually disappeared from public plant breeding, with exception of a few US universities (Ceccarelli and Grando, 2019). In June 2018, we reviewed the websites of CGIAR centres known to have implemented PPB in the past. The results showed that the search for PPB resulted in documents that in the best cases were at least five years old.

As our review revealed no scientific objections to PPB, we came to the conclusion that the major obstacle to institutionalizing PPB, within and outside the CGIAR, lies in the difficulties of sharing power, authority and control in the private and the public sectors.

The way forward: Evolutionary Participatory Plant Breeding (EPPB)

An alternative to PPB is Evolutionary Participatory Plant Breeding (EPPB), which retains many of desirable features of PPB, such as farmers' empowerment, recognition and utilization of indigenous knowledge and breeding for specific adaptation (Suneson, 1956; Murphy *et al.*, 2005; Ceccarelli, 2017). Suneson (1956: 188) defined EPB as a method that 'requires assembly and study of seed stocks with diverse evolutionary origin, recombination by hybridization, bulking of the F_1 progeny and subsequent prolonged mass selection for mass sorting of the progeny in successive natural cropping environments'.

Evolutionary Populations (EPs), formerly known as Composite Crosses (CC), can be made using a large number of parents to generate as much variability as possible, or by careful selection of parents when the EP is made for a specific objective. EPB becomes Participatory when the EPs are planted in farmers' fields and the farmers use these EPs as a source population in conducting selection.

EPPB brings crop genetic diversity back into farmers' fields and into farmers' hands – without necessarily requiring the support of a scientific institution, as EPs can be assembled and managed by the farmers themselves. The EPs are planted and harvested year after year. Through the effect of natural selection and the natural outcrossing (depending on the type of crop), the genetic composition of the EPs changes (Soliman and Allard, 1991; Raggi *et al.*, 2017). Thus, an EP evolves to become progressively better adapted to the environment (soil type, soil fertility, agronomic practices including organic systems, rainfall, temperature etc.) in which is grown, While it is evolving, farmers, or farmers and breeders together, can use it as source for the development of varieties. As climatic conditions vary from one year to the next, the genetic makeup of the population will fluctuate, but if the general tendency is towards a hotter and drier climate, as expected in view of climate changes, the genotypes better adapted to those conditions will gradually become more frequent (Ceccarelli, 2014, 2019).

When an evolutionary population is used by farmers and researchers as a source of genetic diversity from which to select, it is expected that, based on selection theory and assuming similar h^2 and $_p$ (formula 2 earlier), response to selection will increase because of the large population size, thereby leading to greater selection efficiency.

Evolutionary populations of different crops are currently being grown by farmers in Jordan, Ethiopia, Iran, Italy, France, Portugal, UK, and India for cereal crops (maize, barley, bread and durum wheat, and rice), grain legumes (common bean), and horticultural crops (tomato and summer squash). In informal interviews during meeting with scientists, farmers growing these populations have reported higher crop yields and lower levels of weed infestation, disease presence and insect damage as compared to the uniform varieties they used to grow.

Iranian and Italian farmers growing an evolutionary population of bread wheat have begun marketing bread baked from bread-wheat evolutionary population flour. The bread can be safely eaten by gluten-intolerant persons. Anecdotal evidence from farmers growing wheat EPs in France and Italy shows that these EPs not only have greater yield stability: the bread has more aroma and higher quality. Recent experimental evidence has shown that, with evolutionary breeding, it is possible to combine high yield and stability (Raggi *et al.*, 2017). Here it should also be noted that, as EPs are continuously evolving, they cannot be protected by Intellectual Property Rights.

EPPB, being a relatively inexpensive and highly dynamic strategy to adapt crops to various combinations of both abiotic and biotic stresses, has the potential to serve as a suitable method for generating, directly on farmers' hands, the varieties that can feed current and future generations. Because of its ability to adapt crops to the environment, including agronomic practices, EPPB can also be used to breed varieties specifically adapted to organic agriculture.

According to the Commission Implementing Decision of 18 March 2014 pursuant to Council Directive 66/402/EEC (Commission Implementing Decision, 2014), in the EU it is now possible to market experimentally heterogeneous materials of wheat, maize, oats and barley. Originally due to expire in December 2018, this Decision has been recently extended to December 2021, legalizing both marketing and sales of EP seed.

Combining seed saving with evolution returns control of seed production to the farmers. It can produce better and more diversified varieties that can contribute to helping millions of farmers to reduce their dependence on external inputs, as well as reducing vulnerability to disease, insects and climate change. Ultimately this will contribute to food security and food safety for all. As it is simpler to implement and to manage than traditional PPB, EPPB appears particularly well-suited to marginal environments and resource-poor farmers.

While benefiting from advances in molecular genetics, such as molecular markers and genomic selection, participatory and evolutionary plant breeding integrates greater production of more readily available and accessible food with various ecosystem services – including increase in agrobiodiversity, ecosystem maintenance through less use of chemicals, and farmers' intellectual enrichment – while maintaining the evolutionary potential necessary for crops to cope with climate change.

Conclusions

After the original papers by Rhoades and Booth (1982) and Rhoades *et al.* (1986), several workshops[1] have discussed progress and shortcomings in the implementation of the 'farmers first' concept. Some of the questions and comments that emerged during the 2008 workshop – such as 'what changes in paradigmatic assumptions have occurred?' and 'how many participatory research approaches have survived beyond the duration of a specific project?' – seem equally pertinent some ten years later.

It is now recognized that participatory research in general and PPB in particular are not mere technical exercises, but are highly political, touching on consolidated economic interests particularly in the seed, pesticides and food domains. Using the terminology of Kloppenburg (2010), and seeing the advent of commercial (later corporate) breeding as a process of both genetic and epistemic dispossession, PPB can be considered as enabling re-possession of both genetic material and knowledge while at the same time undermining the consolidated interests mentioned above. That makes changing the paradigm on which institutional public breeding is entrenched rather unrealistic.

What appears more feasible is to focus on making progress towards social justice, equity, gender equality and seed/food sovereignty by bringing back to farming communities both genetic diversity *and* updated knowledge on how to manage it, with or without institutional participation. Today's evolutionary–participatory breeding programmes are doing precisely that.

Note

1 Including, but not limited to, one at the University of Sussex in 1987; a second one in 2008, again at the University of Sussex; and one in Oslo in 2017, organized by the Norwegian University of Life Sciences.

References

Ashby JA, 2007. Fostering farmer first methodological innovation: organizational learning and change in international agricultural research. www.future-agricultures. org/farmerfirst/files/D1_Ashby.pdf.

Atlin GN, Cooper M, Bjørnstad Å, 2001. Comparison of formal and participatory breeding approaches using selection theory. *Euphytica* 122, 463–475.

Bassi FM, Bentley AR, Charmet G, Ortiz R, Crossa J, 2017. Breeding schemes for the implementation of genomic selection in wheat *(Triticum spp.)*. *Plant Science* 242, 23–36.

Belay G, 2009. Does client-oriented plant breeding work? *CAB Reviews: Perspectives in Agriculture, Veterinary Science, Nutrition and Natural Resources*, 1–7.

Bishaw Z, Turner M, 2008. Linking participatory plant breeding to the seed supply system. *Euphytica*, 163, 31–44.

Bishaw Z, van Gastel AJG, 2009. Variety release and policy options. In Ceccarelli S, Guimaraes EP, Weltzien E, (eds) *Plant Breeding and Farmer Participation*, 565–588. Rome: FAO.

Bonneuil C, Demeulenaere E, 2007. Une génétique de pair à pair? L'émergence de la sélection participative. In Charvolin F, Micoud A, Nyhart LK, (eds) *Les sciences citoyennes. Vigilance collective et rapport entre profane et scientifique dans les sciences naturalistes*, 122–147. Paris: Editions de l'Aube.

Ceccarelli S, 1994. Specific adaptation and breeding for marginal conditions. *Euphytica* 77, 205–219.

Ceccarelli S, 1996. Adaptation to low/high input cultivation. *Euphytica* 92, 203–214.

Ceccarelli S, 2009. Main stages of a plant breeding programme. In Ceccarelli S, Guimaraes EP, Weltzien E, (eds) *Plant Breeding and Farmer Participation*, 63–74. Rome: FAO.

Ceccarelli S, 2014. GMO, organic agriculture and breeding for sustainability. *Sustainability* 6, 4273–4286.

Ceccarelli S, 2015. Efficiency of plant breeding. *Crop Science* 55, 87–97.

Ceccarelli S, 2017. Increasing plant breeding efficiency through evolutionary–participatory programs. In Pilu R, Gavazzi G, (eds) *More Food: Road to Survival*, 17–40. Sharja, UAE: Bentham Science Publishers.

Ceccarelli S, 2019. Health, seeds, diversity and terraces. In Varotto M, Bonardi L, Tarolli P, Agnoletti M, (eds) *World Terraced Landscapes: History, Environment, Quality of Life*. Proceedings of the III World Meeting on Terraced Landscapes, 6–15 October 2016 (in press).

Ceccarelli S, Grando S, 2009. Participatory Plant Breeding in cereals. In Carena M, (ed) *Cereals*, 395–414. New York: Springer Plant Science.

Ceccarelli S, Grando S, 2019. Participatory Plant Breeding: Who did it, who does it and where? *Experimental Agriculture* 33, 335–344.

Ceccarelli S, Grando S, Tutwiler R, Baha J, Martini AM, Salahieh H, Goodchild A, Michael M, 2000. A methodological study on participatory barley breeding, I: Selection phase. *Euphytica* 111, 91–104.

Ceccarelli S, Grando S, Bailey E, Amri A, El Felah M, Nassif F, Rezgui S, Yahyaoui A, 2001. Farmer participation in barley breeding in Syria, Morocco and Tunisia. *Euphytica* 122, 521–536.

Commission Implementing Decision 2014. *Official Journal of the European Union.* http://eur-lex.europa.eu/legal-content/EN/TXT/?uri=CELEX%3A32014D0150.

Crane TA, 2014. Bringing science and technology studies into agricultural anthropology: Technology development as cultural encounter between farmers and researchers. *Culture, Agriculture, Food and Environment* 36 (1), 45–55.

Dawson JC, Murphy KM, Jones SS, 2008. Decentralized selection and participatory approaches in plant breeding for low-input systems. *Euphytica* 160, 143–154.

De Schutter O. 2014. *Final Report: The Transformative Potential of the Right to Food.* Report of the Special Rapporteur on the Right to Food. United Nations General Assembly, A/HRC/25/57, 1–28.

Desclaux D, Ceccarelli S, Navazio J, Coley M, Trouche G, Aguirre S, Weltzien E, Lançon J, 2012. Centralized or decentralized breeding: the potentials of participatory approaches for low-Input and organic agriculture. In Lammerts van Bueren ET, Myers JR, (eds) *Organic Crop Breeding*, 99–123. Hoboken, NJ: Wiley-Blackwell.

Eyzaguirre PB, Iwanaga M, 1996. *Participatory Plant Breeding.* Proceedings of a workshop on participatory plant breeding, 26–29 July 1995, Wageningen, the Netherlands: International Plant Genetic Resources Institute.

Falconer DS, 1981. *Introduction to Quantitative Genetics*, 2nd edn. London: Longmann.

Fitzgerald D, 1993. Farmers deskilled: Hybrid corn and farmers' work. *Technology and Culture* 34 (2), 324–343.

Gibson RW, Mpembe I, Mwanga, ROM, 2011. Benefits of participatory plant breeding (PPB) as exemplified by the first-ever officially released PPB-bred sweet potato cultivar. *Journal of Agricultural Science* 149 (5), 625–632.

Gyawali S, Sthapit BR, Bhandari B, Bajracharya J, Shrestha PK, Upadhyay MP, Jarvis DI, 2010. Participatory crop improvement and formal release of *Jethobudho* rice landrace in Nepal. *Euphytica* 176, 59–78.

Kloppenburg J, 2010. Impeding dispossession, enabling repossession: Biological open source and the recovery of seed sovereignty. *Journal of Agrarian Change* 10, 367–388.

Laurie SM, Magoro MD, 2008. Evaluation and release of new sweet potato varieties through farmer participatory selection. *African Journal of Agricultural Research* 3 (10), 672–676.

Murphy K, Lammer D, Lyon S, Carter B, Jones SS, 2005. Breeding for organic and low-input farming systems: An evolutionary–participatory breeding method for inbred cereal grains. *Renewable Agriculture and Food Systems*, 20, 48–55.

Omanya GO, Weltzien-Rattunde E, Sogodogo D, Sanogo M, Hanssens N, Guero Y, Zangre R, 2007. Participatory varietal selection with improved pearl millet in West Africa. *Experimental Agriculture* 43, 5–19.

Participatory Research and Gender Analysis (PRGA), 2000. An exchange of experiences from south and south east Asia. International Symposium on Participatory Plant Breeding and Participatory Plant Genetic Resource Enhancement. Pokhara, Nepal, 1–5 May 2000.

Raggi L, Ciancaleoni S, Torricelli R, Terzi V, Ceccarelli S, Negri V, 2017. Evolutionary breeding for sustainable agriculture: Selection and multi-environment evaluation of barley populations and lines. *Field Crops Research* 204, 76–88.

Reguieg M, Labdi M, Benbelkacem A, Hamou M, Maatougui MEH, Grando S, Ceccarelli S, 2013. First experience on participatory barley breeding in Algeria. *Journal of Crop Improvement* 27, 469–486.

Rhoades RE, Booth RH, 1982. Farmer-back-to-farmer: A model for generating acceptable agricultural technology. *Agricultural Administration* 11, 127–137.

Rhoades RE, Horton DE, Booth RH, 1986. Anthropologist, biological scientist and economist: The Three Musketeers or three stooges of farming systems Research? In Jones JR, Wallace BJ, (eds) *Social Sciences and Farming System Research. Methodological Perspectives on Agricultural Development*, 21–40. Boulder, CO: Westview Press.

Schnell FW, 1982. A synoptic study of the methods and categories of plant breeding. *Zeitschrift für Pflanzenzuchtung* 89, 1–18.

Scoones I, Thompson J, Chambers R, 2008. Farmer First revisited: some reflections on the future of the CGIAR, an informal note to the CGIAR Independent Review Team. www.cgiar.org/changemanagement/pdf/farmer_first_revisited.pdf.

Soliman KM, Allard RW, 1991. Grain yield of composite cross populations of barley: Effects of natural selection. *Crop Science* 31, 705–708.

Sperling L, Ashby JA, Smith M, Weltzien E, McGuire S, 2001. A framework for analysing participatory plant breeding approaches and results. *Euphytica* 122 (3), 439–450.

Sthapit BR, Joshi KD, Witcombe JR, 1996. Farmer participatory crop improvement III: Participatory Plant Breeding, a case study for rice in Nepal. *Experimental Agriculture* 32 (4), 479–496.

Suneson CA, 1956. An evolutionary plant breeding method. *Agronomy Journal* 48, 188–191.

Thiele G, Fliert E, Campilan D, 2001. What happened to participatory research at the International Potato Center? *Agriculture and Human Values* 18, 429–446.

Thomas M, Dawson JC, Goldringer I, Bonneuil C, 2011. Seed exchanges, a key to analyze crop diversity dynamics in farmer-led on-farm conservation. *Genetic Resources Crop Evolution* 58, 321–338.

Tripp R, Louwaars N, van der Burg WJ, Virk DS, Witcombe JR, 1997. Alternatives for seed regulatory reform. An analysis of variety testing, variety regulation and seed quality control. *Agricultural Administration (Research and Extension) Network Paper* 69, 1–25. London: Overseas Development Institute.

Tufan HA, Grando S, Meola C, (eds) 2018. *State of the Knowledge for Gender in Breeding: Case Studies for Practitioners*. CGIAR Gender and Breeding Initiative. Working Paper No. 3. Lima: CIP.

Walker TS, 2006. *Participatory Varietal Selection, Participatory Plant Breeding, and Varietal Change*. Washington, DC: World Bank. https://openknowledge.worldbank.org/handle/10986/9182.

Webber HJ, 1908. Plant-breeding for farmers. *Cornell University Bulletin* 251, 289–332.

Weltzien E, Christinck A, 2009. Methodologies for priority setting. In Ceccarelli S, Guimaraes EP, Weltzien E, (eds) *Plant Breeding and Farmer Participation*, 75–106. Rome: FAO.

Witcombe JR, Yadavendra JP, 2014. How much evidence is needed before client-oriented breeding (COB) is institutionalized? Evidence from rice and maize in India. *Field Crops Research* 167, 143–152.

World Bank, 2007. *World Development Report 2008: Agriculture for Development*. https://openknowledge.worldbank.org/handle/10986/5990 License: CC BY 3.0 IGO.

Part IV

Collaborative approaches

International and national legal contexts

16 Participatory plant breeding as a tool for implementing Farmers' Rights and sustainable use under the Plant Treaty

Tone Winge

As the International Treaty on Plant Genetic Resources for Food and Agriculture ('the Plant Treaty'), of 3 November 2001,[1] the stated objectives of which are 'the conservation and sustainable use of plant genetic resources for food and agriculture and the fair and equitable sharing of the benefits arising out of their use' (Article 1.1), inches closer to its 20-year anniversary, the question of implementation progress becomes more and more important. As of August 2019, the Governing Body of the Plant Treaty has met for seven sessions, with an expanding and increasingly complex agenda. Identifying and utilizing key tools and approaches is crucial to continued progress in implementation, especially for provisions where the treaty language lacks precision and/or leaves room for flexibility. Two such provisions in the Plant Treaty are Article 6 on sustainable use of plant genetic resources and Article 9 on Farmers' Rights.

According to Article 6.1, the Contracting Parties of the Plant Treaty 'shall develop and maintain appropriate policy and legal measures that promote the sustainable use of plant genetic resources for food and agriculture'. No definition is provided for the central term 'sustainable use'. However, if seen as related to the much-used term 'sustainable development', which, in line with the Brundtland Report *Our Common Future*, is generally defined as development 'that meets the needs of the present without compromising the ability of future generations to meet their own needs' (World Commission on Environment and Development 1987: 54), it could be defined as use of crop genetic resources that meets the food production needs of the present without compromising the ability of future generations to meet their own food production needs. In the Convention on Biological Diversity (22 May 1992), sustainable use is defined as 'the use of components of biological diversity in a way and at a rate that does not lead to the long-term decline of biological diversity, thereby maintaining its potential to meet the needs and aspirations of present and future generations' (Art. 2). Here it should be noted that for crop genetic diversity, which exists due to human intervention and is dependent on human activity for its maintenance, sustainability is linked to utilization in a different manner than is the case with wild biological diversity.

Article 6.2 lists measures which the sustainable use of crop genetic resources may include. Specifically mentioned in this connection is farmer participation in plant breeding:

> the sustainable use of plant genetic resources for food and agriculture may include such measures as promoting, as appropriate, plant breeding efforts which, with the participation of farmers, particularly in developing countries, strengthen the capacity to develop varieties particularly adapted to social, economic and ecological conditions, including in marginal areas.
>
> (Art. 6.2c)

This represents a strong argument for viewing participatory approaches to plant breeding as a recognized tool for furthering the implementation of Article 6. In this chapter, implementation tools are defined as specific tools that can be implemented nationally.

As for Article 9, it recognizes

> the enormous contribution that the local and indigenous communities and farmers of all regions of the world, particularly those in the centres of origin and crop diversity, have made and will continue to make for the conservation and development of plant genetic resources which constitute the basis of food and agriculture production throughout the world.
>
> (Art. 9.1)

There is no specific mention of participatory plant breeding approaches in the four elements of Farmers' Rights – protection of traditional knowledge (Art. 9.2a), equitable benefit-sharing (Art. 9.2b), participation in national-level related to conservation and use of crop genetic resources (Art. 9.2c), and rights to save, use, exchange, and sell farm-saved seed (Art. 9.3) – but all these can be linked to such approaches. Participatory approaches utilize farmers' traditional knowledge (thereby contributing to protection through use); they bring benefits to farmers, can empower farmers to participate in decision-making, and have the potential to strengthen seed systems and familiarize farmers the legislation and policies that regulate (or limit) these systems, including the rights associated with the use and exchange of farm-saved seed (see also Halewood *et al.*, 2007). Participatory approaches to plant breeding can therefore be seen as a useful tool for implementing all four aspects of Farmers' Rights.

Participatory approaches to plant breeding are here defined as plant breeding efforts where farmers (and potentially other users) collaborate with scientists during key (or all) stages of the plant breeding process. As used here, the term incorporates both formal-led and farmer-led participatory plant breeding programmes. In this context it should be noted that different variants of the term 'participatory plant breeding' are used by practitioners, in the academic literature and in Plant Treaty documents, and a definition or explanation is not always given. Although much has been written about the stages, quality and type of farmer participation in breeding (see Weltzien *et al.*, 2003, Morris and Bellon, 2004) such factors are not necessarily discussed in Plant Treaty meetings and documents.

This chapter uses document analysis to show how participatory approaches to plant breeding are viewed within the Plant Treaty system, and to what extent the Governing Body has recognized them as implementation tools.

Participatory approaches to plant breeding in the Plant Treaty system[2]

Since the Plant Treaty entered into force, the Governing Body has devoted an increasing amount of time and effort to the implementation of Article 6 and Article 9, although not necessarily simultaneously or in linear fashion. This chapter discusses one specific aspect: participatory approaches to plant breeding, first in connection with Article 6 and then with Article 9.

Participatory plant breeding as a tool for implementing Article 6 on sustainable use

Review and analysis of all documents prepared for the Governing Body (working documents, information documents, other documents), and reports from its sessions, held up to and including the seventh session in December 2017, show that participatory approaches to plant breeding have received increased attention and recognition with regard to implementation of Article 6.

At the first session of the Governing Body in 2006, participatory approaches to plant breeding received scant notice;[3] but at the next session, in 2007, there were indications that at least some actors saw such approaches as useful tools for implementation of Article 6.[4] Then, with the third session of the Governing Body, in 2009, official documents[5] devoted greater attention to this topic – not least in the session report, where 'the participation of farmers' (FAO 2009d: para 44) was specifically mentioned in connection with the need to implement Article 6 to enhance capacities for using crop genetic resources through plant breeding. This would appear to indicate growing awareness and acknowledgement on the part of the Governing Body of the role of participatory approaches to plant breeding for sustainable use.

This must also be seen in connection with the time and attention devoted to Article 6 at successive sessions of the Governing Body. Even though implementation of Article 6 had been an agenda item from the first session and onwards, no resolution was adopted on implementation of Article 6 until Resolution 07/2011, at the *fourth* session. Despite the rising attention to participatory approaches to plant breeding, that resolution made no mention of such approaches. However, Resolution 07/2011 did request the Secretary to 'further explore the development of a toolbox on sustainable use' (Para 1), and it established an Ad hoc Technical Committee on Sustainable Use of Plant Genetic Resources for Food and Agriculture (ACSU) (para 7).[6] Further, in the working document on implementation of Article 6, IT/GB-4/11/17 (FAO, 2011a), which dealt specifically with the concept, the justifications for, functions, and elements of such a toolbox, participatory plant breeding was mentioned five times in a very positive manner: for instance, it was lauded for bringing local knowledge and modern science together 'in the most effective pathway to sustainable use' (FAO, 2011a: 10) – surely an indication of growing recognition of participatory plant breeding as an essential tool for promoting the sustainable use of crop genetic resources.

At the fifth session of the Governing Body in 2013, the key results of a stakeholder consultation[7] were presented in the working document on implementation of Article 6, IT/GB-5/13/9 (FAO, 2013a). Included in the short list of main consultation outcomes were 'the need to introduce incentives to promote the sustainable use of plant genetic resources for food and agriculture, and to strengthen capacities for this purpose, including local plant breeding capacity, by promoting participatory plant breeding and supporting small breeder organizations and seed companies' (FAO 2013a: 5) – indicating the importance given to participatory approaches to plant breeding in connection with sustainable use by a majority of consulted stakeholders. Indeed, according to the summary and synthesis provided in the information document IT/GB-5/13/Inf.7 (FAO, 2013c), a majority of those participating in the consultation saw participatory plant breeding as a key strategy to enhance local breeding capacity (FAO, 2013c: 8) – and the need for increased funding was specifically mentioned.

Participatory plant breeding was also included among the best practices and measures to improve the sustainable use of crop genetic resources mentioned in the submissions from Contracting Parties and other relevant institutions (FAO, 2013b). However, although the discussions of the Governing Body on implementation of Article 6 resulted in Resolution 07/2013, participatory approaches to plant breeding were not referred to, directly or indirectly, in the text of that resolution.

In 2015, prior to the sixth session of the Governing Body, an online stakeholder consultation was held in order to collect views relevant to developing a toolbox on sustainable use. Participatory plant breeding/ variety selection programmes were one of the 12 predefined areas of sustainable use in this consultation – indicative of the recognition given to such approaches within the context of Article 6. Of the 289 survey responses from respondents in 109 countries, 122 mentioned an involvement or interest in participatory plant breeding or participatory varietal selection: most of these responses came from the public sector. The survey also indicated that national policies for sustainable use did not support farmers sufficiently, and that participatory plant breeding and participatory varietal selection were among the measures in need of supportive policies and funding (FAO, 2015b).

At the sixth session of the Governing Body in October 2015, a new resolution on the implementation of Article 6 on sustainable use was adopted: Resolution 4/2015. Unlike its predecessor, this resolution made reference to participatory approaches to plant breeding in connection with the assessment of farmers' needs and the 'identification of possible means to address those needs also through participatory approaches in the context of the Programme of Work' (Resolution 4/2015: Art. 4e).

This resolution also reconvened the Ad hoc Committee on Sustainable Use, which met later in October 2016 to continue discussions on developing the toolbox. At this meeting, participatory plant breeding was identified as one of three priority areas (along with 'adding value to and sustaining the use of landraces/farmers' varieties' and 'seed systems') recommended for inclusion in the first phase of the toolbox on sustainable use (FAO, 2017h) – clear recognition of participatory plant breeding as a tool for implementation of Article 6.

A toolbox prototype has since been developed, and is currently hosted by the Secretariat on the Plant Treaty website.[8] The prototype addresses the priority areas identified by the ACSU through the webpage on 'sustaining local crop diversity', which is divided into 'strengthening seed systems', 'enhancing crop diversity for local needs', and 'promoting local crop diversity'. Participatory plant breeding is included under 'enhancing crop diversity for local needs' (Plant Treaty, 2017), which perhaps does not accord it as prominent a place as might have been expected. However, its inclusion in the prototype does imbue it with importance for implementation of Article 6.

Then, at the seventh session of the Governing Body in November 2017, a new resolution on implementation of Article 6, Resolution 6/2017, marked a further step forward for participatory approaches to plant breeding as tools for implementing Article 6. The Secretary (working in collaboration with other stakeholders, and subject to the availability of financial resources) was requested to 'organize regional capacity-building workshops on topics such as participatory plant breeding, community seed bank development, sustainable biodiverse production systems and promoting the value of farmers' varieties' (Resolution 6/2017: para. 2). Regardless of whether such workshops are actually held, this represents indisputable recognition of the role that participatory plant breeding can play for the sustainable use of crop genetic resources.[9]

Additionally, the working document on implementation of the programme of work on sustainable use mentions participatory plant breeding as a suggested topic for regional capacity-building workshops to be organized by a possible joint programme on biodiversity in agriculture for sustainable use of crop genetic resources (FAO, 2017a). Further, according to the report of the Compliance Committee to the seventh session of the Governing Body, 11 out of the 14 Contracting Parties that had submitted reports 'reported promoting plant breeding efforts, with the participation of farmers, which strengthen the capacity to develop varieties particularly adapted to social, economic and ecological conditions, including in marginal areas' (FAO, 2017c: 7). This may indicate that participatory approaches to plant breeding are now being applied by many Contracting Parties in their implementation efforts, but with so few reports received it is too early to draw firm conclusions here.

Participatory plant breeding as a tool for implementing Article 9 on Farmers' Rights

Also for Article 9 on Farmers' Rights, review and analysis of relevant documents and the reports from the seven Governing Body sessions that have been held as of December 2017, show that participatory approaches to plant breeding are increasingly being recognized.

At the first session of the Governing Body in 2006, very little attention was paid to participatory approaches to plant breeding in connection with Farmers' Rights – perhaps not so surprising, as implementation of Article 9 on Farmers' Rights was not even on the agenda at this first session.[10] However, at the informal international consultation on Farmers' Rights held in 2007, 10 of 18 presentations addressed PPB/PVS in some way (Norwegian Ministry of Agriculture and Food, 2007); seen together, these references indicate a shared understanding among certain key figures involved in discussions of Farmers' Rights as to the positive role such breeding efforts can play for implementation, especially of Article 9.2 (b) on benefit sharing.

Then, at the second session of the Governing Body, one of the three resolutions adopted concerned farmers' rights: Resolution 2/2007. It encouraged Contracting Parties to submit views on and experiences with implementation of Article 9 – but there was no reference to participatory approaches to plant breeding.

The submissions received in response to Resolution 2/2007 were presented in connection with the third session of the Governing Body in 2009. Here, participatory plant breeding was mentioned in three of the 17 submissions. One of these concerned an online conference on 'Options for Farmers' Rights' in which 55 invited participants took part in three discussion rounds; one point agreed upon by all participants was that farmers can be breeders. In addition, 'most participants also agreed that participatory plant breeding can form a major instrument in improving the maintenance' of crop genetic resources on-farm (FAO, 2009b: 8). Adding weight to these results were the findings from an international stakeholder survey on Farmers' Rights conducted in 2005. This survey found participatory plant breeding to be one of the non-monetary benefits most frequently mentioned by respondents (FAO, 2009c). It outlined three categories of benefit-sharing measures relevant to Farmers' Rights, and included participatory plant breeding as one example of measures in the 'reward and support' category (FAO, 2009c) – further strengthening the focus on participatory plant breeding as a tool for benefit-sharing.

Another resolution on the implementation of Article 9 on Farmers' Rights, Resolution 6/2009, which focused *inter alia* on the submission and collection of views and experiences, was adopted at the third session of the Governing Body. Again there was no mention of participatory approaches to plant breeding.

However, the results of the e-mail based survey of the 2010 Global Consultations on Farmers' Rights, conducted from July to September 2010, further confirmed participatory plant breeding as an important measure for non-monetary benefit-sharing related to Farmers' Rights. This survey showed that participatory plant breeding projects were among the five most commonly used benefit-sharing measures – and, perhaps more importantly, were one of the two[11] most favourably viewed by respondents. Moreover, participatory plant breeding was mentioned when respondents were asked to provide recommendations to the Governing Body (Andersen and Winge, 2011).

At the consultation conference of the 2010 Global Consultations, participatory plant breeding received attention particularly in connection with equitable benefit-sharing, but it was also recognized that such projects can contribute to the protection of traditional knowledge.

In response to Resolution 6/2009, ten submissions were received: three of these, all from organizations, made reference to participatory plant breeding. Of these three, the submission from the International Institute for Environment and Development went into greatest empirical detail regarding the use of participatory plant breeding. It noted that participatory plant breeding has been central to Chinese efforts to protect traditional knowledge and promote benefit-sharing, and that also some province-level governments in China have come to support these approaches (FAO, 2011b).[12]

At the fourth session of the Governing Body, another resolution on Farmers' Rights was adopted, Resolution 6/2011, out of altogether nine resolutions. This resolution similarly encouraged the submission of views and experiences, but did not go into much detail regarding implementation measures, and there was no direct reference to participatory approaches to plant breeding as a tool for implementation. However, this resolution did encourage Contracting Parties to 'closely relate the realization of Farmers' Rights' (Resolution 6/2011: para 8) with the implementation of Article 6 (and 5), giving particular emphasis to 'the measures in Articles 5.1(c and d), and 6.2 (c, d, e, f, and g)' (ibid.). As the participation of farmers in certain plant breeding efforts is specifically mentioned in 6.2.c, this represents important recognition of the centrality of such participation for the realization of Farmers' Rights.

This resolution led to 14 submissions prior to the fifth session of the Governing Body. With regard to participatory plant breeding, once again, such approaches were highlighted only in submissions from relevant organizations, and not from Contracting Parties. In this round, four submissions mentioned participatory plant breeding. Some merely noted it briefly in connection with Farmers' Rights in general;[13] however, the submission from the Norway-based Development Fund[14] addressed its potential role for the various elements of Farmers' Rights, by including participatory plant breeding among the tools for promoting and protecting traditional knowledge, and for empowering farmers' involvement in decision-making (FAO, 2013d). Further, this submission argued that the Benefit-Sharing Fund of the Treaty should finance participatory plant breeding programmes to enable farmers to take part in indirect benefit-sharing.

The fifth session of the Governing Body adopted Resolution 8/2013 on implementation of Article 9 on Farmers' Rights. In its invitation to Contracting Parties to 'consider developing national action plans for the implementation of article 9' (Resolution 8/2013: para 5), this resolution used almost the same phrase concerning Articles 5 and 6 as the previous resolution, suggesting that such plans should be 'in line with' the implementation of these two articles, and in particular the measures listed before, including that in 6.2.c. In addition to further strengthening the relationship between these articles, and implementation efforts, this constitutes another indirect reference to farmer participation in some plant breeding endeavours.

This new resolution also reiterated many of the same elements as previous resolutions. However, instead of again inviting Contracting Parties and relevant organizations to submit views and experiences, it requested the Secretary to 'review the knowledge, views, experiences and best practices that have been submitted' (Resolution 8/2013: para 1).

Prior to the sixth session, the Secretariat reviewed all submissions. This review included participatory plant breeding among the positive experiences connected to the sharing of benefits and protection of traditional knowledge. In addition, 'participatory plant breeding and participative research in farmers' fields and under the control of farmers' organizations are considered as further positive approaches to the realization of Farmers' Rights' (FAO, 2015a: 6) in connection with 9.3 and farm-saved seed. Thus, the review explicitly linked participatory approaches to plant breeding to three of the four elements of Farmers' Rights. However, there are considerably more Contracting Parties than submissions received;[15] participatory approaches to plant breeding may well be central to the implementation efforts in even more Contracting Parties, and in more ways, than indicated by these submissions and this review.

The resolution adopted at the sixth session of the Governing Body, Resolution 5/2015, expands upon and adds to the content of previous resolutions on implementation of Article 9 – *inter alia* by requesting the Secretary to prepare a study on lessons learned (subject to available resources). Again, however, there is little focus on specific implementation measures, and therefore no mention of participatory approaches to plant breeding.[16]

To gather views on and experiences from the implementation of Farmers' Rights and to provide additional data for the study on lessons learned, as requested in Resolution 5/2015, in June–August 2016 the Secretariat conducted an online consultation on implementation of Farmers' Rights.[17] Again, participatory plant breeding was regarded quite favourably: in fact, community seed banks were the only other measure seen as having positive effects by a higher number of respondents. As for the prevalence of participatory plant-breeding projects and programmes, however, there were more respondents who replied that their countries did not have any such measures, than those confirming that their countries did (FAO, 2017d).

Another global consultation[18] on Farmers' Rights, with an impressive 95 participants from 37 countries (Plant Treaty, 2016), was organized in Bali, Indonesia, in September 2016. Here, 5 of the 27 presentations outlining measures and progress were wholly or partly about participatory plant breeding. One of these in particular made clear the potential outreach of participatory approaches to plant breeding: it highlighted the plans of the Cambodian government to include participatory plant breeding in its strategic programme on rice management and development (Sabran and Torheim, 2017). Such expansion strategies showcase the strength of the participatory approaches utilized, and also indicate one possible way forward for participatory plant breeding and participatory varietal selection to reach more farmers.

Draft recommendations to the Governing Body were also discussed at the Bali consultation, resulting in the co-chairs drawing up a summary of the recommendations. Included in this summary was a call for Contracting Parties to 'promote participatory approaches such as community seed banks, community biodiversity registries, participatory plant breeding and seed fairs as tools for realizing Farmers' Rights' (Andersen and Correa, 2016: para 10).

Then, at the seventh session of the Governing Body (2017), a resolution adopted on the implementation of Farmers' Rights represented an important step forward for the negotiation process. The Contracting Parties agreed on the establishment of an Ad hoc Technical Expert Group on Farmers' Rights, and, for the first time, specifically referred to participatory plant breeding as an implementation tool for the realization of Farmers' Rights. Contracting Parties are invited to 'promote sustainable biodiverse production systems and facilitate participatory approaches such as community seed banks, community biodiversity registries, participatory plant breeding and seed fairs as tools for realizing Farmer's Rights' (Resolution 7/2017: para 3).[19] As this also was the first resolution on Farmers' Rights to list specific implementation tools, this means that participatory plant breeding was included the first time any such tools were mentioned.

This followed on another noteworthy acknowledgement concerning the role of participatory approaches to plant breeding: in the Educational Module on Farmers' Rights.[20] This module was finalized after a round of expert peer reviews and in consultation with the Bureau of the Plant Treaty (FAO, 2017b). It is organized into lessons; the third and final lesson presents a range of activities implemented by stakeholders – and participatory approaches to plant breeding are mentioned as among the key approaches. Participatory plant breeding and/or participatory varietal selection are recognized in connection with three of the elements of Farmers' Rights – protection of traditional knowledge, benefit-sharing, and participation in decision-making at the national level. Moreover, 3 of the 21 cases in the section of lesson 3 providing practical examples concern or prominently feature participatory approaches to plant breeding. One is even introduced in connection with the rights of farmers to use, exchange, and sell farm-saved seed. Additionally, farmer–scientist collaboration and participatory approaches are listed among the 'important features of projects that have contributed to the realization of Farmers' Rights' (FAO, 2017i: 91). If utilized by substantial numbers and a broad range of stakeholders and decision-makers, this module might lead to more widespread use of participatory approaches to plant breeding in implementing Farmers' Rights.[21]

However, as there are a great many projects that use and lay claim to the term 'participatory plant breeding,' and there are many ways of approaching and organizing such plant breeding, we must ask: if the goal is to contribute to the realization of Farmers' Rights, what does that mean for the organization of participatory plant breeding? Aspects worth considering include: which stages farmers take part in; how much actual influence they have; how representative they are (socio-economic spectrum, gender, religion, etc.), how they are 'selected' and the consequences of this; the types and importance of benefits achieved for participating farmers and wider communities (empowerment, new knowledge, higher or more stable incomes, improved livelihoods, etc.); whether participation results in greater awareness of and knowledge about legislation and policy related to the utilization of and distribution of seed and propagating material; the effects on farmers' seed systems – and possible reform efforts if effects are found to be negative; and whether traditional knowledge is applied and/or documented. If participation in plant breeding is viewed along a continuum, and not as being something that is only either/or, it can also be argued that plant breeding in general is important to Farmers' Rights, but that it is crucial to consider whose interests are being promoted in various breeding programmes: for instance, those of farmers, other end-users – or perhaps commercial breeding companies? As the discussions surrounding participatory plant breeding as a tool for implementing Farmers' Rights become more complex and detailed, such aspects might be seen moving to the forefront.

Concluding remarks

Participatory approaches to plant breeding are indeed gaining recognition as important and useful tools for implementing Plant Treaty Article 6 on sustainable use and Article 9 on Farmers' Rights. The seventh session of the Governing Body in 2017 resulted in progress for the implementation process concerning both articles, and explicit recognition of participatory plant breeding. For Article 6 it is perhaps the inclusion of participatory plant breeding in the first version of the toolbox that best illustrates this recognition; with regard to Article 9, participatory plant breeding was included as one of four specific implementation tools – the very first time any such were mentioned in a resolution on Farmers' Rights.

Also other important developments demonstrate the rising status of participatory approaches. For Article 6 on sustainable use, the need to introduce incentives for participatory plant breeding was among the elements highlighted when the results of a stakeholder consultation on the implementation of Article 6 were presented in 2013. These results also revealed that such plant breeding efforts were seen as a key strategy for enhancing local breeding capacity. Further, the results of the 2015 stakeholder consultation concerning views on the development of a toolbox, where participatory plant breeding was included as 1 of 12 pre-defined areas of sustainable use, were important. Participatory plant breeding received most attention from the public sector, accompanied by recognition of the need for funding and supportive policies. A further milestone was reached in 2016 in connection with the work of the ACSU, when participatory plant breeding was proposed as one of three priority areas for the planned toolbox. Most recently, the inclusion of participatory plant breeding among the four topics suggested for the regional capacity-building workshops requested in Resolution 6/2017 represented an important breakthrough.

The implementation of Article 9 on Farmers' Rights can be said to have attracted more controversy, and more academic and activist attention, than Article 6. Key developments with regard to participatory approaches to plant breeding include the consensus that became clear in the online conference 'Options for Farmers' Rights' regarding farmers as be breeders, and the 2005 international stakeholder survey where participatory plant breeding was among the most frequently mentioned non-monetary benefits. In addition, the 2010 Global Consultations revealed that participatory approaches to plant breeding were among the five most widely employed benefit-sharing measures, and also one of the two most favourably viewed. This high standing was also evident from the online consultation conducted by the Secretariat in 2016, where participatory plant breeding was one of the most positively viewed implementation tools, in that context second only to community seed banks.

Although this chapter has shown that participatory approaches to plant breeding are often regarded as benefit-sharing measures, and are discussed largely in connection with them, they are also relevant for and are recognized in connection with the other elements of Farmers' Rights. This was illustrated prior to the sixth session of the Governing Body, when the Secretariat's review of the submissions received to date linked participatory plant breeding to three[22] of the four elements of Farmers' Rights; and again when the Educational Module on Farmers' Rights referred to participatory approaches to plant breeding in connection with both benefit-sharing and protection of traditional knowledge, as well as participation in decision-making.

That participatory approaches to plant breeding are seen as central to the implementation of both Article 6 and Article 9 of the Plant Treaty adds to their usefulness: this double importance makes them an implementation tool that connects the two articles and their implementation processes.

There are good reasons to expect that participatory approaches to plant breeding will be recognized as increasingly important for promoting Article 6 on sustainable use and Article 9 on Farmers' Rights over the next 20 years. Perhaps also the process and dialogue concerning them within the Plant Treaty system will become more nuanced. There are many ways of approaching and organizing farmer participation in crop development – and these different approaches might also have different consequences for the sustainable use of crop genetic resources and the realization of Farmers' Rights.

Notes

1 The Plant Treaty entered into force on 29 June 2004; as of May 2019 it had 146 Contracting Parties.
2 This review is based on a study conducted under the project 'From base broadening to enhancing crop adaptation to climate change: a preparatory study for the farmer evaluation activity in the project "Adapting Agriculture to Climate Change: Collecting, Protecting and Preparing Crop Wild Relatives"', funded by the Crop Trust.
3 There was no mention of such approaches in the working document on sustainable use, or in the report.
4 A new publication on participatory breeding and farmers' participation in varietal development processes was highlighted in the information document IT/GB-2/07/Inf.8.1 (FAO, 2007) on the relevant work of FAO for sustainable use (in para 17).
5 Reference is made to the participation of farmers in plant breeding, including participatory plant breeding, in the working document on the implementation of Article 6, as well as in the submissions regarding sustainable use (annexed in the information document IT/GB-3/09/Inf.5.Add.1 (FAO, 2009a) from the Centers of the Consultative Group on International Agricultural Research (CGIAR) and SEARICE.
6 The establishment of this committee also made explicit, and strengthened, the link between Article 6 and Article 9, as Resolution 07/2011 used the terminology 'taking into account Resolution 6/2011 on Farmers' Rights' (ibid.) in connection with the establishment of the committee; and that resolution on Farmers' Rights requested the Secretary to compile submissions received in response to the resolution and reports from regional workshops 'for consideration by' the ACSU (Resolution 2/2011, para 6).
7 This consultation used an online questionnaire developed by the Ad hoc Technical Committee on Sustainable Use of Plant Genetic Resources for Food and Agriculture (ACSU) established by Resolution 07/2011; it sought to collect views on possible elements of a programme of work on sustainable use of plant genetic resources for food and agriculture.
8 See: www.fao.org/plant-treaty/tools/toolbox-for-sustainable-use/overview/en/.
9 This resolution also contained some of the same elements as the previous resolution on sustainable use, including the reference to 'the assessment of needs of local farmers and other relevant local stakeholders and identification of possible means to address those needs, including through participatory approaches in the context of the Programme of Work' (Resolution 6/2017: para 2).
10 However, under 'other business', Norway proposed that their implementation be included as an agenda item at the Governing Body's second session (FAO, 2006).
11 The other concerned community seed banks.

12 The other submissions came from the Centre for Sustainable Development (CENESTA), an Iranian NGO, which referred to 'support from participatory plant breeding' as one of the seven aspects of Farmers' Rights in Iran (FAO, 2011b: 11), and from the international Via Campesina movement.

13 The 'Declaration of Szeged' from the European forum 'Let's Liberate Diversity' included a call to promote participatory plant breeding. The submission from the Nepalese NGO LI-BIRD briefly mentioned participatory plant breeding as one of the practices it supports and promotes (FAO, 2013d).

14 This submission can be seen as interlinked with the submission from ASOCUCH, based in Guatemala, which works the Development Fund in Central America, including on participatory plant breeding.

15 Of the 114 Parties to the Treaty (see www.fao.org/fileadmin/user_upload/legal/docs/033s–e.pdf), only 17 Contracting Parties (and 17 relevant organizations) had submitted their views and experiences regarding Farmers' Rights (FAO, 2015a).

16 In its reiteration of the invitation to Contracting Parties to consider developing national action plans, this resolution also uses the phrase 'in line with the implementation of Articles 5 and 6' (Resolution 5/2015: para 2), but without referring to the sub-articles or their measures.

17 An updated version of the questionnaire from the e-mail based survey of the 2010 Global Consultations was used, and a total of 166 respondents from 53 countries took part. By contrast, the original 2010 survey had involved 61 responses from 36 countries.

18 Prior to, and in preparation for, this consultation, a small informal consultation on Farmers' Rights was held in May 2016, by the Secretariat of the Plant Treaty and the Quaker United Nations Office, which gathered ten experts and country representatives (QUNO, 2016). However, no reference was made to participatory approaches to plant breeding. In addition, a stakeholder consultation on Farmers' Rights in Africa was organized in June 2016, with 59 participants. Here, participatory plant breeding was taken up in connection with the rights to participate in decision-making at the national level, and was recognized as a relevant measure for non-monetary benefit-sharing. The need for awareness-raising on the importance of such breeding efforts was also noted (CTDT, 2016).

19 The main difference between this phrasing and that of the co-chairs' summary from the Bali consultations is that the resolution uses the term 'facilitate' instead of 'promote', perhaps because the latter was seen putting too much pressure on the Contracting Parties, so it was easier to reach consensus on the former.

20 The finalization of this module was highlighted in the working document on the implementation of Farmers' Rights (FAO, 2017b).

21 Participatory approaches to plant breeding were also briefly referred to in the proceedings of the symposium held on 26 October 26 in Geneva, on possible interrelations between the Plant Treaty and the International Convention for the Protection of New Varieties of Plants (UPOV), attended by more than 140 participants (FAO, 2017e). In addition, a document with supplementary information on plant-breeding impacts, non-monetary benefit-sharing and contributions to Farmers' Rights for the CGIAR Report (FAO, 2017f) and a report from the Sowing Diversity = Harvesting Security programme (FAO, 2017g) both mention participatory plant breeding as a tool.

22 Benefit-sharing, protection of traditional knowledge and rights related to farm-saved seed.

References

Articles, chapters, reports and books

Andersen R, Winge T, 2011. *The 2010 Global Consultations on Farmers' Rights: Results from an Email-based Survey.* FNI Report 2/2011. Lysaker, Norway: The Fridtjof Nansen Institute. www.fni.no/getfile.php/131678-1469868948/Filer/Publikasjoner/FNI-R0211.pdf.

Halewood M, Deupmann P, Sthapit B, Vernooy R, Ceccarelli S, 2007. *Participatory Plant Breeding to Promote Farmers' Rights.* Rome: Bioversity International. www.bioversityinternational.org/fileadmin/_migrated/uploads/tx_news/Participatory_plant_breeding_to_promote_Farmers__Rights_1254.pdf.

Morris ML, Bellon MR, 2004. Participatory plant breeding research: Opportunities and challenges for the international crop improvement system. *Euphytica* 136, 21–35.

Norwegian Ministry of Agriculture and Food 2007. *Informal International Consultation on Farmers' Rights*, 18–20 September 2007, Lusaka, Zambia. Report from meeting co-hosted by the Norwegian Ministry of Agriculture and Food; the Fridtjof Nansen Institute, Norway; and the Zambia Agriculture Research Institute of the Ministry of Agriculture, Food and Fisheries. Oslo: Norwegian Ministry of Agriculture and Food. www.regjeringen.no/globalassets/upload/lmd/vedlegg/brosjyrer_veiledere_rapporter/lusakarapporten.pdf.

QUNO, 2016. *Summary Report of the Stakeholder Consultation on Implementation of Farmers' Rights.* 28 May 2016, Geneva, Switzerland. Quaker United Nations Office and the International Treaty on Plant Genetic Resources for Food and Agriculture. www.quno.org/sites/default/files/resources/Farmers%27%20Rights%20Consultation%20-%20Summary%20Report%20submitted.pdf.

Sabran M, Batta Torheim S-I, 2017. *The Global Consultation on Farmers' Rights 2016: Summary of Presentations and Discussions.* 27–30 September 2016, Bali, Indonesia. Indonesian Agency for Agricultural Research and Development (IAARD) and Norwegian Ministry of Agriculture and Food. www.fao.org/plant-treaty/meetings/meetings-detail/en/c/414974/.

Weltzien E, Smith ME, Meitzner LS, Sperling L, 2003. *Technical and Institutional Issues in Participatory Plant Breeding – from the Perspective of Formal Plant Breeding: A Global Analysis of Issues, Results, and Current Experience.* PPB Monograph No. 1. Cali, Colombia: International Center for Tropical Agriculture (CIAT). http://ciat-library.ciat.cgiar.org/Articulos_Ciat/Technical_Institutional_Issues_Participatory_Plant_Breeding_Perspective_of_Formal.pdf.

Documents of the Governing Body of the Plant Treaty, and the Food and Agriculture Organization of the United Nations (FAO)

FAO, 2006. *Report of the Governing Body of the International Treaty on Plant Genetic Resources for Food and Agriculture.* IT/GB-1/06/Report. Rome: FAO. www.fao.org/3/a-be210e.pdf.

FAO, 2007. *Information received from Relevant Organizations concerning the Implementation of Article 6.* IT/GB-2/07/Inf.8.1. Rome: FAO. www.fao.org/3/a-be154e.pdf.

FAO, 2009a. *Compilation of Submissions sent by Contracting Parties, Other Governments, and Relevant Institutions and Organizations on the Implementation of Article 6.* IT/GB-3/09/Inf.5.Add.1. Rome: FAO. www.fao.org/3/a-be093e.pdf.

FAO, 2009b. *Collection of Views and Experiences Submitted by Contracting Parties and Other Relevant Organizations on the Implementation of Article 9.* IT/GB-3/09/ Inf.6.Add.2. Rome: FAO. www.fao.org/3/a-be096e.pdf.

FAO, 2009c. *Collection of Views and Experiences Submitted by Contracting Parties and Other Relevant Organizations on the Implementation of Article 9.* IT/GB-3/09/ Inf.6.Add.3. Rome: FAO. www.fao.org/3/a-be097e.pdf.

FAO, 2009d. *Report of the Governing body of the International Treaty on Plant Genetic Resources for Food and Agriculture.* IT/GB-3/09/Report. Rome: FAO. www.fao. org/3/a-be112e.pdf.

FAO, 2011a. *Implementation of Article 6.* IT/GB-4/11/17. Rome: FAO. www.fao. org/3/a-be136e.pdf.

FAO, 2011b *Compilation of views and experiences on the Implementation of Farmer's Rights submitted by Contracting Parties and relevant organizations.* IT/GB-4/11/Inf.6. Rome: FAO. www.fao.org/3/a-be075e.pdf.

FAO, 2013a, *Implementation of the Article 6 Sustainable Use of Plant genetic Resources for Food and Agriculture.* IT/GB-5/13/9. Rome: FAO. www.fao.org/3/a-be567e.pdf.

FAO, 2013b. *Compilation and Analysis of Submissions sent by Contracting Parties, Other governments and Relevant Institutions and Organizations on the Implementation of Article 6.* IT/GB-5/13/Inf.6. Rome: FAO. www.fao.org/3/a-be544e.pdf.

FAO, 2013c. *Synthesis of the Outcomes of the Stakeholder Consultation on the Implementation of Article 6.* IT/GB-5/13/Inf.7. Rome: FAO. www.fao.org/3/a-be001e.pdf.

FAO, 2013d. *Compilation of Submissions by Contracting Parties and other Relevant Organizations, and the Reports of Regional Workshops on the Implementation of Article 9.* IT/ GB-5/13/Inf.8. Rome: FAO. www.fao.org/3/a-be547e.pdf.

FAO, 2015a. *Report and Review of Submissions on the Implementation of Article 9, Farmers' Rights.* IT/GB-6/15/13. Rome: FAO. www.fao.org/3/a-mo440e.pdf.

FAO, 2015b. *Development of a Toolbox for Sustainable Use of Plant Genetic Resources for Food and Agriculture.* IT/GB-6/15/Inf.3. Rome: FAO. www.fao.org/3/a-bb352e. pdf.

FAO, 2017a, *Implementation of the Programme of Work on Sustainable Use of Plant Genetic Resources for Food and Agriculture.* IT/GB-7/17/16. Rome: FAO. www.fao.org/3/a-mu386e.pdf.

FAO, 2017b. *Report on Implementation of Article 9, Farmers' Rights.* IT/GB-7/17/17. Rome: FAO. www.fao.org/3/a-mu391e.pdf.

FAO, 2017c. *Report of the Compliance Committee.* IT/GB-7/17/18. Rome: FAO. www.fao.org/3/a-mt576e.pdf.

FAO, 2017d. *Results of the online consultation to gather views and needs on the implementation of Farmers' Rights.* IT/GB-7/17/Inf.11. Rome: FAO. www.fao.org/3/a-bs783e. pdf.

FAO, 2017e. *Proceedings of the Symposium on Possible Interrelations between the International Treaty on Plant Genetic Resources for Food and Agriculture and the International Convention for the Protection of New Varieties of Plants.* IT/GB-7/17/Inf.14. Rome: FAO. www.fao.org/3/a-bs781e.pdf.

FAO, 2017f. *Supplementary Information for CGIAR Report: Plant Breeding Impacts, Non-monetary Benefit-sharing and contributions to Farmers' Rights.* IT/GB-7/17/Inf.20. Rome: FAO. www.fao.org/3/a-bs785e.pdf.

FAO, 2017g. *Submission from the Governments of Bhutan and Zimbabwe containing the Sowing Diversity – Harvesting Security Programme Report.* IT/GB-7/17/Circ.2. Rome: FAO. www.fao.org/3/a-bs791e.pdf.

FAO, 2017h. *Report of the Third Meeting of the* Ad hoc *Technical Committee on Sustainable Use.* IT/GB-7/17/Inf. 9. (IT/ACSU-3/16/Report). Rome: FAO. www.fao. org/3/a-bt106e.pdf.

FAO, 2017i. Farmers' Rights. Rome: FAO. www.fao.org/3/I7820EN/i7820en.pdf.

Governing Body Resolutions, the Plant Treaty, FAO

Resolution 2/2007 on Farmers' Rights. www.fao.org/3/a-be008e.pdf.

Resolution 6/2009. Implementation of Article 9, Farmers' Rights. www.fao.org/3/a-be080e.pdf.

Resolution 6/2011. Implementation of Article 9, Farmers' Rights. www.fao.org/3/a-be456e.pdf.

Resolution 7/2011. Implementation of Article 6, Sustainable use of plant genetic resources. www.fao.org/3/a-be457e.pdf.

Resolution 8/2013. Implementation of Article 9, Farmers' Rights. www.fao.org/3/a-be600e.pdf.

Resolution 4/2015. Implementation of Article 6, Sustainable use of Plant Genetic Resources for Food and Agriculture. www.fao.org/3/a-bl143e.pdf.

Resolution 5/2015. Implementation of Article 9, Farmers' Rights. www.fao.org/3/a-bl144e.pdf.

Resolution 6/2017. Implementation of Article 6, Sustainable Use of Plant Genetic Resources for Food and Agriculture. www.fao.org/3/a-mv086e.pdf.

Resolution 7/2017. Implementation of Article 9, Farmers' Rights. www.fao.org/3/a-mv102e.pdf.

Other documents

Andersen R, Correa C, 2016 *Co-chairs' Summary of Recommendations to the Governing Body of the International Treaty on Plant Genetic Resources for Food and Agriculture.* Global Consultations on Farmers' Rights, Bali, Indonesia, 27–30 September.

CTDT, 2016. *Stakeholders' Consultation on Farmers' Rights.* African Position Paper, 27–29 June, HICC Hotel, Harare, Zimbabwe. Community Technology Development Trust (CTDT). www.fao.org/3/a-bq550e.pdf.

Plant Treaty, 2016. *Bali Global Consultation: shares views and experiences to strengthen the implementation of Farmers' Rights.* www.fao.org/plant-treaty/news/news-detail/en/c/445335/.

Plant Treaty, 2017. *Toolbox for Sustainable Use of PGRFA: Sustaining local crop diversity.* www.fao.org/plant-treaty/tools/toolbox-for-sustainable-use/sustaining-local-crop-diversity/en/. Accessed 20 December 2017.

World Commission on Environment and Development, 1987. *Our Common Future: Report of the World Commission on Environment and Development.* A/42/427. New York: United Nations. www.sswm.info/sites/default/files/reference_attachments/UN%20WCED%201987%20Brundtland%20Report.pdf.

Treaties

Convention on Biological Diversity (CBD), 22 May 1992. www.cbd.int/doc/legal/cbd-en.pdf.

International Treaty on Plant Genetic Resources for Food and Agriculture (Plant Treaty), 3 November 2001. www.fao.org/3/a-i0510e.pdf.

17 Funding participatory plant breeding

Outlook and future challenges

Álvaro Toledo[1]

Introduction

In the course of the two first decades of the twenty-first century, participatory plant breeding (PPB) has managed to establish and refine its methodologies, and has been disseminated around the world. We now need to reflect on how PPB may contribute to tackling the many challenges that will face agricultural development in the years to come. The new development agenda presents new opportunities for support to PPB – but there are also major challenges on the horizon.

Agricultural biodiversity will be instrumental in coping with the challenges to be dealt with in the coming decades. Sustainable management of agricultural diversity can contribute to the diversification of agricultural systems, and to making national food and agriculture sectors more sustainable. As a vehicle for enabling farmers' access to and use of plant diversity to improve their livelihoods, PPB has a central role to play in connection with family farming. It will be an important tool for development, in centres of crop origin or diversity, and in farming systems that are heavily dependent on natural resources, vulnerable to climate change, and characterized by rural poverty.

How can we fund participatory plant breeding in the next decade? It can be argued that this is not just a 'million dollar question', but a multi-million dollar one. This chapter starts by discussing the various plant genepools that are available for advancing the development agenda, and the potentials regarding PPB. It presents the new development agenda, noting the intertwined challenges facing agriculture, and then turns to how PPB relates to the Funding Strategy of the International Treaty on Plant Genetic Resources for Food and Agriculture (the Plant Treaty), which is currently being updated. PPB has featured strongly in the Plant Treaty's Benefit-sharing Fund, with an estimated USD 10.5 million invested thus far in programmes for participatory characterization, evaluation, variety selection or breeding. The chapter concludes with some remarks on challenges and opportunities that the PPB community may wish to discuss in further advancing its agenda and strengthening its role in implementation of the Plant Treaty.

Sustaining the plants that feed the world

The total diversity of food plants available for making progress in the realization of the UN Sustainable Development Goals can be divided into four genepools: *in situ* crop wild relatives, on-farm landraces, *ex situ* materials, and breeding materials. Genetic sequence information on crop diversity is becoming increasingly relevant: indeed, it has been postulated that access to genetic information may come to replace the need for access to genetic material in the future. In any case, this total diversity has been fragmented into genepools by the practices and mind-sets of various research and development groups. Within the PGRFA (Plant Genetic Resources for Food and Agriculture) community these genepools are seen as a continuum involving ongoing contributions from one to the other; however, a divide has emerged in recent decades, due to various policies and regulations arising from the agricultural, environmental, research and trade sectors.

In this chapter I consider PPB as containing a wide range of approaches that include participatory variety selection, participatory improvement or restoration of local varieties. While many PPB practitioners focus on the on farm genepool, others work mainly with the commercial breeding material. Many initiatives start by using the diversity available within local seed systems and then go on to integrate, at the local level, new genetic diversity arriving in the form of material accessed from genebanks or pre-breeding material or varieties from breeders in public institutions, even commercial varieties available in national seed markets. In recent years, PPB has been integrated into larger initiatives for on-farm management of PGRFA that include the establishment and maintenance of community seedbanks; the multiplication, dissemination and even commercialization of seeds at local level; and the strengthening of local markets for biodiversity-rich products.

The real potential of PPB lies in its use as a bridging device for connecting various genepools for the benefit of small-scale farmers, and for empowering farming communities.

Agricultural biodiversity, participatory approaches and the new development agenda

The Agenda for Sustainable Development (2030 Agenda), including the 17 Sustainable Development Goals (SDGs), was adopted by the United Nations in late 2017. It presents a set of new global objectives intended to shape national development plans and guide the actions of the international community until 2030. Unlike the earlier Millennium Development Goals,[2] the 2030 Agenda and SDGs are relevant, and apply, to all countries.

From ending poverty and hunger to responding to climate change and sustaining natural resources, food and agriculture lies at the heart of the 2030 Agenda. Emphasis here is placed on the role played by agricultural biodiversity, as it is critical to the sustainable production of nutritious and abundant food and to adapting agriculture to global challenges, such as climate change and growing populations. Two SDG targets directly concern the three objectives of the Plant Treaty:

- Target 2.5: By 2020, maintain genetic diversity of seeds, cultivated plants, farmed and domesticated animals and their related wild species, including through soundly managed and diversified seed and plant banks at national, regional and international levels, and promote access to and fair and equitable sharing of benefits arising from the utilization of genetic resources and associated traditional knowledge, as internationally agreed;
- In launching the 2030 Agenda, as a guiding principle, countries pledged that no one would be left behind in the collective efforts to achieve the Agenda. There is a clear emphasis on prioritizing the rural poor in sustainable development: they are amongst the most vulnerable, but are also critical agents of change. Implementation of Agenda 2030 focuses on the poorest, most vulnerable and those who have lagged furthest behind. It is people-centred, gender-sensitive, and human rights-oriented.

PPB has specific contributions to make to the Agenda 2030. Managing and sustainably deploying plant genetic diversity is one of the key options available to vulnerable farmers seeking to increase the resilience of the production systems and secure their livelihoods. Increased attention and support are needed if farmers are to reap the full benefits of genetic diversity. The key role of women in managing biodiversity needs to be further recognized and strengthened.

All this calls for a strengthened role for PPB in the agenda for sustainable rural development. In 2015, three Rome-based UN agencies published a joint report that presented estimates on the investments required to eradicate poverty and hunger sustainably by 2030 (FAO, IFAD and WFP, 2015). The report specifically considered how to eliminate poverty and hunger through a combination of investments in social protection and pro-poor sustainable production. It examined the specific role of PGRFA conservation and improvement in enabling and incentivizing sustainable smallholder agriculture, with estimates of the additional average annual investments in PGRFA needed to maximize contributions to achieving global food security and the eradication of rural poverty.

The 2015 report concluded that USD 977 million per year of additional rural investments in developing areas would be required in activities related to 'Preservation/improvement of crop genetic resources' for sustainably ending hunger by 2030. It focused on activities strictly supporting crop breeding for yield increases to set this target. Although this financial target has not been endorsed by the international community (donor countries in particular), it indicates the order of magnitude of the financing needed if plant genetic resource management is to contribute to achievement of the SDGs.

Further, the report stressed that governments and the international community will need to build on approaches that have already proven effective, combining three important elements: (1) promote immediate access to food and nutrition-related services that directly address food insecurity; (2) create opportunities for the poor and hungry to improve their livelihoods; (3) increase the sustainability of food systems by conserving natural resources and adopting sustainable agricultural practices.

Participatory plant breeding has been shown to be an effective approach that meets these three goals in an integrated manner. Therefore, PPB programmes, in conjunction with other strategies to increase seed security and access to locally adapted plant resources should be at the core of additional investments to support realization of the SDGs.

PPB within the funding strategy: who funds PPB at the international level?

Article 18 of the Plant Treaty provides for a Funding Strategy to ensure Treaty implementation through multiple funding sources that include contributions from Contracting Parties, in particular at the national level, and multilateral and bilateral funds and programmes, as well as the contributions from stakeholders and user-based incomes arising from the Treaty's Multilateral System for Access and Benefit-sharing ('monetary benefit-sharing'). The Funding Strategy calls for the establishment of a funding target to mobilize funding for priority activities, plans and programmes under the Treaty.

In spring 2019, the Governing Body of the Plant Treaty set about updating the Funding Strategy, including establishing funding targets for the next five years for the overall Funding Strategy and the target for the Benefit-sharing Fund. While none of these targets will focus solely on participatory plant breeding, they will lay the basis for the PGRFA community, including those working on PPB, to work together to mobilize funding towards Treaty implementation.

The information provided in this section is taken from an analysis of selected funding mechanisms active in the past five years in the wider area of on-farm management (Moeller, 2018). The analysis indicates that PPB receives support because it links in with the otherwise highly differing priorities among donors – for instance, through its contribution to biodiversity conservation or to rural poverty alleviation, or the development of new local markets or demonstrating the value of PGRFA for climate change adaptation. Three funding mechanisms that have been active are the Global Environment Facility (GEF), the International Fund for Agricultural Development (IFAD) and the Benefit-sharing Fund of the International Treaty. Various CGIAR Centres, Bioversity International and ICARDA in particular, have supported the implementation of projects from these and other funding bodies; and FAO has become increasingly involved in the implementation of GEF projects on agro-biodiversity. Other key global programmes include that of Oxfam Novib, which works with donor partners such as Sweden. A brief summary of GEF's and IFAD's support for agrobiodiversity conservation and sustainable use is presented in the following paragraphs.

The Global Environment Facility has promoted agro-biodiversity projects since the 1990s. It works through four-year replenishment cycles. GEF6 finished in June 2018 and had a total programme budget of USD 4,309 million, of which 29.2 per cent or USD 1,296 million in the Biodiversity Focal Area. The GEF6 strategic programme, 'Securing Agriculture's Future: Sustainable use of plant and animal genetic resources', supported the in-situ conservation and sustainable use, through farmer management, of plant genetic resources in Vavilov Centres of Diversity. The programme allocated USD 37.7 million in grants during GEF-6, which also attracted an additional USD 325.9 million in co-financing. Almost all these allocations have been for on-farm management of plant genetic resources, including PPB. Also GEF7 (1 July 2018–30 June 2022) features a strategic programme for support to agrobiodiversity.

The IFAD has actively promoted the implementation of PPB through several recent grants. A very rough estimate of overall funding provided to various Treaty areas (Moeller, 2018) shows that, in 2017, between USD 19.68 and 32.8 million went to on-farm management and sustainable use (combined); IFAD established partnerships through these grants with many of the leading PPB actors, including Oxfam-Novib and Bioversity International. These are not robust figures, but give an indication of the orders of magnitude involved. The IFAD Strategic Framework 2016–2025 (IFAD, 2016) offers ample opportunities for further support to PPB, especially through its thematic focus on climate change, adaptability, nutrition security, environmental sustainability and access to natural resources and agricultural technologies and production services.

Many bilateral donors are involved in agriculture and agro-biodiversity. While assessing the bilateral funding flow that supports agro-biodiversity is a complex undertaking with many pitfalls, it has become easier to find accurate information on what is done with aid, development assistance and other international finance flows at the level of individual countries and activities. The OECD's Creditor Reporting System (CRS) is an authoritative source of activity-level data. Filtering CRS data by seven relevant sectors (agricultural development, food crop production, industrial crops/export crops, alternative agricultural development, agricultural extension, agricultural education/training, agricultural research), provides a total of USD 7,253.551 million in aid flows for over 7,000 projects between 2013 and 2016 (between USD 1.5 and 2 billion a year). Although it is currently not possible to filter these data to identify the bilateral funding specifically targeting plant genetic resources for food and agriculture, let alone for PPB, this can be expected to change in the next years. Improving the monitoring of funding flows towards Treaty implementation is likely to be a component of the reviewed Funding Strategy of the Governing Body. Such monitoring should allow PPB practitioners and others to find key donor partners at regional and national levels.

Participatory plant breeding and the benefit-sharing fund

The Funding Strategy of the Plant Treaty includes the Benefit-sharing Fund. This is a mechanism established by the Governing Body to receive and utilize the financial resources that accrue to it for purposes of implementing the Treaty, including the monetary benefits arising from the Multilateral System of Access and Benefit-sharing and voluntary contributions from Contracting Parties, the private sector, non-governmental organizations and other sources. The Fund is under the direct control of the Treaty's Governing Body.

Since 2009, the Benefit-sharing Fund has supported projects to enable adaptation to climate change, food security and on-farm conservation of crop diversity. Current projects contribute to the implementation of the Plant Treaty and the 2030 Agenda for Sustainable Development – Sustainable Development Goals (SDGs) 2.5 and 15.6 in particular.

To date, the Benefit-sharing Fund has invested more than USD 20 million in 61 projects in 55 developing countries over three project cycles (FAO, 2017).[3] As of September 2018, a total of USD 27.4 million in contributions had been received and pledged – from traditional donors (Norway, Italy, the European Commission, Australia), developing countries such as Indonesia, seed-sector institutions such as the European Seed Association, the Groupement National Interprofessional des Semences, the International Seed Federation or Syngenta, and others such as IFAD (Moeller, 2018). The first payment to the Benefit-sharing Fund, corresponding to 0.77 per cent of seed sales of ten new varieties of vegetables derived from germplasm accessed under the Plant Treaty's Standard Material Transfer Agreement (SMTA), was made in June 2018.[4]

Box 17.1 The benefit-sharing fund in brief

The Governing Body has used available funds to play a catalytic role in international cooperation in the following areas:

1 Information exchange, technology transfer and capacity-building

- Projects support the training of a new generation of scientists and technical experts on plant genetic resources in the developing world.
- Activities increase capacity and expertise in the areas of germplasm evaluation and characterization, phenotyping and genotyping, genetic base broadening, plant breeding, data management and use of new technologies.
- Over 29,000 researchers and local partners have been trained through the Treaty's Benefit-sharing Fund projects.

2 Managing and conserving plant genetic resources on-farm

- Farmers have taken the lead in the organization of surveys, seed fairs, community biodiversity registers, training and capacity building, participatory variety selection, plant breeding and establishment of community seed banks;
- 1,000,000 people, mostly small-holder farmers, have directly or indirectly benefited from the activities conducted.

3 The sustainable use of plant genetic resources

- Diversification of crop production, genetic enhancement and broadening the crop genetic base has featured in many projects, for more sustainable agricultural production.
- More than 8,000 accessions have been characterized and evaluated, to increase the relevance of germplasm held *ex situ* and on farm for breeding purposes.
- Benefit-sharing Fund projects have helped to identify and disseminate drought-tolerant rice (India), flood-resistant rice (Indonesia) and drought-tolerant sorghum (Tanzania).

The Benefit-sharing Fund promotes the establishment of multi-level partnerships involving a wide range of institutions. The over-350 partnering institutions involved in the second and third project cycles include universities, institutes for biodiversity conservation, international organizations, governmental and non-governmental organizations, donor agencies, genebanks and national and international research institutes. These partnership arrangements enable each organization to add value to the others within the project, ensure a science-based approach that is relevant for local realities, and foster a sense of local ownership and accountability.

Participatory plant breeding has featured prominently in the Benefit-sharing Fund grant portfolio so far, with 27 PPB projects in three first project cycles – 44 per cent of the projects funded (as estimated by the Treaty Secretariat). This prominence is due the fact that PPB supports realization of the three current Benefit-sharing Fund priorities: (1) on-farm management; (2) sustainable use and (3) information exchange, capacity-building and technology transfer. The focus of the second to fourth project cycles was on supporting vulnerable communities to increase their food security and capacity to adapt to climate-changes. The strong call by many of the Plant Treaty's Contracting Parties for the co-development of varieties and technologies is also supportive of the collaborative activities established by farmers and breeders. The Secretariat estimates than USD 10.5 million has been invested in PPB in the first three project cycles, with more than 100 institutions involved in various phases of PPB. Targeted crops have included beans, wheat, maize, potato, sorghum, finger millet, pearl millet, cowpeas, bambara groundnuts, rice, cassava and aroids (*Araceae*).

Benefit-sharing Fund partners characterize the overall strategy used in conducting PPB as follows:

- participatory and inclusive: farmers interact with breeders jointly to evaluate germplasm for individuation of preferable/adaptive traits of local importance.
- multi-sectoral: involves a wide range of institutions with various types of expertise and complementary know-how.
- empowering: farmers' capacities are enhanced to breed locally adapted varieties.
- community-based: projects work directly with targeted rural communities, to ensure that interventions are based on local needs, collective strengths and shared resources (FAO, 2017; FAO, 2017b).

The main benefits arising from PPB initiatives reported by Benefit-sharing Fund partners include greater access and availability of genetic resources, new locally adapted varieties generated, increased on-farm diversity, and improved seed security (FAO, 2017; FAO, 2017b). Box 17.2 shows the key findings of the independent evaluation report of the second Benefit-sharing Fund project cycle (FAO, 2017). The evaluation team did not examine PPB programmes funded in isolation, so the box provides a summary of findings relevant to these programmes.

Box 17.2

Strengths and weaknesses of programs undertaking Participatory Plant Breeding through the Benefit-sharing Fund – summary arising from the Independent Evaluation of the Benefit-sharing Fund, second project cycle.

Relevance:
- The actions funded facilitated closer ties between farmers and genebanks through field trials designed to identify and register accessions of potential new local varieties, as well as test their tolerance to abiotic and biotic stresses, with the aim of increasing PGRFA in the multilateral system.
- The actions funded focused on ensuring food security – but there was insufficient attention to other relevant aspects for PGRFA use, such as improving household nutrition or facilitating income generation.

Effectiveness:
- Participatory field trials in the actions funded resulted in the provision of new accessions of PGRFA to gene banks and research institutions and encouraged farmers to exchange information, thereby increasing their sense of ownership of PGRFA.
- Training leads farmers in seed selection, variety selection and plant breeding in four of the five case studies undertaken by the Evaluation facilitated

an increase in productivity and production over traditional practices and in some cases increased on-farm crop diversity by more than planned.

- The increased production arising from newly adapted PGRFA has not only enhanced food security: it has, in some cases, enabled farmers to generate an income from the sale of surpluses.
- Internal monitoring and reporting of the actions funded did not include the collection of reliable statistics on PGRFA performance to facilitate participatory analysis of results or identify lessons learned and best practices to guide planning for the next planting season.
- The duration of the actions funded was limited to two years: however, the vast majority of farmers of seed-bearing and tuber crops stated that at least three or five growing seasons, respectively, were needed to fully develop capacity in areas such as seed quality control and safe storage.

Efficiency:
- Most of the actions funded represented good value for money because they were able to convert limited resources (an average of USD 221,000) into positive outputs that demonstrated the important role played by PGRFA in safeguarding food security in vulnerable farming communities.

Sustainability:
- Farmers have remained highly committed to conserving newly adopted PGRFA in their farms since 2014, especially where they have witnessed increased levels of productivity and production in the 2014–2015 and 2015–2016 growing seasons.
- Even where abiotic or biotic stresses have reduced productivity and production levels of PGRFA, most farmers kept as much seed as possible for the next growing season, to reduce their dependency on external inputs.
- Information exchange on PGRFA has continued within most beneficiary farming communities visited by the Evaluation. In some cases this has increased the availability of PGRFA to sustain crop diversity in farmers' plots.
- PGRFA are unlikely to be conserved in the medium to long term in cases where farmers have not been able to adopt adequate seed-quality and storage controls, supported by agricultural extension services to supervise such developments.
- The general lack of capacity among farmers to compare and contrast the production costs of PGRFA in relation to traditional staple crops grown has reduced the opportunities for expanding the production of PGRFA and showcasing it to decision-makers.

Benefit-sharing Fund support to PPB is important, as it provides good example to showcase how farmers invest at the individual and community levels in the management of PGRFA, and how knowledge from farmers and scientists can be applied for achieving a common goal. PPB programmes combine Benefit-sharing Fund funding with the delivery of non-monetary benefit-sharing called for in the Treaty in the form of access to PGRFA, capacity-building, information sharing and transfer of technology. PPB also helps to bring materials from the Multilateral System to farming communities. Participatory plant breeding assists implementation of various provisions of the Treaty in an integrated manner.

Outlook for PPB funding: challenges and opportunities

In the past two decades, PPB programmes have increasingly demonstrated their positive impact on the livelihoods of small-scale farmers and the relevance of PPB to addressing the intertwined challenges of biodiversity conservation, food insecurity and climate-change adaptation. There is in principle a strong case for financial support in the next decade to PPB for sustainable development. However, there are also weaknesses and challenges to be dealt with if PPB programmes are to be upscaled and outscaled in the next years. This final section offers some reflection on challenges and opportunities for funding PPB programmes in the future.

Visualizing impact and sharing lessons on upscaling and outscaling

Other chapters in this book describe the various PPB methodologies and record the initiatives undertaken for crops in many regions. In the next decade, greater attention should be paid to assembling and disseminating evidence on the impact of PPB programmes around the world for the livelihoods of family farmers. This information exists, but has remained within the remit of the donor–executing partner bilateral relationship – not a sustainable practice at a time when donor priorities are shifting rapidly. Additionally, greater emphasis is needed on analysing lessons learned from upscaling and outscaling PPB initiatives, as these have often involved automatically reproducing the methodologies and practices from in one area to another.

Integrating PPB in larger programmes

The complex and interconnected nature of the UN SDGs requires an integrated and interlinked approach, and the 2030 Agenda emphasizes the importance of partnerships in this regard. This integrated approach has an effect on how multilateral and bilateral agencies, as well as national decision-makers, will provide funding in future, with greater emphasis on larger programmes within several areas supported at the same time. This will require the PPB practitioners to establish new alliances and partnerships, not only within the immediate realm of plant-genetic resource conservation and use but also, more importantly, with others working to promote sustainable agriculture, climate-change response and sustainable diets and nutrition.

One opportunity for lesson-learning would be to assess how some recent larger programmes on sustainable agriculture have dealt with PPB, or with agro-biodiversity management more generally. The GEF Integrated Approach

Pilot (IAP) 'Fostering Sustainability and Resilience for Food Security in Sub-Saharan Africa' could be one such programme. The GEF has been a leading donor partner supporting agro-biodiversity management, so its IAP offers excellent opportunities for understanding how such integration could be undertaken, as well as the current limitations. Lessons learned could be brought back to the community of PPB practitioners.

Mobilizing climate change financing

The biggest change currently underway in the global funding landscape is the rise of international donor finance for climate change ('climate finance') which has mushroomed over the two past decades and is predicted to increase. In 2000, climate finance was estimated to provide USD 560 million to developing countries. By 2016, this had grown to almost USD 20 billon – a 40-fold increase – and is expected to reach USD 100 billion by 2020. By comparison, total official development assistance (ODA) was USD 72 billion in 2000 and USD 145 billion in 2016, or a twofold increase in this period (all figures from Johnston, 2018).

Agriculture and ecosystem-based approaches feature prominently in numerous country National Adaptation Plans (NAPAs) and Nationally Determined Contributions (NDCs). An analysis of NDCs submitted by 189 countries as of 29 July 2016 showed that 34 countries note the sustainable use of plant genetic diversity (FAO, 2019), mainly in connecting with enabling crops to respond to stresses caused by climate change: drought, flood, soil salinization, pest and diseases as well as shorter crop-cycles. Mention is also made of the development of conservation, both *ex situ* and on farm, as is the importance of preserving traditional knowledge of breeding, R&D in crop varieties, and the adoption of climate-resilient crops from other regions.

All PPB initiatives that the Benefit-sharing Fund has financially supported have had a strong focus on ensuring the adaptation of farming systems to the effects of climate change. Such linkage to climate change is not uncommon for other agro-biodiversity projects funded by others, such as GEF.

The experience gained from PPB projects funded by the Benefit-sharing Fund and others, together with the focus on agriculture in many national planning tools for climate change, provides a strong basis for approaching several climate–change financing institutions that have not yet supported work on PGRFA. These include the Green Climate Fund, the Adaptation Fund, and bilateral programmes like the International Climate Initiative (IKI) of the German Federal Ministry for the Environment, Nature Conservation and Nuclear Safety (BMU).

Mainstreaming into national policy and budgets

Funding is cited as a key constraint to better use of crop genetic diversity. The FAO 2010 State of the World's on PGRFA states:

[p]lant breeding, seed systems and associated research are all expensive and require a long-term commitment of financial, physical and human resources. Success, for both the public and private sectors, is greatly dependent on government support through appropriate policies as well as funds.

(FAO, 2010)

This is especially true for PPB, which generally targets farming systems not central to other breeding programmes. This FAO report also points out that, although some progress has been achieved with regard to integrating PPB in national breeding strategies, this area still requires attention.

Promotion of and support to PPB has relied extensively on funding from international organizations and donors. The goal for the next decade must be to transit to a more diversified approach based on ensuring the local sustainability of the initiatives, together with better national policy and budgetary support.

This is especially relevant in countries that are major centres of plant domestication and diversification and where rural communities still manage agro-biodiversity in farmer fields. Unfortunately, the World Bank considers many of these as Middle Income Countries, entitled in principle less support from multilateral and bilateral programmes in the future. However, we can note some on-going efforts and experiences on which to build.

In Mexico, CONABIO and other institutions are currently developing an initiative for the use of genetic diversity of domesticated plants and their wild relatives to strengthen food and forestry production in a socially fair and environmentally friendly way (Mastretta-Yanes *et al.*, 2018). This initiative promotes shifting official policies to better support smallholder needs and adaptation to local conditions, not least through participatory breeding together with other tools like functional genetics or documenting and sharing campesino-to-campesino experiences. There is a strong focus on mainstreaming agro-biodiversity into official policies and planning, also within national development plans, sectoral programmes related to the environment, farming development, social development and certain indigenous peoples, and related budgets.

In Ecuador, the Instituto Nacional of Investigaciones Agrarias (INIAP) is promoting the establishment of a national fund for research on agrobiodiversity, seeds and sustainable agriculture (Cesar Tapia, personal communication). The aim is to ensure that a certain percentage of the agricultural GDP will be invested in agrobiodiversity, based on the recognition that the sustainability of the entire agricultural sector relies on biodiversity. Although this initiative has not yet been approved, it represents one of the most innovative ways recently explored at the national level to ensure funding for activities such as PPB.

A community of practice to enable PPB growth?

This final section has offered a menu of options that the PPB community at large, including farmer organizations, local non-governmental organizations, like-minded donors, national research institutions and ministries of agriculture, may wish to explore during the next decade.

The options discussed above would profit from the establishment of community of practice that could bring together the various initiatives and key institutions to share lessons learned and disseminate success stories, as well as discussing how to deal with barriers and failures. This community of practice could make the knowledge gained available to global forums such as the Governing Body of the International Treaty, thereby providing a valuable global platform to build awareness and support the further growth of successful approaches to PGRFA conservation and use. This will need to be aligned with ongoing efforts to review the Treaty's Funding Strategy. The aim must be to strengthen the linkages between various funding sources and partners, also by pursuing collaborative planning and co-spending opportunities and identifying the channels suitable for such linkage.

Notes

1 This study reflects the technical opinions of its author, not necessarily those of the FAO, or the Secretariat of the International Treaty on Plant Genetic Resources for Food and Agriculture in particular.
2 The MDGs were agreed by all countries and all countries were committed to their achievement. However, the MDGs focused on action in developing countries, so developed countries were not required to achieve much within their national boundaries.
3 The Fourth Call for Benefit-sharing Fund (Benefit-sharing Fund-4) project proposals was launched in December 2017.
4 www.fao.org/plant-treaty/news/news-detail/en/c/1143273/.

References

FAO, 2010. *Second Report on the State of the World's Plant Genetic Resources for Food and Agriculture.* Rome: FAO.

FAO, 2015. *Final Report on the Execution of the Second Project Cycle of the Benefit-sharing Fund.* Rome: FAO. www.fao.org/3/a-bb364e.pdf.

FAO, 2017. *Evaluation of the Benefit-sharing Fund Second Project Cycle, International Treaty on Plant Genetic Resources for Food and Agriculture.* Rome: FAO. www.fao.org/3/a-bd706e.pdf.

FAO, 2017b. *Report on the Benefit-sharing Fund: 2016–2017.* Rome: FAO. www.fao.org/3/a-bs793e.pdf

FAO, 2019. *Assessment of the Role of Genetic Resources for Food and Agriculture for Climate Change Adaptation and Mitigation.* Rome: FAO www.fao.org/3/my589en/my589en.pdf.

FAO, IFAD, WFP. 2015. *Achieving Zero Hunger: The Critical Role of Investments in Social Protection and Agriculture.* Rome: FAO.

IFAD, 2016. *IFAD Strategic Framework 2016–2025: Enabling Inclusive and Sustainable Rural Transformation.* Rome: IFAD. www.ifad.org/documents/38714170/40237917/IFAD+Strategic+Framework+2016-2025/d43eed79-c827-4ae8-b043-09e65977e22d.

Johnston J, 2018. *Climate Finance and the ITPGRFA Review.* Rome: FAO www.fao.org/3/CA1038EN/ca1038en.pdf.

Mastretta-Yanes A, Acevedo Gasman F, Burgeff C, Cano Ramírez M, Piñero D, Sarukhán J, 2018. An initiative for the study and use of genetic diversity of domesticated plants and their wild relatives. *Frontiers in Plant Science* 9, 209. doi: 10.3389/fpls.2018.00209.

Moeller NI, 2018. *Report on Progress: Matrix of Funding Tools Analysis.* Rome: FAO. www.fao.org/3/CA1024EN/ca1024en.pdf. Accessed 20 March 2019.

18 Seed laws

Bottlenecks and opportunities for participatory plant breeding

Bram De Jonge, Gigi Manicad, Andrew Mushita,
Normita G. Ignacio, Alejandro Argumedo and
Bert Visser

Introduction

Seed laws address a fundamental challenge: at the time of purchase, farmers cannot reliably assess the quality and the identity (variety) of seed. Seed laws are intended to protect the farmer by establishing a legal obligation for the seller to guarantee the quality and identity of seed, by means of standardized inspection and testing procedures.

Seed laws are typically intended to regulate and systemize variety development and subsequent seed multiplication by seed companies, research organizations and governmental institutions – collectively referred to as the formal seed sector. Many developing-country legislators have drawn on legislation in developed countries where there are well-established seed sectors with public research organizations and private seed companies catering for the needs of mostly large-scale, commercial farmers. By contrast, seed systems in many developing countries are predominantly farmer-managed, with smallholder farmers obtaining seed though informal channels of exchange and trade in farm-saved seed. And even though seed laws have not necessarily been designed with the intention of affecting the smallholder farmer-based seed sector, studies find that they often do so (ACB, 2018; GRAIN, 2005 and 2015; Halewood, 2016; Herpers *et al.*, 2017b; Wattnem, 2016).

This chapter examines current seed laws in developing countries around the world, analysing the bottlenecks and opportunities for Participatory Plant Breeding (PPB). It draws on work and research undertaken under the 'Sowing Diversity = Harvesting Security' (SD = HS) programme.[1] Information and experiences have been gathered from the programme countries (Laos, Myanmar, Peru, Vietnam and Zimbabwe), including from farmers, researchers and policymakers (Visser, 2017). Initial findings have been validated through national and regional policy workshops. Further, the chapter builds on information from other studies and academic publications in the field.

The first section of this chapter sets the stage by presenting the key concepts: farmers' seed systems and participatory plant breeding, seed policies, laws and formal seed production. The second section offers a concise overview of current seed policies and laws in the developing world, noting the (lack of) recognition and support for farmers' seed systems or PPB. Examples from countries in Africa, Asia and Latin America are provided, and ongoing initiatives towards the regional harmonization of seed laws are briefly discussed. Presenting a description of alternative measures applied by some countries and organizations, the third section indicates possible policy options for supporting farmers' seed systems and accommodating PPB.

Setting the stage

Farmers' seed systems and Participatory Plant Breeding

Seed systems in many developing countries are predominantly farmer-managed (Richards *et al.*, 2009). These systems are based on collection, selection, crossing, testing, multiplication and storage of seeds and vegetative propagation materials by local farmers, without formal oversight or quality control, relying instead on trust and social networks, not least among the women. In farmers' seed systems, widening of the gene pool and testing of new materials take place continuously, through introgression of wild material into successfully cultivated material, and introduction and hybridization with improved materials from the formal sector.

Learning by farmers about the management of genetic diversity may or may not involve external parties. In PPB,[2] farmers collaborate with scientists in setting selection or breeding goals, and following up with the development of better-adapted varieties. Mutual learning in which both farmers and breeders or extensionists are involved plays a major role throughout the process. In the absence of support by external parties, farmers acquire planting material from various sources, including gifts, purchases, and accidental introgression from neighbouring populations. Farmers may – whether consciously or unintentionally – select new varieties and traits in response to local environmental conditions and cultural considerations, and then pass these on through their social and economic networks.

PPB generates new varieties that are deliberately bred and maintained by active seed selection on-farm. Almekinders *et al.* (2007) describe how well-adapted cultivars developed through PPB may be disseminated by seed production and distribution, without encountering major bottlenecks. In farmers' seed systems, varieties are shared amongst smallholder farmers within and between communities, through local (barter) markets, personal exchange between neighbours and family members, and through local and national seed fairs. In contrast, the provision of quality seed from the formal sector to smallholder farmers may be constrained by poor access to product information, or problems of affordability, infrastructure and market conditions. These constraints may affect women farmers in particular, as their access to materials and information is often restricted by discrimination pertaining to their gender, class and ethnicity. In addition, the public and private breeding sectors rarely cater to the specific needs and preferences of women farmers (ADB, 2013; Oxfam Novib *et al.*, 2015). In developing countries, informal markets remain the most important sources of seed for smallholder farmers for most food crops, except perhaps for maize and vegetables (McGuire and Sperling, 2016). Field studies show diverse trends in the functioning of these informal markets, which have unrealized potentials for delivering a wider range of higher-quality seed (Sperling and McGuire, 2010).

The interplay between farmers' seed systems and the formal, often commercial, seed sector varies among countries and regions, and among crops. Both systems are important: farmers' seed systems offer seeds that exhibit high levels of diversity and are well adapted to local conditions, which can help to cope with climate change; formal seed systems offer seeds that may be of higher quality or have important traits concerning yield and resistance. Seed from the formal sector is often not readily accessible to small-scale farmers, but it may be absorbed into informal seed systems if offered for sale in local markets by seed retailers or farmers and crossed with local farmers' varieties to better suit farmers' needs and preferences.

The challenge for policymakers is to create policies and laws that support both formal and farmers' seed systems where they are most effective (Louwaars and de Boef, 2012). Or as stated in the FAO Guide for National Seed Policy Formulation:

> The availability of, and access to, quality seeds of a diverse range of adapted crop varieties is essential for achieving food and livelihood security and for eradicating hunger, especially in developing countries. Strengthening both formal and informal seed systems is therefore an integral part of the sustainable use of plant genetic resources for food and agriculture.
>
> (FAO, 2015: 1)

Seed policies, laws and formal seed production

A national seed policy provides the general objectives and framework for the seed sector, while legislation creates the means to enforce particular standards and procedures, especially those relating to seed quality. By 'formalizing' seed production and transfer, seed policies and laws aim to protect consumers (here: the farmers) who otherwise cannot assess the quality and variety of seed in the marketplace. By doing so, seed legislation aims to strengthen the national seed sector and economy. Formal seed production requires means of assuring quality standards and mechanisms for coordinating functions. Formal seed production, in the private and public sectors, involves a systematic process of variety evaluation and seed multiplication in which seeds fall into various classes recognized in the law (breeder/pre-basic, foundation/basic, and certified seed). In the farmer-managed system of saving and exchanging seeds, similar processes of variety evaluation and selection may be undertaken, but usually informally and without supervision.

Seed laws commonly include procedures and standards for:

- variety release systems, registering only varieties of proven value to be made available to farmers through the formal seed system;
- seed certification, which aims to monitor and guarantee varietal identity and purity throughout the seed chain, and which requires registration of seed producers;
- seed quality control, which checks on additional seed characteristics such as viability and seed health.

Regarding the release of new varieties, most countries apply standards for distinctiveness, uniformity and stability (DUS). A DUS trial is an evaluation to determine whether a candidate variety meets the criteria set for that particular crop concerning its distinctness from any known variety, uniformity in its main characteristics, and stability in repeated multiplication cycles. In addition to DUS testing, several countries also require testing for value for cultivation and use (VCU) before a candidate variety can be released. VCU testing, commonly conducted in national performance trials, is carried out to establish if the cultivation of a newly developed variety offers substantial benefits compared to existing local or standard varieties. These tests are usually executed over the course of two years or two seasons, at a prescribed minimum number of sites.

Seed certification regulations aim to guarantee seed quality by providing minimum standards of genetic purity, physical purity and germination rates. All these control mechanisms add costs to the seed-production process and, ultimately, to the price of seed in the formal market. Policymakers must consider complicated trade-offs between stricter measures for quality control and the need to encourage the multiplication and distribution of lower-cost seed (Rohrbach *et al.*, 2003). If large quantities of seed are produced in a small area, inspection costs are manageable, but they may become prohibitive if seed production is dispersed across widely distributed groups of small-scale farmers. The higher the number of varieties that need to be certified, the greater the expense.

Some countries have responded to these difficulties by relaxing their inspection requirements: stricter certification standards are demanded for only a few commercial (export) crops. Countries may have also different standards for different classes of seed, trying to balance the need to protect farmers with the need to avoid overburdening producers. Such measures can be particularly important for PPB projects, as these are often small-scale and/or involve commercially less important crops. Before we delve more deeply into policy options that can accommodate PPB, the next section provides a concise overview of current seed policies and laws, and their recognition and support for farmers' seed systems and PPB.

Current seed policies and legislation in the developing world: how do they relate to PPB?

Here we begin with a brief look at some current *seed policies* in developing countries, and then turn to the more specific provisions included in national *seed laws*, with examples from Africa, Asia and Latin America.[3] Finally, we mention ongoing initiatives aimed at *regional harmonization* of seed laws.

Seed policies

A seed policy is a statement of principles which guide government actions for seed-sector development, and which explain the general structure of the seed sector and the roles and responsibilities of relevant stakeholders. The function of a seed policy is to establish the overall framework for all other instruments that govern the seed sector, such as the seed law and regulations.

Few developing countries have national seed policies in place (FAO, 2015). To provide assistance in developing such policies, the FAO Commission on Plant Genetic Resources for Food and Agriculture has developed a Voluntary Guide for National Seed Policy Formulation (2015), which aims to assist countries in formulating policies that can help to create an enabling environment for seed-sector development. Recognizing that formal seed sector institutions (e.g. seed companies, NARS) and informal farmer seed systems often operate side by side, the Guide argues that seed policy should strive at strengthening both seed systems as well as their interlinkages:

> National seed policies may recognize the informal sector's important role and promote support in appropriate areas such as extension, training schemes for farmers, community seed banks, germplasm conservation, and seed quality control, or even promote official recognition of some of these activities. The role of women in these various functions should be given particular attention.
>
> (FAO, 2015: 32)

Recent studies indicate that most developing-country seed policies aim at promoting the distribution and use of formal-sector seed varieties, with scant reference to or recognition of the role of farmer-managed seed supply (GRAIN, 2015; Halewood, 2016; Herpers *et al.*, 2017b; Wattnem, 2016). Where the 'informal sector' is mentioned, the connotations are often negative. For example, Ghana's seed policy states that 'due to the dominance of small-scale holders, the use of quality seed is very much limited. If not checked, this trend will lead to a continuous diminishing of agricultural productivity and compromise the cherished national goal of food security' (Ghana, 2013: 13). The typical remedy for countering this weakness is to strengthen seed provision by the formal sector. As stated in India's seed policy:

To meet the Nation's food security needs, it is important to make available to Indian farmers a wide range of seeds of superior quality, in adequate quantity on a timely basis. Public Sector Seed Institutions will be encouraged to enhance production of seed towards meeting the objective of food and nutritional security.

(India, 2002: Art. 2.2)

However, some seed policies do invite farmers to participate in formal breeding processes. According to Section 10.2.3 of Ghana's seed policy (2013), 'Scientists and farmers will be encouraged and supported to test and release popular local landraces as official varieties', implying that these varieties must fulfil the standard DUS and VCU requirements as prescribed by the Ghanaian seed law and its regulations (Ghana, 2014: Art 5.4). Similarly, the seed policy of Niger restricts seed production to varieties that have fulfilled DUS and VCU testing, while also encouraging farmers to contribute to breeding and selection of new varieties, stating that successful farmers will be rewarded (Niger, 2012: 10). One promising draft seed policy comes from Uganda: it aims to 'enhance the production of quality seed within the informal system', *inter alia* by providing 'for the listing of traditional and participatory bred varieties' (Uganda, 2016: 12), and strengthening 'participatory variety selection to enhance adoption of new improved varieties' (ibid.: 10).

Seed laws

Seed laws establish the standards and procedures that govern seed production and marketing. The laws and regulations set the institutional framework for enforcement by defining prohibitions and obligations regarding the marketing of seeds, and may stipulate registration for seeds and sellers and set other quality requirements (FAO, 2015). By determining who can produce and sell seeds of what varieties and under which conditions, seed laws can strongly impact the functioning of farmers' seed systems and PPB projects.

Farmers' seed systems are characterized by the use, exchange and local trade of farm-saved seed. Yet, despite being common to smallholder farmers throughout the developing world (IFAD and UNEP, 2013; McGuire and Sperling, 2016), these practices are permitted by the seed laws of very few developing countries. A recent study of the seed laws in 35 African countries finds that for the specified crops to which these laws apply,[4] 23 countries forbid the trade and exchange of unregulated seed (Herpers et al., 2017a): seed that is sold or exchanged must be from a variety that is listed in the national catalogue and/or must be certified or have its quality declared. For example, 'No seed shall be offered for sale unless it is certified in accordance to these regulations' (Herpers et al., 2017b); 'Only varieties that have been approved for release and notified and included in the variety list may be sold' (Malawi, 2005: Art. 5). The definition of 'sell' includes 'distribution or give away' (Tanzania 2003: Art. 2) and 'exchange or barter or to offer' (Malawi, 2005: Art. 2).

For seed to be formally allowed to be sold in the market, seed laws usually require the registration of the seed producer and seller and the listing of the variety for sale, as well as the certification of seed lots. Anyone wishing to register as a seed grower and seller must normally be able to demonstrate certain educational qualifications, and to show proof of access to seed processing and storage facilities. Such requirements make it difficult for small-scale farmers to register as official seed producers and/or sellers, and can create legal uncertainty for farmers or farming communities taking their first steps towards more organized seed production.

Farmers and partner organizations involved in PPB in the SD=HS project have indicated that it is unclear as to which requirements farmers must comply with, especially once a farming community starts producing seed in greater quantities. The Seeds Act of Zimbabwe, for example, stipulates that the obligation to register 'shall not apply to the sale of seed which is grown by any farmer and sold by him to a person for use as seed by such person' (i.e. another farmer) (Zimbabwe, 2001: Ch. 19:13). This provision would appear conducive to the exchange and sale of farm-saved seed among smallholder farmers. However, it is not clear whether, and up to what level of production, this exemption covers small-scale farmers in the UMP and Tsholotsho districts who have established community seed banks where they save, use, exchange and sell farm-saved seeds. Some of these farming communities are now part of the Champion Farmer Seed Cooperative, which has hundreds of farmer members and fulfils all registration requirements, producing certified seed for maize, sorghum, pearl millet and groundnuts (Oxfam Novib *et al.*, 2017). Still, even when farmers can meet the registration and certification requirements, many feel that these are unnecessary when social structures can control the quality of the seed offered. Moreover, compulsory seed certification has been reported to act as a disincentive for local seed companies to invest in low-value, non-hybrid seeds (e.g. sorghum, millet, soybean and groundnut) due to the time and costs involved (Mutonhori and Muchati, 2013).

The above points concern possible challenges for farmer seed-producers seeking to comply with seed *certification* procedures. However, most countries demand that varieties from which certified seed may be produced must first be officially *released*. The Ethiopian Seed Proclamation, for example, defines the 'release' of a variety as the permission whereby seed of the registered variety 'can be multiplied, produced or supplied to [the] domestic market' (Ethiopia, 2013: Art. 2.17). Moreover, variety release is commonly associated with ownership or recognition of origin, which may have consequences for breeders' rights and aspects of access and benefit-sharing.

Under most of the seed laws reviewed, new varieties must be tested for DUS and VCU and evaluated by a committee before being released and entered in a national catalogue of varieties. Such tests are usually executed over the course of several years or seasons, and at a prescribed minimum number of sites. The procedure is expensive and time-consuming, often beyond the possibilities of smallholder farmers. Indeed, in the SD=HS programme countries,[5] legal requirements make it unrealistic for smallholder farmers to register new farmers' varieties. In the SD=HS programme, many of the farmer breeders are women, who are likely to have further constraints as to cash and time, as well as other obstacles to participating in the complex and male-dominated formalities of varietal release (Oxfam Novib *et al.*, 2015). In Vietnam, some small-scale farmers involved in the PPB projects of the SD=HS projects have succeeded in registering new farmers' variety, but only with the assistance of programme partners such as the Mekong Development Institute and the financial resources of the programme. Mr Nguyen Can Tinh registered the first farmer-bred variety (HD1) in the country in 2010 after a procedure that took five years, first with VCU and DUS tests on the local level and then, after approval, followed by large-scale production tests. Registration fees alone cost a minimum of USD 625 per variety, with the farmer having to shoulder the costs the multi-location trials in addition (SEARICE *et al.*, 2013). Besides these financial constraints and the long and complex process, many farmers have difficulties in meeting requirements such as technical capability in documenting and analysing the characteristics of their rice varieties.

The registration process is cumbersome and expensive. Moreover, many farmers' varieties do not fulfil standard DUS requirements, as such varieties are characterized by relatively greater genetic and phenotypic diversity – characteristics that may make them more resilient and adaptable to local agro-ecological conditions. Farmers' varietal selection practices maintain this diversity, whereas the varieties listed in the official catalogues are often those that do well 'on average' (in multi-locations and under high-input conditions), and may not be the varieties most preferred by farmers who operate in more challenging or less-typical environments. Of the 35 African countries whose seed legislation has been analysed, only four countries indicated that they will register landraces or farmers' varieties under alternative or looser criteria,[6] and only Benin has an actual list for farmers' varieties in place (Herpers *et al.*, 2017a; Vodouhe and Halewood, 2016).

Regional harmonization processes

Regional organizations and trading blocs in Africa and Asia have been developing regional seed legislation aimed at improving the seed sector by harmonizing national seed laws, to reduce the costs and time associated with seed trade within the region. In Africa, these processes are executed by the Common Market for Eastern and Southern Africa (COMESA, 19 member countries), the Southern African Development Community (SADC, 15 member countries), and the Economic Community of West African States (ECOWAS, 15 member countries);[7] in Asia, such regional harmonization has been proposed in the Association of Southeast Asian Nations (ASEAN, 10 member countries).

The aim of regional harmonization is to promote and speed up the distribution of new varieties among countries in the region. In the COMESA region, for example, a new variety needs to be released only in two COMESA member states to trigger inclusion in the COMESA Variety Catalogue, which then allows the variety to be legally commercialized in all 20 COMESA countries (COMESA, 2014). In addition, regional legislation aims to harmonize phytosanitary procedures and seed certification standards and labels for some key crops in the region. The main objective is to

> encourage investment in seed business in the COMESA Member States, to enhance access to new and existing varieties in the COMESA Member States, and to stimulate the breeding and availability of seed varieties resulting in increased crop variety choices for all farmers.
>
> (COMESA, n.d.)

Thus, current harmonization processes focus predominantly on the formal seed sector – more specifically, the commercial seed sector (GRAIN, 2005; 2015). With respect to farmers' seed systems, ECOWAS regulations state: '[this] shall not be applicable to freely used farm grains and seeds' (ECOWAS 2008: Art. 3.2), with farm grain and seed defined as 'any seed or grain produced by a farm meant for the personal use of the farmer and not destined for the market' (Art.1). Herpers *et al.* (2017a) interpret this to mean that farm-saved seed may be freely used by farmers and exchanged among themselves, but not sold in the market. They also observe that SADC and COMESA member countries retain the possibility, but not the obligation, to endorse such exemptions for farmers in their national legislation; and that in all three African regions, national governments may set up variety lists that include farmer varieties. However, Herpers and colleagues (2017a: 12) further note that, by paying so little attention to farmers' seed systems, these regional initiatives 'represent lost opportunities to promote seed system integration', with legal support for farmers' seed systems and PPB being dependent on voluntary efforts at the national level, without the benefit of encouragement from actors who support regionalization.

Thus we must conclude that the bureaucratic procedures that regulate the certification of seed producers and seed-lots and/or the registration of new varieties may create extra hurdles for PPB projects. The main reason is that the set requirements are usually not conducive to the needs and realities of the small-scale farmers involved, who, for example, may not meet the educational standards required for registering as seed producers or cannot afford the expense and time involved in multi-year, multi-location testing of their new varieties.

Policy options that recognize and promote PPB and farmers' seed systems

Despite the very limited recognition and support for farmers' seed systems and PPB at the regional level and in most national seed laws and policies, some countries have implemented alternative measures. Here we offer some examples and options that countries can apply to accommodate PPB and support farmers' seed systems, based on these experiences. We focus on the role of exemptions, the opportunities for registering farmers' varieties, and the introduction of an alternative seed class like Quality Declared Seed.

Exemptions

One way to ensure that a country's seed legislation does not obstruct the use, exchange and local trade of farm-saved seed is to include a general exemption for farmers' seed systems in the national seed law. The Ethiopian Seed Proclamation, for example, notes that it 'may not be applicable to (a) the use of farm-saved seed by any person; (b) the exchange or sale of farm-saved seed among smallholder farmers or agro-pastoralists' (Ethiopia, 2013: Art. 3.2). Another example can be found in India's draft Seeds Bill, which is said to hold that a

> farmer can sow, exchange or sell his farm seeds and planting material without having to conform to the prescribed minimum limits of germination, physical purity and genetic purity (as required by registered seeds). However, farmers cannot sell any seed under a brand name.
>
> (PRS, 2010: 1)

The downside of legal exemptions is that their exact scope is seldom clarified. In Vietnam, for example, the Seed Ordinance provides that seed production and seed business involving varieties belonging to major crops will be managed strictly, and producers must fulfil specified requirements (Vietnam, 2004: Art. 4). However, it also provides a small-holder exemption:

> households or individuals who produce and trade in major crops and do not belong to a person that has to register for business do not have to obey regulations stipulated (...), but must ensure the quality of plant variety and environmental sanitation [identity and phytosanitary condition] according to regulations of the law on plant protection and quarantine, the law on environmental protection and the law on fishery.
>
> (Art. 36.3)

Together with Decision No. 35/2008/QD-BNN from 2008, which applies to farm households, cooperatives and other organizations that produce, circulate and transact 'farm household plant varieties' (Dinh and Kinh, 2016: 373), this legislation has created space for PPB projects in Vietnam to produce and sell uncertified seed of new and unregistered varieties in the marketplace, as long as the variety performs well and the seeds are of good quality. However, because the relevant legal provisions are not clearly defined, this practice of selling uncertified seed (of any variety) is tolerated by the provincial authorities on a limited scale and within certain geo-graphic boundaries, and cannot be practised on the national level (SEARICE personal communication, in Visser, 2017).

One way to avoid legal uncertainty is by specifying the boundaries of an exemption in the law or its bylaws, which can more readily be updated in line with changing needs and circumstances. The seed law in Myanmar, for example, exempts seed control on traditional farmers' activities: 'the provi-sions contained in this Law shall not apply to the following facts [sic]: distri-bution and sale of seed produced by any peasant by himself to another peasant' (Myanmar, 2011: Art. 31). To establish greater clarity as to the cat-egory of farmers who fall within the scope of this exemption, the government may provide a definition of 'peasant' in the regulation to implement the seed law. Further, the law defines 'commercial distribution' as the distribution of seed above a certain weight or volume threshold to be determined by the National Seed Committee (Art. 2). By establishing such flexibility in their national legislation, governments can create space to set standards better suited to the (changing) needs and opportunities of farmer seed enterprises. Further, according to Article 32, 'peasants and seed researchers who produce seed in co-operation with the departments, services under the Ministry shall be exempted from obtaining license under this law'. This exemption may stimu-late PPB projects, as they may now avoid the otherwise compulsory licensing of seed-lots as well as sellers.

Registration of farmers' varieties

As noted, of all African countries, only Benin maintains a list in which farmers' varieties are registered. There are in fact three distinct lists of varieties in the Benin catalogue: List A comprises released varieties that must be tested for DUS and VCU in order to be registered; List B consists of varieties tested for DUS only, which can be multiplied, but only for export; and List C com-prises traditional or local varieties that must be tested for VCU in order to be registered (Herpers *et al.*, 2017a). As the DUS requirements are not included in List C, landraces and new varieties developed through PPB methods that are more heterogeneous can still be registered and their seed multiplied and sold at the national level. Smallholder farmers may appreciate heterogeneity in existing and newly developed varieties, as this can make the crop more resilient and adjustable to changing environmental conditions, so this legal space is very much needed.

In this context, it is interesting to note that the European Union, which has probably the most strictly regulated seed sector in the world, has developed mechanisms to allow for the registration of varieties that do not fulfil standard DUS requirements. These include conservation varieties (EU, 2008), amateur varieties, preservation mixtures (EU, 2010), and populations (EU, 2014). With respect to the latter, the Commission implementing the decision has stated that the general seed law is to prevent the marketing of seed not belonging to a variety. However, new research in the Union on plant reproductive material that does not fulfil the variety definition as regards uniformity, shows that there could be benefits of using this diverse material, in particular with regards to organic production or in low input agriculture for example to reduce the spread of diseases. To allow seed from those populations to be marketed, it would be necessary to (...) add the possibility to market seed which does not fulfil the requirements concerning varietal aspects.

(EU, 2014: Articles 1–3)

Another example can be found in Peru's National Register for Native Potatoes (Peru, 2008). In its implementing guidelines, the Peruvian National Agricultural Research Institute (INIA specifies that the objectives are to register and recognize the genetic diversity and variability of the Peruvian native potato cultivars that originated in Andean communities and have been developed and conserved by generations of farmers. By establishing a national database with passport data, morphological and agronomic information, the Register further aims to promote inter-institutional collaboration and contribute to developing tools to identify developers of these native varieties and prevent acts of bio-piracy. Small-scale farmers or individuals may request INIA to register their native potato varieties. They must comply with certain conditions, which in practice can be met with the technical assistance of INIA and other institutions, including NGOs. This can serve to facilitate the registration and recognition of new varieties and their developers in PPB projects (Visser, 2017).

An alternative seed class

To support and facilitate the production, use and marketing of seed of farmers' varieties resulting from PPB, which are generally maintained in small-scale systems and contribute to wider diversity in farmers' fields, the government may establish an alternative seed class. This could be regulated by a separate quality assurance mechanism, adjusted to the needs and capacities of small-scale farmers to promote farmers' seed production while ensuring high seed quality.

One example of such an alternative seed classification is Quality Declared Seed (QDS). FAO introduced this concept in 1993, noting that many countries lacked the expertise and infrastructure necessary to certify all seed-lots offered in the market; further, that not all producers would be able to meet the strict requirements set for commercial seed enterprise, but minimum standards of genetic purity, physical purity and germination rates were still relevant. Aimed at providing 'seed standards for a wide range of crop species and agro-ecologies for the development of the agricultural sector', systems based on this concept give major responsibilities to seed producers and dealers (FAO, 2006: vii).

The concept was revised in 2003–2006 for better accommodation of local varieties. For a variety developed through PPB, for example, the following requirements have been proposed: (a) a statement giving the origin of the variety; (b) data obtained during the farmer evaluation process; (c) a description of the main characteristics that distinguish the variety from other varieties; (d) a statement defining the agro-ecological zone for which the variety is suited, and (e) a statement indicating the procedures to be followed for maintaining the variety (FAO, 2006: 13).

Whereas the recognition of QDS created some legal space for the registration of farmers' varieties, for which seed may be produced only under the QDS system, most countries that have implemented QDS use it merely as a simpler certification system for seed production of formally registered varieties. Tanzania introduced a QDS system in 2000. The system is characterized by fewer inspections, as only a portion of the fields are randomly selected for inspection, and the inspections are conducted mainly by local agricultural officers. A small percentage of the seeds inspected are sent to a central laboratory for testing of purity and germination rates. The reduced testing and travel costs make this QDS system cheaper than full certification, with estimated costs being 11 per cent of the retail price, compared to 25 per cent for full certified seed (Gildemacher *et al.*, 2017: 7). The introduction of a QDS system in Uganda has shown that QDS can meet similar or even better quality standards than certified seed (Gildemacher *et al.*, 2017). Gildemacher and colleagues hold that

> the essential elements that distinguish QDS from certified seed systems are not the quality standards. (…) What should distinguish a QDS system from a full certification system is that it is basic, simple and easily accessible, which will also automatically make it cheaper.
>
> (2017: 8)

However, the same study also notes several challenges to the QDS system in general, and in Tanzania in particular. There may be limited capacity at the local/regional level for laboratory testing; and the labelling and testing requirements add costs and delays that, although lower/less than with full certification, can still obstruct the production and delivery of affordable quality seed at the right time. This is confirmed by Haug and Hella (2013), who concluded that small-scale farmers in Tanzania still make relatively little use of QDS, as they consider it to be either too expensive or of too little added value.

Another alternative certification system has been developed by the International Federation of Organic Agriculture Movements (IFOAM), and is known as the participatory guarantee system (Fonseca, n.d.). This system emphasizes food sovereignty, food security and food safety, and aims to be appropriate to the realities of small farmers and enterprises. It certifies producers on the basis of active participation of stakeholders and is built on a foundation of trust (e.g. farmers' pledges), social networks, co-responsibility, knowledge exchange and decentralized decision-making.[8] Such systems are still rare, but they offer guidelines that can be useful in a wider context.

Lack of substantial implementation of the QDS concept, and mixed reports from the few cases where QDS has been introduced, warrant more in-depth study of its strengths and weaknesses. To support PPB projects, countries may give consideration to elements of the QDS principles, to facilitate the registration of small-scale farmers as seed producers and sellers and to include farmers' varieties in the national catalogue, while offering farming communities a wider diversity of affordable, quality seeds.

Conclusions

This chapter has presented an overview of current seed policies and laws in developing countries in Africa, Asia and Latin America, drawing on experiences from the SD=HS programme and the relevant literature. We find that most developing-country seed policies and laws fail to recognize the importance of farmers' seed systems – which form a major part of the seed sector in virtually all developing countries – and that they do not support participatory breeding practices. Instead, these seed laws and related regional harmonization processes aim primarily at regulating and stimulating the formal sector – more specifically, the commercial seed sector.

That seed laws neglect farmers' seed systems and the diverse needs and realities of smallholder farmers involved in breeding and seed production can result in serious bottlenecks for PPB projects. For example, smallholder farmers may not legally exchange farm-saved seed amongst themselves; they may not register as seed producers if they cannot show evidence of formal education in this field, or if they lack access to the required processing and storage facilitates; they may not be able to get their seed certified because of the standards required;[9] or because their seed is from a farmer variety, which they cannot register due to strict DUS requirements or the high registration fees as well as the time and costs involved in multi-location testing. These complicated procedures may also serve to further discriminate against women, who tend to have less access to cash and may experience obstacles to participating in male-dominated formal institutions and procedures.

Some countries and organizations have devised remedies to counter these bottlenecks – for example, by including a general exemption for farmers' seed systems in their seed laws; by limiting the set of (commercial export) crops for which stricter certification standards apply; by offering the possibility of registering farmer varieties; or by establishing an alternative seed class like QDS, aimed at protecting farmers by setting certain standards for seed quality while not overburdening the seed producers or the administrators who implement the certification system. Such alternative approaches have potential for supporting and facilitating PPB projects; however, for further development and upscaling, more research, impact assessments and investment will be needed.

Notes

1 This programme is coordinated by Oxfam Novib; see www.sdhsprogram.org/. The SD=HS programme has been generously funded by the Swedish International Development Cooperation Agency, the Dutch National Postcode Lottery and the International Fund for Agricultural Development.
2 In this chapter, the term 'participatory plant breeding' (PPB) is used in a wide sense, encompassing not only crossing, but also selection in either segregating populations (PPB *sensu stricto*), comparative evaluation between stable varieties (PVS), and participatory variety enhancement (restoration of traditional varieties: PVE).
3 National implementing regulations were, in general, not consulted for this research.
4 The seed laws of most countries regulate only the crops most important in terms of food security and trade value.
5 Laos, Myanmar, Peru, Vietnam and Zimbabwe; see www.sdhsprogram.org/country/.
6 Benin, Malawi, Niger and Uganda (in draft seed policy).
7 See Herpers *et al.* (2017b) for more on the specific (binding and non-binding) legislation applicable in the three regional trade organizations in Africa.
8 For more information on how the PGS works, its guidelines and examples, see www.ifoam.bio/en/organic-policy-guarantee/participatory-guarantee-systems-pgs.
9 It must be acknowledged that, overall, the SD=HS farming communities in Laos, Vietnam and Zimbabwe noted no problems in meeting certification standards. On the contrary, they are confident and proud of being able to meet the same standards as commercial producers (Visser, 2017: 29).

References

ACB, 2018. A review of participatory plant breeding and lessons for African seed and food sovereignty movements. Melville, South Africa: ACBIO. www.acbio.org.za/en/review-participatory-plant-bre0eding-and-lessons-african-seed-and-food-sovereignty-movements.

ADB, 2013. Gender equality and food security: Women's empowerment as a tool against hunger. Mandaluyong City, Philippines: Asian Development Bank. www.fao.org/wairdocs/ar259e/ar259e.pdf.

Almekinders CJM, Thiele G, Danial DL, 2007. Can cultivars from participatory plant breeding improve seed provision to small-scale farmers? *Euphytica* 153, 363–372.

COMESA, 2014. COMESA Seed Trade Harmonization Regulations, 2014. https://varietycatalogue.comesa.int/web/legislationregional.

COMESA, (n.d.). COMESA Plant Variety Catalogue website. https://varietycatalogue.comesa.int/web/.

Dinh NV, Kinh NN, 2016. Commentary on the regulation on production management of farm households' plant varieties in Vietnam. In Halewood, M, (ed) *Farmers' Crop Varieties and Farmers' Rights: Challenges in Taxonomy and Law*, 373–380. New York: Routledge.

ECOWAS, 2008. Regulation C/REG. 4/05/2008 on Harmonization of the Rules Governing Quality Control, Certification and Marketing of Plant Seeds and Seedlings in ECOWAS Region. www.coraf.org/wasp2013/wp-content/uploads/2013/07/Regulation-seed-ECOWAS-signed-E-NG.pdf.

Ethiopia, 2013. Ethiopian Seed Proclamation No. 782/2013. https://chilot.files.wordpress.com/2014/09/proclamation-no-782-2013-seed-proclamation.pdf.

EU, 2008. Commission Directive 2008/62/EC of 20 June 2008. https://eur-lex.europa.eu/LexUriServ/LexUriServ.do?uri=OJ:L:2008:162:0013:0019:EN:PDF.

EU, 2010. Commission Directive 2010/60/EU of 30 August 2010. https://eur-lex.europa.eu/LexUriServ/LexUriServ.do?uri=OJ:L:2010:228:0010:0014:EN:PDF.

EU, 2014. Commission Implementing Decision of 18 March 2014. https://eur-lex.europa.eu/legal-content/EN/TXT/?uri=CELEX%3A32014D0150.

FAO, 2006. Quality declared seed system. FAO Plant Production and Protection Paper No. 185. www.fao.org/docrep/pdf/009/a0503e/a0503e00.pdf.

FAO, 2015. Voluntary guide for national seed policy formulation. FAO Commission on Genetic Resources for Food and Agriculture. www.fao.org/3/a-i4916e.pdf.

Fonseca MF (n.d.). Alternative certification and a network conformity assessment approach. International Federation of Organic Agriculture Movements. www.ifoam.bio/sites/default/files/page/files/alternativecertificationandanetworkconformityassessmentapproach.pdf.

Ghana, 2013. National Seed Policy of Ghana. https://agriknowledge.org/downloads/xk81jk41k.

Ghana, 2014. Seeds (Certification and Standards) Regulations.

Gildemacher P, Kleijn W, Ndung'u D, Kapran I, Yogo J, Laizer R, Kadeoua A, Karanja D, Simbashizubwoba C, Ntamavukiro A, Niangado O, Oyee P, Chebet AN, Marandu E, Gitu G, Walsh S, Kugbei S, 2017. Effective seed quality assurance. ISSD Africa Synthesis paper. *KIT Working Papers* 2017–2. www.issdseed.org/sites/default/files/case/issd_africa_twg1_sp2_seed_quality_assurance_170412.pdf.

Grain, 2005. Africa's seed laws: Red carpet for the corporations. Seedling July – 2005. www.grain.org/article/entries/540-africa-s-seeds-laws-red-carpet-for-corporations.

Grain, 2015. Seed laws that criminalise farmers. www.grain.org/article/entries/5142-seed-laws-that-criminalise-farmers-resistance-and-fightback.

Halewood, M, (ed) 2016. *Farmers' Crop Varieties and Farmers' Rights. Challenges in Taxonomy and Law.* New York: Routledge. www.bioversityinternational.org/fileadmin/user_upload/Farmers_crop_varieties-Halewood.pdf.

Haug R, Hella J, 2013. The art of balancing food security: Securing availability and affordability of food in Tanzania. *Food Security* 5, 415–426.

Herpers S, Vodouhe R, Halewood M, De Jonge B, 2017a. The support for farmer-led seed systems in African seed laws. ISSD Africa synthesis paper. *KIT Working Papers* 2017–9. www.issdseed.org/thematic-working-group-3-matching-global-commitments-national-realities.

Herpers S, Vodouhe R, Halewood M, De Jonge B, 2017b. The support for farmer-led seed systems in African seed laws. Full report. www.issdseed.org/sites/default/files/case/issd_african_seed_laws_full_report_final.pdf.

India, 2002. National Seeds Policy. https://nsai.co.in/editor/fmanage/userfiles/Rules/National%20Seed%20Policy,2002.pdf.

International Fund for Agricultural Development (IFAD), United Nations Environment Programme (UNEP), 2013. Smallholders, food security and the environment. http://wedocs.unep.org/handle/20.500.11822/8127.

Louwaars NP, de Boef WS, 2012. Integrated seed sector development in Africa: A conceptual framework for creating coherence between practices, programs, and policies. *Journal of Crop Improvement*, 26, 39–59.

McGuire S, Sperling L, 2016. Seed systems smallholder farmers use. *Food Security* 8 (1), 179–195. DOI 10.1007/s12571-015-0528-8.

Malawi, 2005. Seeds Act 1996 (amended 2005). www.wipo.int/wipolex/en/details.jsp?id=11276.

Mutonhori S, Muchati J, 2013. A situational analysis of the scope for smallholder seed development and marketing. Working Paper, Ruzivo Trust. Harare, Zimbabwe.

Myanmar, 2011. The Seed Law. www.myanmartradeportal.gov.mm/kcfinder/upload/files/The%20Seed%20Law-English.pdf.

Niger, 2012. Politique semencière nationale. www.reca-niger.org/IMG/pdf/Document_de_Politique_Semenciere_du_Niger.pdf.

Oxfam Novib, ANDES, CTDT, SEARICE, CGN-WUR, 2015. From lessons to practice and impact: Scaling up pathways in people's biodiversity management. *Briefing Note*. The Hague: Oxfam Novib. www.sdhsprogram.org/publications/publication-three/.

Oxfam Novib, ANDES, CTDT, SEARICE, 2017. The power to exercise choice: Implementing Farmers' Rights to eradicate poverty and adapt to climate change. *SD = HS Briefing Note* 3. The Hague: Oxfam Novib. www.sdhsprogram.org/publications/the-power-to-exercice-choice-implementing-farmersrights-to-eradicate-poverty-and-adapt-to-climate-change-briefing-note/.

Peru. 2008. Crean el Registro Nacional de la Papa Nativa Peruana – RNPNP. RESOLUCIÓN MINISTERIAL N° 0533–2008-AG. www.inia.gob.pe/wp-content/uploads/NormasSustantivas/N_18_Resolucion_Ministerial_N_533-2008-AG.pdf.

PRS (India). 2010. Official Amendments to the Seeds Bill, 2004. www.prsindia.org/uploads/media/Note%20on%20official%20Amendments%20in%20Seeds%20Bill%202004.pdf.

Richards P, de Bruin-Hoekzema M, Hughes SG, Kudadlie-Freeman C, Offei SW, Struik PC, 2009. Seed systems for African food security: Linking molecular genetic analysis and cultivator knowledge in West Africa. *International Journal of Technology Management* 45, 197–214.

Rohrbach DD, Minde IJ, Howard J, 2003. Looking beyond national boundaries: Regional harmonization of seed policies, laws and regulations. *Food Policy* 28, 317–333.

SEARICE, Utviklingsfondet, Oxfam Novib, 2013. Farmer-bred varieties: Finding their place in the seed supply system of Vietnam. The case of the HD1 variety. SEARICE, Philippines. www.sdhsprogram.org/publications/farmer-bred-varieties-finding-their-place-in-the-seed-supply-system-of-vietnam-the-case-of-hd1-variety/.

Sperling S, McGuire S, 2010. Understanding and strengthening informal seed markets. *Experimental Agriculture* 46, 119–136.

Tanzania, 2003. The Seeds Act. https://members.wto.org/crnattachments/2017/SPS/TZA/17_3213_02_e.pdf.

Uganda, 2016. Draft National Seed Policy.

Vietnam, 2004. Seed Ordinance. www.wipo.int/edocs/lexdocs/laws/en/vn/vn053en.pdf.

Visser B, 2017. The impact of national seed laws on the functioning of small-scale seed systems: A country case study. The Hague: Oxfam Novib. www.sdhsprogram.org/publications/the-impact-of-national-seed-laws-on-the-functioning-of-small-scale-seed-systems-a-country-case-study/.

Vodouhe RS, Halewood M, 2016. Commentary on the registration of traditional varieties in Benin. In Halewood, M (ed) *Farmers' Crop Varieties and Farmers' Rights: Challenges in Taxonomy and Law*. New York: Routledge.

Wattnem T, 2016. Seed laws, certification and standardization: outlawing informal seed systems in the Global South. *Journal of Peasant Studies* 43 (4), 850–867.

Zimbabwe, 2001. Seeds Act of Zimbabwe. www.wipo.int/wipolex/en/text.jsp?file_id=214681.

19 Participatory plant breeding and *sui generis* plant variety protection

Daniele Manzella and Selim Louafi[1]

Introduction

This chapter examines the role of *sui generis* plant variety protection (PVP) regimes for the products of participatory plant breeding (PPB). Within PVP, *sui generis* refers to means of protection that are alternatives to patenting.

From a literature review of PPB's scope of action, and the normative definition of *sui generis* PVP and its interpretation, we select the legal constructs relevant to the operation of PPB, to examine whether and how *sui generis* PVP can incentivize PPB in practice. The results of this analysis indicate the need to investigate other incentives, beyond the legal realm of IP, for sustaining and upscaling PPB in modern agricultural research.

Context and forms of PPB

PPB encompasses a wide range of forms of collaboration between farmers and breeders, as noted throughout this book (see also Sperling *et al.*, 2001). The historical context of PPB is situated within the decreasing role of public vis-à-vis private breeding and the consequent need to develop partnerships that can deliver suitable crop varieties to small-scale farmers (Visser and De Jonge, 2016). Findings in the literature are mixed regarding impacts of PPB. While some scholars call for further evidence of the impacts of PPB (Almekinders and Elings, 2001), others are more positive as to the role played by production-oriented PPB in poverty alleviation in marginal areas and resource-poor crop improvement programmes (Walker, 2006).

As highlighted in previous chapters, PPB has been variously defined. However, the common value proposition in all forms of PPB resides in the exploitation of the comparative advantages of two systems of plant breeding: those of farmers, and those of professional breeders. PPB involves farmers, while maintaining the science of plant breeding (Ceccarelli, 2016). PPB is held to augment the production of cultivars by farmer and the generation of famers' varieties, especially in collaborative ventures where farmers play a leading role and public breeders provide a support base (Salazar *et al.*, 2006). PPB has also been deemed conducive to the protection of crop diversity in diverse environmental conditions – more so than conventional plant breeding, which seeks uniformity (Mendum and Glenna, 2010). PBB is thus highly relevant to other global problems such as climate change and famine (Ceccarelli, 2016). Within these different value propositions, PPB involves a compromise between utilitarian and fairness-oriented views of collaboration; it implies balancing the interests of different stakeholders, and relates to the effectiveness of legal measures for safeguarding those interests (Visser and De Jonge, 2016).

It has been argued that the increased need for PPB has come in conjunction with the shift from considering plant genetic resources as the common heritage of humankind, to being subject to intellectual property (IP) rights, national sovereignty and access and benefit-sharing (ABS) regulations. IP rights and ABS laws create limitations to the accessibility of plant breeding inputs and technologies that PPB could ease by creating decentralized spaces for famer–breeder collaboration (Aoki, 2009). The teleology of PPB is production-oriented, centred on generating plant varieties for, and by, farmers. This is not necessarily limited to the selection of modern cultivars: it may apply also to the results of deliberate crosses between farmers' materials, or between breeders' and farmers' materials. We agree with Salazar and colleagues (2006) on their identification of areas where effective legal measures for PPB are needed: access to source germplasm for improvement; recognition of collective innovation; and keeping the resulting germplasm available for further improvement. We examine these three areas below.

Normative definition of *sui generis* PVP, evolution of PVP debates

The UPOV (International Union for the Protection of New Varieties of Plants) Convention established PVP as an international IP rights regime, specific to the protection of plant varieties. Originally concluded in 1961 after urgent requests for the protection of agricultural and horticultural innovations by industry associations (Blakeney, 2012), the Convention was revised in 1978 and in 1991. The UPOV Convention establishes plant breeders' rights as a means of protecting plant varieties that are 'new' as regards novelty on the market, and are morphologically and genetically distinct, uniform and stable. In the implementing legislation of UPOV member-countries, exclusivity of the rights-holder typically applies to the following acts performed on the propagating material of the plant variety: production or reproduction; conditioning for propagation; offer for sale and sale; import and export; and stocking for all of the previous purposes. The specificity of these exclusivities increases the ability of the rights-holder to exclude access without permission (Lesser, 2012). In the 1991 revision of the Convention, exceptions to exclusivity were foreseen for acts performed privately and for non-commercial purposes, for experimental purposes, or for the purpose of breeding other plant varieties, except for varieties essentially derived from the protected variety: this is *the breeder's exemption* (Art. 15(1)). UPOV member-countries may also introduce a seed-saving exception in their domestic legislation – *the farmers' privilege* – but exclusively from the harvest that farmers obtain by planting a protected variety, for propagation on their own holdings, and within reasonable limits and subject to the safeguarding of the legitimate interests of the breeder (Article 15(2)). In the 1978 version of the Convention, the farmers' privilege was broader, as the breeder's right did not extend to acts that farmers may perform in order to save, use or exchange seeds.

The use of the term *sui generis* in relation to PVP comes from the World Trade Organization (WTO) Agreement on Trade Related Aspects of Intellectual Property Rights (TRIPS), whose Article 27.3b obliges WTO member-countries, as a minimum standard of harmonization, to provide for the protection of plant varieties either by patents or by an effective *sui generis* system, or by any combination of the two systems. The TRIPS Agreement does not specify which *sui generis* system WTO members shall provide in order to meet its requirements, but most of them have implemented legislation in line with the UPOV Convention.

The literature has examined in detail the various options offered under TRIPS, including to exclude plants from patentability and establish a *sui generis* system; not exclude plants and establish co-existence with a *sui generis* regime; exclude only plant varieties from patentability (Leskien and Flitner, 1997).[2] The merits and deficiencies of the UPOV Convention have been studied in order to assess the opportunities and disadvantages for developing countries considering joining UPOV with the dual objective of incentivizing the seed industry and maintaining food security (Rangnekar, 2001). In a food-security perspective, PVP is intertwined with questions of sustainable development and biodiversity management, and cannot be addressed in isolation from those broader societal objectives. Blakeney (2012) has discussed the interrelation of UPOV with other international regimes dealing with plant genetic resources – the Convention on Biological Diversity (CBD) and the International Treaty on Plant Genetic Resources for Food and Agriculture (Plant Treaty) – with a focus on food security. Rangnekar (2001) has explored the *sui generis* flexibilities of TRIPS as regards crop coverage, scope and conditions for protection.

Sui generis PVP and measures to safeguard the interests of PPB

Drawing on the body of policy and legal analysis outlined above, we review measures to safeguard three sets of interests revolving around PPB: access to source germplasm for improvement; recognition of collective innovation; and keeping the resulting germplasm available for further improvement.

Access to source germplasm for improvement

A PPB project may access germplasm from various sources – local, national, international; and differing legal conditions may, or may not, be attached to such access, depending on the providers' management and control of the source germplasm. Such legal conditions may have been created as part of national efforts to implement the CBD/Plant Treaty and TRIPS/UPOV proprietary regimes. The complex interplay of such regimes may be a limiting factor for PPB (see Bjørnstad and Westengen, this volume). Attempts to integrate the TRIPS IP harmonization standards with the CBD requirements for access and benefit-sharing (ABS) at the global level have stalled; as a result, the overall international legal framework integrates the different models of agricultural innovation and management found in IP systems and biodiversity frameworks only partially. Despite the lack of global synchronization, *sui generis* continues to be interpreted as creating a legal space for the exercise of sovereign rights through the prior informed consent and benefit-sharing requirements of the CBD in UPOV-based domestic legislation (Blakeney, 2012). India's Protection of Plant Varieties and Farmers' Rights Act (2001) utilizes the legal space of *sui generis* PVP for the dual objective of revisiting the scope and criteria for protection of established IP in order to protect extant varieties and farmers' varieties, and of integrating biodiversity protection into IP regimes through disclosure requirements (Dang and Goel, 2009).

A pertinent access-issue is how to ensure freedom to operate for PPB, i.e. to ensure that the legal status of the material pooled into a PPB project allows it to be used for project purposes. Important here is whether and how the applicable domestic legislation regulates access, use and benefit-sharing of germplasm that the PPB programme plans to use. Sovereign or ownership rights may concern the genetic materials for PPB and also possibly traditional knowledge embodied in traditional varieties. This is a typical upstream restriction that may affect *all* phases of PPB, up to the diffusion of the material resulting from PPB among farming communities.

Materials pooled into a PPB programme may come from various sources: from public national collections and breeding programmes of CGIAR (Consultative Group on International Agricultural Research) from farmers' fields, or from other private owners (Mancini *et al.*, 2017). Materials in public collections are normally not subject to IP rights, but CBD rules may apply. Materials in CGIAR collections and breeding programmes are available either under Plant Treaty rules or under CBD rules. Germplasm may be exchanged from farmer to farmer, for instance through community seed banks (African Centre for Biodiversity, 2018). If there are IP rights on private material, the holder of those rights normally waives them for the purposes of the PPB programme. Should IP rights not be surrendered, PPB programmes can benefit from the breeder's exemption from PVP titles whereby, in the case of patent-protected material, the consent of the title-holder must be sought.

A formal agreement must be negotiated on a case-by-case basis with the provider of the resources that are subject to national systems implementing the CBD and the Nagoya Protocol. The Plant Treaty has standardized ABS conditions for several agricultural crops (the 64 crops and forages listed in its Annex 1), as expressed in the standard material transfer agreement (SMTA) which is used for all transfers under the Plant Treaty's multilateral system of access and benefit-sharing. A negotiated material transfer agreement or an SMTA may also be required for the subsequent use of PPB results.

The SMTA provides useful incentives for PPB. It reduces the transaction costs of bilateral negotiations. It prohibits IP protection on germplasm, in the form received with the SMTA, which protects the source material of PPB from appropriation. It exempts recipients from having to track germplasm at the level of individual accessions, which simplifies the identification of source material throughout the breeding programme. Significantly, the CGIAR Centers which share germplasm under the Plant Treaty apply the SMTA to breeding lines, and not only to genebank accessions (CGIAR, 2013). However, the SMTA entails several possible disadvantages – some common to any system of regulated access, and others specific to the Plant Treaty system. Like any material transfer agreement, the SMTA may over-formalize access, through a contractual instrument perhaps not necessary in trust-based collaborative endeavours. The SMTA extends this contractual approach beyond the first-access transaction: once the accessed germplasm, even when improved, is transferred to another entity (e.g. a partner institution in the PPB programme), a subsequent SMTA must be signed. SMTA benefit-sharing obligations, including the royalty-based scheme triggered by the commercialization of a resulting product under certain conditions, may well be less suitable in the typical PPB context, which is collaborative and not geared towards commercial exploitation of the results. However, the multilateral nature of the Plant Treaty system precludes any simplification or relaxation of SMTA terms and conditions to meet the needs of specific PPB projects.

Recognition of collective innovation and modalities of participation

Sui generis PVP may recognize farmers' innovations in the form of farmers' varieties. This recognition has important equity and functional dimensions. In terms of equity, it puts farmers on a par with formal breeders, who have benefited from a formal, well-enforced and tailor-made system of protection of their business interests. From a functional standpoint, prevention against misappropriation and compensation may create material incentives for farmers to continue maintaining and augmenting the genetic patrimony of traditional plant varieties. Essential here is the PPB methodology of collaboration, whereby the object of protection (the plant variety) is generated. This gives rise to the question of how IP protection relates to the collective nature of farmers' breeding and the similarly collective nature of PPB collaborations, and whether and how the role of farmers in PPB programmes should be recognized by assigning proprietary rights.

Viewed in theoretical terms, the misfit between the communitarian/collective logic of farmer innovation, and the assigning of property to an individual person or legal entity, has been debated, regardless of whether the form of property would mirror the protection offered to professional breeders. Culturally, peasant farmers are generally not characterized by possessive individualism over genetic resources: they need to adapt rapidly to environmental risks by maximizing the benefits of exchange (Brush, 1998). Moreover, PPB involves whole communities, and innovations are the result of a collective process and collective action. Property rights, even in collective form, do not satisfactorily capture the value of biodiversity, due to the public-goods nature of many of the benefits that farmers generate (Eyzaguirre and Dennis, 2007).

From a practical viewpoint, the main legal problem is how to apply PVP criteria to protect PPB-produced varieties, which are somewhere between farmers' varieties and formal sector bred varieties. PPB products will often be more distinct, uniform and stable than farmers' varieties that have not been subjected to PPB techniques (especially for outbreeding crops, less so for clonally propagated ones). However, they will often be less distinct, uniform or stable than modern PVP-protected varieties created through the breeding programmes of the formal sector (Salazar *et al.*, 2006). In general, traditional farmers do not value uniformity in their selection, so the stability of the variety over ensuing generations is also lower.

Through the adaptation or rejection of UPOV-derived PVP criteria (novelty, distinctiveness, uniformity and stability), normative recognition has been given to farmers' varieties. India's Act of 2001 utilizes the *sui generis* flexibility to exclude PVP criteria vis-à-vis farmers' varieties of which there is public knowledge. The Malaysian Protection of New Plant Varieties Act of 2004 has followed a similar route by introducing 'identifiability' of farmers' varieties. As examined in the literature, different requirements for protection are also coupled with different rights conferred (Correa, 2016). Obviously, in cases where PPB produces a variety that fulfils UPOV criteria for protection, adaptation of the criteria for protection is not relevant.

Another legal problem, less noted by academics and legislators, is how to cater for collective innovation through joint ownership or rights assignment. In principle, joint ownership of the IP title, as applicable to PVP, could support the recognition of collective innovation stemming from PPB programmes, including the role of farmers in PVP-protected innovation. Anyone – an individual or a legal entity – who is the owner, breeder, developer or discoverer of a unique cultivar of a crop is entitled to apply for PVP, in accordance with the UPOV Convention and the laws of implementing nations (Dodds *et al.*, 2007). PVP is open to joint ownership. IP joint ownership frequently results from collaborative innovation in projects where IP subject-matter has been generated jointly by partners, and exactly how the work has been shared is difficult to determine (European IPR Helpdesk,

2015). Normally, IP joint ownership is contractually agreed, but national laws will determine the default option if no agreement exists. The agreement involves allocating shares between joint owners on the basis of contributions made to the joint endeavour, and establishing the rights at the disposal of the co-owner (e.g. use or exploitation); it may also provide mutually restrictive conditions. There has been abundant legal analysis concerning various points related to joint ownership, such as the distinction between joint inventor and executor, the possibility of separating inventorship from ownership by formally assigning rights, and the duty of notification among co-inventors or co-owners (McGee, 2012).

PPB may involve various forms of participation: contractual, consultative, collaborative and collegial, and 'farmer-led' can emerge as distinct types (African Centre for Biodiversity, 2018). There may be a wide range of modes of participation by farmers and breeders, depending on the stage of the breeding process, geographical location, and the design and management of germplasm evaluation (Bellon and Morris, 2002). These multiple forms of participation may animate all, or only some, phases of a PPB programme – such as selection of material for crossing, and testing – involving various forms of intellectual or material contributions. Under each of those modalities, various actors – e.g. one or more farming communities, a national institute of agricultural research, a state university – may be involved, or not (INRA, 2015). Collaborative research agreements in agriculture normally contain a list of the materials to be provided by each party to the project (to acknowledge each party's ownership), and provisions as to the ownership of any new properties discovered during the project (Steinbock, 2012). However, this form of contractualizing the collaborative relationship assumes that contributors and materials are defined. In PPB programmes, a binary categorization (farmer/breeder) and a definition of materials entering the PPB programme may not capture the diversity in collaboration and inputs. This can make it extremely challenging to assess the impact of participation, in all the relevant forms of participation, in order to attribute joint inventorship and/or ownership of the title and ultimately of the 'final product' that is subject to PVP protection.

One approach could be to bring the farmers who have participated in PPB under the umbrella of a representative legal entity (Salazar *et al.*, 2006). This option can be pursued in UPOV-style application forms that may be filed by an agent, or by an applicant who, not being the breeder of the variety, is to declare the details of his right to make the application (Blakeney, 2012). However, given the fluidity of collaborative modalities that typically characterize PPB programmes, and aside from practical expedients that can facilitate the filing of applications or the assigning of rights, joint IP ownership appears to be more a solution for value appropriation than a legal tool for managing PPB relations among contributors towards value creation. In PPB programmes, stakeholders are highly interdependent. The discourse around PVP, similarly to other forms of IP, has addressed interdependency among innovation stakeholders as to societal sustainability through the prism of binary relations between the sole inventor and the society in question, delimiting the sphere of what can be subject to appropriation and in return for what benefit to that society. It has only marginally addressed the pluralist, social dimension of the invention in-the-making – which is the most salient feature of PPB.

Availability of germplasm for further improvement

Continuous availability of the germplasm generated by PPB programmes is seen as a value of PPB (Salazar *et al.*, 2006). An unintended negative consequence of property recognition could be to obstruct the subsequent improvement of PPB-generated material. Here UPOV-style PVP offers two important principles to consider: the breeder's exemption, and the farmers' privilege. The former is an essential component of the ad hoc system of protection for the breeding industry, which differentiates it from patent systems. Even though the exemption was not specifically intended to protect farmer's interests, famers may have some interest in applying it in PPB projects, for instance to breed new varieties that are useful to them at lower costs.

As to the farmers' privilege, it does not encompass the entirety of traditional seed practices among farmers. According to such practices, seed saving is not only for sowing and re-sowing but also for exchange and sales of seeds. Those practices have found partial international recognition in the Plant Treaty, as one element of Farmers' Rights. According to Article 9 of the Plant Treaty, nothing in its provisions should be interpreted as limiting any rights that famers have to save, use, exchange and sell seed and propagating material, as appropriate and subject to national law. Under the 1978 version of the UPOV Convention, the realization of Farmers' Rights is facilitated by the broader definition of the farmers' privilege which excludes only the sale of saved seeds (Correa, 2017).

In order to guarantee the continuous availability of material for breeding, two forms of co-existence of norms are advocated: co-existence of patents and plant breeder's rights, with the introduction of a breeder's exemption in patent law to mirror the exemption in plant breeders' rights (Prifti, 2017); and co-existence of breeders' and farmers' rights, as in India's Protection of Plant Varieties and Farmers' Rights Act (Tuhairwe, 2017). However, these debates are largely exogenous to PPB dynamics. The debate over the exemption for breeders revolves around achieving fairness and stimulating innovation, with a focus on the formal breeding sector, in the context of the current division between small- and large-scale breeding companies – *not* as an incentive to farmer–breeder collaboration (Visser and De Jonge, 2016). From the perspective of traditional farmers, the farmers' privilege in the 1991 version of UPOV does not fully cater to the practices of seed exchange among farmers, which may be a prerequisite for some PPB programmes but could be compatible with UPOV legislation only in countries that adhere to the 1978 Convention. Farmers' rights have the potential to fill this gap; however, specification is needed in domestic legislation, not necessarily as regards PVP.

The multilateral system of the Plant Treaty can provide another solution to ensure that the PPB-generated material remains available for further research and breeding. Such material can be made available with the SMTA by its developers, on a voluntary basis. Examples of this practice include PPB projects funded by the Benefit-sharing Fund of the Plant Treaty, and from the global project on crop wild relatives led by the Global Crop Diversity Trust (see Toledo, this volume).

Attempts have been made to adapt the open-source model to seeds, so as to keep access to germplasm open. The US-based Open Source Seed Initiative (OSSI) has devised and implemented a protected commons mechanism for seeds through a chain of bilateral commitments, in the form of a pledge not to restrict use of the material or derivatives thereof (Luby *et al.*, 2016). An open-source model for plant genetic resources is based on the dual role of farmers, as both users and developers, borrowing from software terminology (Aoki, 2009). When OSSI was established, it was believed that the 'viral' effect would suit the dynamics of PPB, where material is continuously generated in a collaborative fashion and remains freely accessible (ibid.). The underlying model of free and continuous access to germplasm that has animated OSSI has inspired PPB projects where farmers, working in collaboration with plant scientists, are empowered to improve seeds and develop new varieties adapted to various agricultural conditions (African Centre for Biodiversity, 2018; Kotschi and Horneburg, 2018; Moeller and Pedersen, 2018).

A critical review of *sui generis* PVP in light of PPB instances

> Although the design of *sui generis* regimes for the protection of plant varieties that do not apply the UPOV model has been on the agenda of many developing countries, nongovernmental organizations and academics for at least 20 years, little progress has been made in finding solutions to the complex conceptual and technical problems that are involved. Despite the experiences in a few developing countries, there is little evidence about what such regimes have achieved. Indeed, reliable models that can be followed do not seem to exist yet, and considerable work is still necessary to design a national regime that effectively addresses the needs of farming communities in a particular national context.
>
> (Correa, 2016: 155)

This chapter has reviewed some legal arguments arising from *sui generis* PVP that can apply to PPB. However, the discourse has remained largely hypothetical, even though PPB is certainly not a newcomer in agricultural research and a considerable body of policy literature on the topic exists. There is no evidence of *sui generis* PVP having being applied systematically to sustain PPB – and, as we argue below, the potential for doing so may be limited. More relevant here may be the legal analysis of other, non-proprietary, normative categories applicable to PPB processes and products (e.g. those belonging to seed legislation – see De Jonge *et al.*, this volume). A conclusive body of evidence justifying the promotion of *sui generis* IP protection for PPB is yet to emerge.

Our legal analysis of existing PPB tensions between formalization and flexibility in access, collective innovation and individual appropriation for farmers' varieties, exclusivity and incremental innovation in relation to continuous access to germplasm, links to the conclusion reached by Correa (see p. 302) made in relation to *sui generis* IP for farmers' varieties, and leads to an additional consideration. The lack of impact of *sui generis* IP regimes that is due to the difficulties involved in deviating from standard IP rules to tackle technical, administrative and political challenges, may be one expression of a more fundamental discord between IP in general on plant variety (even under the broadest configuration of *sui generis*) and PPB.

The type of IP protection examined here is not fully suited to collective, social systems of innovation in agriculture. IP is designed to reward the final innovator in return for a benefit to society in general, and not so much those who have collaborated on the innovation. Solutions for incentivizing PPB may need to be sought outside the IP system, as PPB is more about increasing collaboration across a wide range of actors, than providing individual, property-based incentives in return for the development of a unique final product. Indeed, PPB may well be seen as a socially binding system of collaborative innovation that can operate on the basis of other mechanisms, such as reputation or reciprocity, and not on the legal protection of an end product. In this sense, PPB would benefit from being considered not only as an alternative breeding methodology to generate innovation in the form of new varieties, but also as an organizational innovation that can generate societal outcomes in the form of new collaborations or increased capacity among farmers.

Conclusions

PVP offers protection to a defined output, a biological object, that is generated through breeding. However, PPB cannot be assessed solely in terms of new varieties produced and how to protect them. PPB brings other benefits than just the final 'product'. Notably, farmers' empowerment and building organizational capacity are recognized as social benefits of PPB (African Centre for Biodiversity, 2018; Almekinders and Elings, 2001). The configuration of PPB initiatives within the logic of open source seed is promising, as it focuses on community management for the creation and sharing of intellectual and cultural resources that are not necessarily incorporated into biological material.

Seeds cannot be viewed solely as biological material: they must be appraised in their cultural, societal and political contexts and meanings.

Experiences with *sui generis* PVP demonstrate how fitting biological or utilitarian criteria to a *multi-dimensional* object is not only complex but, in some respects, misleading. In order to incentivize PPB, it may be necessary to identify and deal with the socio-economic factors that obstruct the upscaling of PPB (Mendum and Glenna, 2010).

Despite the property-focused narrative that animates PVP, *sui generis* PVP gives rise to a wide range of policy questions related to the institutionalization of collaboration, involving cultural, social and political dimensions which are not captured by biological or utilitarian criteria for recognition of innovation. Seen through the prism of *sui generis* PVP, the dilemmas of fitting IP criteria to PPB (what resources does PPB use, who is the innovator, what is the resulting object of PPB?) challenge the whole IP view of resources as being produced through individual entitlements and utilized through market mechanisms.

In PPB, it is difficult to determine the upstream and downstream creators on which the conventional approach to IP is based (Frischmann *et al.*, 2014). Nor is there a clear dividing line between private rights and the public domain – another foundational concept of functionalist IP theory. These binary framings seem less applicable to PPB situations, where farmers and breeders often collaborate in seeking solutions to their own problems, innovating on the expectation that the returns will be for their own uses, without worrying about possible free-riding. Indeed, property rights may not be an essential driver of PPB innovation. PPB should be explored as a form of user innovation in agriculture that needs to be governed through communal institutional arrangements for the co-production of knowledge in the relevant cultural contexts.

Notes

1 The opinions expressed in this chapter are those of the authors. They do not purport to reflect the opinions or views of the institutions with which the authors are affiliated.
2 This is the solution that the European Patent Convention has embraced. However, the exclusion is interpreted restrictively, and patents with claims covering one or more plant varieties have been granted, provided that the invention is not technically limited to an individual plant variety, and essential biological processes are not the exclusive method of reproducing the invention.

References

African Centre for Biodiversity, 2018. *A Review of Participatory Plant Breeding and Lessons for African Seed and Food Sovereignty Movements.* Johannesburg: ACB.

Almekinders CJM, Elings A, 2001. Collaboration of farmers and breeders: Participatory crop improvement in perspective, *Euphytica* 122, 425–438.

Aoki K, 2009. Free seeds, not free beer: Participatory Plant Breeding, open source seeds, and acknowledging user innovation in agriculture, *Fordham Law Review* 77 (5), 2275–2310.

Bellon MR, Morris ML, 2002. Linking global and local approaches to agricultural technology development: The role of participatory plant breeding research in the CGIAR, *CIMMYT Economics Working Paper 02–03*. Mexico, D.F.

Blakeney M, 2012. Plant variety protection, international agricultural research, and exchange of germplasm: legal aspects of sui generis and patent regimes. In Krattiger A, with Mahoney RT, Nelsen L, Thomson JA, Bennett AB, Satyanarayana K, Graff GD, Fernandez C, Kowalski SP, (eds) *Intellectual Property Management in Health and Agricultural Innovation: A Handbook of Best Practices.* Oxford: MIHR, Oxford/ Davis, CA: PIPRA.

Brush SB, 1998. Bio-cooperation and the benefits of crop genetic resources: The case of Mexican maize. *World Development* 26 (5), 755–766.

Ceccarelli S, 2016. Participatory barley breeding in Syria: Policy bottlenecks and responses. In Halewood M, (ed) *Farmers' Crop Varieties and Farmers' Rights: Challenges in Taxonomy and Law.* London: Earthscan.

CGIAR, 2013. *Implementation Guidelines for the CGIAR IA Principles for the Management of Intellectual Assets.* Montpellier: CGIAR.

Correa C, 2016. Sui generis protection for farmers' varieties. In Halewood M, (ed) *Farmers' Crop Varieties and Farmers' Rights: Challenges in Taxonomy and Law.* London: Earthscan.

Correa C, 2017. *Implementing Farmers' Rights Relating to Seeds.* Research paper, March 2017, South Centre, Geneva.

Dang R, Goel C, 2009. Sui generis plant variety protection: The Indian perspective, *American Journal of Economics and Business Administration* 1 (4), 303–312.

Dodds J, Krattiger A, Kowalski SP, 2012. Plants, germplasm, genebanks and intellectual property. In Krattiger A, Krattiger A, with Mahoney RT, Nelsen L, Thomson JA, Bennett AB, Satyanarayana K, Graff GD, Fernandez C, Kowalski SP, (eds) *Intellectual Property Management in Health and Agricultural Innovation: A Handbook of Best Practices.* Oxford: MIHR/Davis, CA: PIPRA.

European IPR Helpdesk, 2015. *Fact Sheet: IP Joint Ownership*, www.iprhelpdesk. Accessed 29 April 2018.

Eyzaguire P, Dennis E, 2007. The impacts of collective action and property rights on plant genetic resources, *World Development* 35 (9), 1489–1498.

Frischmann BM, Madison MJ, Strandburg KJ, 2014. Governing knowledge commons. In Frischmann BM, Madison MJ, Strandburg KJ, (eds) *Governing Knowledge Commons.* New York: Oxford University Press.

INRA, 2015. *Need for IP Rights in Public Pre-Breeding Research*, International Workshop 3 January 2015, Montpellier.

Kotschi J, Horneburg B, 2018. The open source seed licence: A novel approach to safeguarding access to plant germplasm, *PLoS Biology* 16 (10), e3000023.

Leskien D, Flitner M, 1997. Intellectual property rights and plant genetic resources: Options for a sui Generis system, *Issues in Genetic Resources* no. 6, Rome: IPGRI.

Lesser W, 2012. Plant breeders' rights: An introduction. In Krattiger A, Krattiger A, with Mahoney RT, Nelsen L, Thomson JA, Bennett AB, Satyanarayana K, Graff GD, Fernandez C, Kowalski SP, (eds) *Intellectual Property Management in Health and Agricultural Innovation: A Handbook of Best Practices.* Oxford: MIHR/Davis, CA: PIPRA.

Luby C, Kloppenburg J, Goldman L, Ortiz R, 2016. Open source plant breeding and the Open Source Seed Initiative, *Plant Breeding Reviews*, 271–298.

Mancini C, Kidane YG, Mengistu DK, Melfa and Workaye Farmer Community, Pè ME, Fadda C, Dell'Acqua M, 2017. Joining smallholder farmers' traditional knowledge with metric traits to select better varieties of Ethiopian wheat, *Scientific Reports* 7, 9120.

McGee D, 2012. Invention disclosures and the role of inventors, In Krattiger A, Krattiger A, with Mahoney RT, Nelsen L, Thomson JA, Bennett AB, Satyanarayana K, Graff GD, Fernandez C, Kowalski SP, (eds) *Intellectual Property Management in Health and Agricultural Innovation: A Handbook of Best Practices*. Oxford: MIHR/Davis, CA: PIPRA.

Mendum R, Glenna LL, 2010. Socio-economic obstacles to establishing a Participatory Plant Breeding program for organic growers in the United States, *Sustainability* 2, 73–91.

Moeller NI, Pedersen JM, 2018. *Open Source Seed Networking: Towards a Global Community of Seed Commons. A Progress Report.* The Hague: HIVOS.

Prifti V, 2017. The breeder's exception to patent rights as a new type of research exception, *Rights and Science* 2017, 109–116.

Rangnekar D, 2001. *Access to Genetic Resources, Gene-based Inventions and Agriculture*, Commission on Intellectual Property Rights, Study Paper 3a, London.

Salazar R, Louwaars NP, Visser B, 2006. *On Protecting Farmers' New Varieties: New Approaches to Rights on Collective Innovation in Plant Genetic Resources*, CAPRI Working Paper 45, Washington, DC: IFPRI.

Sperling L, Ashby JA, Smith ME, Weltzien E, McGuire S, 2001. A framework for analyzing participatory plant breeding approaches and results, *Euphytica* 122, 439–450.

Steinbock, MB, 2012. How to draft a collaborative research agreement. In Krattiger A, Krattiger A, with Mahoney RT, Nelsen L, Thomson JA, Bennett AB, Satyanarayana K, Graff GD, Fernandez C, Kowalski SP, (eds) *Intellectual Property Management in Health and Agricultural Innovation: A Handbook of Best Practices*. Oxford: MIHR/Davis, CA: PIPRA.

Tuhairwe H, 2017. Farmers' rights and plant variety protection in Uganda: Considerations and opportunities, *Journal of Intellectual Property Law and Practice* 12 (12), 1004–1011.

Visser B, De Jonge B, 2016. Intellectual property rights in plant breeding and the impact on agricultural innovation, CTA, http://knowledge.cta/int/. Accessed 29 April 2018.

Walker TS, (2006). *Participatory Varietal Selection, Participatory Plant Breeding, and Varietal Change*, background paper for the World Development Report 2008, World Bank.

20 The straitjacket of plant breeding

Can it be eased?

Åsmund Bjørnstad and Ola Tveitereid Westengen

Introduction

Access to genetic resources is crucial in any programme aimed at crop improvement, not least the participatory programmes often undertaken outside formal breeding institutions that have access to a wide range of germplasm. Exchange and utilization of germplasm has been the foundation for all crop-related innovation since crops were first domesticated. Breeders and other actors involved in crop improvement lament that access to genetic resources from other organizations, especially those in other countries, is now restricted or completely blocked.

By definition, national access and benefit-sharing (ABS) regulations in line with international biodiversity laws and intellectual property rights (IPRs) under national and international law impose restrictions on the use of genetic resources. But do these restrictions limit the possibilities of breeding programmes to develop useful varieties to such an extent as to outweigh the positive effects of the rewards they are meant to create? Scientists affiliated with breeding programmes for climate adaptation of African agriculture have indicated that there is indeed reason for concern:

> If such programmes are to deliver rapid climate change adaptation, it is critical that they have access to elite germplasm from other regions already experiencing the "future climate". Such access has become increasingly restricted, due to the enforcement of strong forms of intellectual property protection on plant varieties in some jurisdictions and to restrictions imposed by some on the exchange or export of germplasm considered to be proprietary or part of a national genetic patrimony.
>
> (Atlin *et al.*, 2017)

These restrictions comprise what Ghijsen has called the 'straitjacket' on plant breeding, which limits plant breeders' freedom to operate (Ghijsen, 2009). As Visser *et al.* (this volume) and Manzella and Louafi (this volume) show, this straitjacket applies to the impact of ABS and IPR policies and contractual restrictions also on participatory breeding programmes.

Khoury *et al.* (2016) found that the average nation depends on plant genetic resources (PGR) originating abroad (outside that country) for two-thirds of its food supply. Further improvement and adaption of these crops will be difficult without continued access to genetic resources from genepools existing in other geographical areas and jurisdictions. To promote food security, exchange of PGR used in breeding *and* effective dissemination of new adapted varieties are needed (Burke *et al.*, 2009; Challinor *et al.*, 2016).

The relevance of IPR legislation and ABS regimes for access to diversity for adaptation to climate change is obvious, as both regulate access. Too often, however, the potentially negative impact of only a single regulation is considered at a time. Breeders accepting IPRs complain about unreasonable ABS regulations, whereas critics of IPRs see ABS regulations as providing legitimate protection against 'biopiracy'. The historical dynamic between IPRs protecting ownership of new varieties and ABS legislation protecting countries' sovereign right to genetic resources has been referred to as the 'seed wars' (see Berg and Westengen, this volume). In the words of PGR legal experts Halewood, Lopez Noriega and Louafi, since the 1980s there has been a 'race-to-the-bottom scenario' in terms of establishing forms of private control over PGR (Halewood *et al.*, 2012).

In this chapter, we first present an overview of the major legislation affecting access to breeding material. Second, we explore the empirical evidence for the claim that the two types of legislation limit breeders' freedom to operate. We then present a case study of the global wheat-rust epidemic that illustrates the importance of access to genetic resources in the face of global crop production challenges. In conclusion, we discuss some promising new pathways for improving exchange and benefit-sharing.

Legal frameworks affecting access

The Convention on Biological Diversity (CBD) has now been in effect for more than 25 years.[1] The CBD has a threefold objective: the conservation of biodiversity, the sustainable use of its components, and the fair and equitable sharing of the benefits arising from genetic resources. Operationalizing the third objective has proved difficult, and it was not until 2014 that the Nagoya Protocol on ABS entered into force. With 198 parties[2] (the USA is a notable exception), the CBD is among the most widely ratified of all international conventions. By 2019, the Nagoya Protocol had been ratified by 116 countries. Under the CBD and Nagoya, access is a matter of bilateral agreements between 'provider' states and 'users', private or public.

As of 2019, the International Treaty on Plant Genetic Resources for Food and Agriculture (the Plant Treaty) has been in force for 15 years. It links 'facilitated access' (the 'A' of ABS) to a list of 64 of some of the most important food and forage crops through a multilateral system of access and benefit-sharing 'in harmony with the Convention on Biological Diversity, for sustainable agriculture and food security' (Article 1). The formal exchange is effected through signing a standard material transfer agreement (SMTA). The 'BS' (benefit-sharing) part of ABS is provided by guaranteeing other users' access on these same terms, through various forms of non-monetary benefit-sharing, such as exchange of information, access to and transfer of technology, and capacity-building. There is also monetary benefit-sharing in the form of mandatory financial contributions based on sales of seeds of new PGRFA products that are derived from materials accessed through the multilateral system. These payments are made to the Plant Treaty's international Benefit-sharing Fund (BSF), which supports projects and programmes for the benefit of farmers and local communities in developing countries and countries with economies in transition who work towards maintaining and increasing the use of genetic resources for food and agriculture.[3] Payments are mandatory only when the new PGRFA products (e.g. plant varieties) are not available for others for research or breeding (as when they are subject to patent protection regulations).

Intellectual property protections for plant varieties are promoted through the Agreement on Trade-Related Aspects of Intellectual Property Rights (TRIPs) under the World Trade Organization (WTO). TRIPs entered into force in 1995 and is binding for all 162 WTO member-states. Member-states are obliged to 'provide for the protection of plant varieties either by patents or by an effective *sui generis* system or by any combination thereof' (Article 27.3(b)). Access to patented material requires a bilateral agreement between a lessor (owner) and a lessee (the party obtaining access). Ideally, patented inventions should be available for 'pure' research; and some EU countries allow free and unconditional use of patented material in breeding, up to the release of varieties (the 'Breeding Exception'). In contrast, in the USA, utility patents are issued on genes as well as plant varieties.

The term *sui generis* means 'of its own kind, unique' and in most countries such regulation refers to plant breeder's rights (PBR) or plant variety protection (PVP) (see Manzella and Louafi, this volume). The most common model for PVP is that of the Union for the Protection of New Varieties of Plants (UPOV). The first UPOV Convention was established in 1961 as an attempt to harmonize PVP, an experiment that has now provided 48 years of data. As a 'soft' intellectual property system for plant varieties, it protects plant varieties that are distinct, uniform and stable (DUS), restricting the right to market them to the owner/breeder, but allowing others access to them for further breeding. This 'breeder's exemption' is an explicit feature of the system to promote breeding progress (Gouache, 2015). Farmers may save and use seed of protected varieties for their own use (the 'farmer's privilege'). The UPOV 1991 Convention reduced the farmers' privilege to a national option, and only for crops where the harvested product is to be used as propagation material. Countries joining UPOV after 1991 are required to adopt the stringent UPOV91 convention, while countries that joined before 1991 are allowed to follow earlier versions. China and Brazil are parties to UPOV78, and therefore have the flexibility to develop national laws which allow their farmers to propagate GM varieties of soybeans and cotton without first seeking permission of the rights holders. Patents and PVP are commonly valid for 20 to 30 years before the variety becomes freely accessible for others to use in their breeding programmes.

Commercial varieties are also increasingly protected by contract law, in addition to PVPs or patents. A 'bag tag' on the seed bag states that opening the bag constitutes agreement to the terms of the license which includes not saving or replanting seed. Thus the farmer does not buy or own the seed, but merely licenses its use (Kloppenburg, 2014). Such 'technology agreements' effectively add a layer of protection to patents and PVPs, specifying that the seeds can be used 'for planting a commercial crop only in a single season'; they do not allow the buyer to 'save any crop produced from this seed for replanting, or supply saved seeds to anyone for replanting' (Janis and Kesan, 2002: 1163). In the USA, new varieties of major crops like maize and soybeans are usually under both patent and contract law; corporate IPR holders regularly check for breaches such as illegal seed saving (Howard, 2015). Such arrangements negate the UPOV farmer's privilege and breeder's exemption.

As experiments in ABS and IPR, these legal instruments entered into force 15 years (the Plant Treaty), 24 years (TRIPS), 25 years (CBD) and 48 years (UPOV) ago. The accumulated years of experience since can be expected to yield data on how they promote or prevent the sustainable use of plant genetic resources for food and agriculture (PGRFA).

Effects of ABS and IPR legislation on access

ABS under CBD and the Plant Treaty

In June 2018, an open letter published in *Science* signed by 177 biologists from 35 countries held that restricted access due to CBD and its Nagoya Protocol is thwarting basic biodiversity research. According to the signatories, this hampers not only field studies, but also access to digital sequence information (Prathapan *et al.*, 2018). Interestingly, the average country signs only 2.05 commercial ABS agreements per year, 'suggesting a lack of demand for genetic resources by users, as well as restrictive procedures for access, as factors for poor performance' (Prathapan *et al.*, 2018: 1405). This may apply to access for domestic scientists as well. The letter contrasts this with the Plant Treaty, noting its multilateral system as a model for escaping from the impasse caused by the Nagoya Protocol.

The Nagoya Protocol applies to Annex 1 crops also when the country in question is not a party to the Plant Treaty. A case in point concerns the nitrogen-fixing maize recently discovered in Mexico, where a US company signed an ABS contract in line with the Nagoya Protocol (Van Deynze *et al.*, 2018). There are many anecdotal accounts of how the CBD and the Nagoya Protocol have hampered access to genetic resources, but limited direct evidence is available. The reason could be that when research has been denied permission to collect genetic resources *in situ* or has failed to obtain an export permit, this is hardly likely to yield publishable results. Or it could be that researchers find alternative ways to obtain material, for example by requesting genetic resources from international genebanks operating under Plant Treaty or other open access regimes. However, one may question the ability of these genebanks to acquire new diversity through collecting under the current regime. Fowler and Hodgkin (2004) report that the CGIAR saw a 94 per cent drop in numbers of accessions acquired or collected annually between 1985 and 1999 – the period when the CBD entered into force.

A key purpose of the Plant Treaty was to regulate but maintain the open access policy practised by the CGIAR centres over the past 50 years. It is widely recognized that this was (literally) seminal in the Green Revolution, and has been so since. As the name implies, the Plant Treaty concerns plant genetic resources for food and agriculture (PGRFA). The term PGRFA has been defined as 'seeds, plants, and plant parts useful in crop breeding, research, or conservation for their genetic attributes' (Fowler and Hodgkin, 2004). This definition includes crop wild relatives (CWR), traditional varieties and landraces, as well as breeding lines and improved varieties.

A commonly held view underpinning the demand for ABS is that there is a global asymmetry in patterns of exchange between the less developed countries of the South and the more advanced industrial nations of the North (Kloppenburg, 2005). This view is reflected in the writings of legal scholars concerned with operationalizing ABS under CBD and Nagoya: 'The idea behind ABS is that the world's most biodiverse regions, usually located in developing countries and dubbed "providers", shall partake in the benefits created by "users", located more traditionally in richer economies' (Tvedt, 2017: 21). However, empirical evidence on PGRFA reveals that developing and developed countries alike are *net* recipients of genebank accessions. Galluzi *et al.* (2016) analysed distribution data from seven CGIAR genebanks from the period 1985–2009. They compared the number of accessions 'provided' according to where the accessions were collected originally, with the number of accessions received by the same countries. For example, while the seven CG genebanks distributed 48,635 accessions from the top provider India, over the same period India itself received 115,849 accessions from the same genebanks (Galluzzi *et al.*, 2016). The study also reported that the majority of global exchanges were South–South, which reinforced the message about interdependency between all countries as regards access to genetic resources. Further, several other assessments of distribution data have shown that most recipient institutions of genebank accessions are public (national agricultural research organizations and universities), whereas the private sector is a small recipient group (Fowler *et al.*, 2001; Galluzzi *et al.*, 2016; Westengen *et al.*, 2018).

Galluzi *et al.* (2016) reported an annual average of 40,000 samples distributed from the seven CG genebanks assessed in the period 1985–2009.[4] In line with this, Westengen *et al.* (2018) reported that the 10 CG genebanks assessed in that study distributed 40,000 samples in 2015. In comparison, the US National Plant Germplasm System in 2012 distributed more than 300,000 samples (Heisey and Day-Rubenstein, 2015), the German Institute of Plant Genetics (IPK) 4,400, the Centre for Genetic Resources in the Netherlands (CGN) 2,500 and the Vavilov Institute in Russia (VIR) 6,400 samples (Galluzzi *et al.*, 2016).

The relatively stable CG distribution figures suggest that the introduction of the SMTA of the Plant Treaty has not had a significant effect on access to these international collections, because the Centers were providing facilitated access to them under the 1994 CGIAR-In Trust Agreements and long before in the international nurseries. It is rather the dearth of distribution figures from various other genebanks that gives reason for concern. The FAO has registered 1,750 genebanks around the world (FAO, 2010), but many of these do not distribute – even organizations located in countries that are contracting parties to the Plant Treaty (Bjørnstad *et al.*, 2013; FAO, 2014; Fowler and Hodgkin, 2004). Bjørnstad *et al.* (2013) tested the concept of 'Facilitated Access' by sending seed requests to 121 genebanks in Contracting Party countries. This resulted in only 44 seed shipments; while 54 countries never responded, and the process halted for various reasons in 23 before access was provided. The results of this experiment indicate that the Plant Treaty, with its principle of facilitated access, still has some way to go before it can be said to be a fully functional regime for ABS. The reasons behind the low seed return can be many, but there were some cases of apparent non-willingness to comply with the multilateral system. India, Ethiopia and Turkey subjected the SMTA access to their national biodiversity laws (Bjørnstad *et al.*, unpublished). Turkey provided a modified 'Material Transfer Agreement' obliging the Recipient to relate to the AARI National Seed Bank under Turkish law, without any reference to the Plant Treaty, the governing body or the multilateral system.[5] Ethiopia required a research proposal, an official letter from the university and a letter from the CBD Focal Point in the country from which the request was made.[6]

The fact that not all countries that are parties to the Plant Treaty comply with its rules can hardly be held against the Plant Treaty itself. Rather, the key role of the genebanks that do play by the rules of the multilateral system indicates that this indeed is a functional regime for access.

The effect of IPRs on access

A survey of public plant breeders in the USA published in 2017 found that almost two-thirds of the breeders thought IPRs on material they received restricted their freedom to operate, either 'strongly' or 'somewhat' (Shelton and Tracey, 2017). The study quotes a respondent saying

> the change in the last 20 years has been dramatic. In the 1990s, I could send a postcard to a research director or president of a major company

and receive seed (with no or minimal restrictions) by return mail. Today such seed is simply unavailable.

(Shelton and Tracy, 2017: 1833)

The authors also note that there are particularly strict restrictions on access to genetic resources in those crops where genetically engineered varieties are on the market – with soybean as an example of a crop with very limited access to genetic resources, even more so since it is not on the Annex 1 list and thus not in the multilateral system of the Plant Treaty. The same survey showed that IPR restrictions on material received from public breeding programmes were almost as strict as those on private ones. The explanation cited was the need for public programmes to generate revenues following substantial funding cuts for public breeding since the 1980 Bayh–Dole Act that permits and encourages public IPRs on innovations (Coffman *et al.*, 2003).

A few studies have tried to quantify the impact of IPRs on utilization. Mikel and Dudley (2006) traced the origin of 908 inbred maize lines and found that much of the protected lines originated from 7 progenitor inbred lines. Since elite inbreds are the most commonly used genetic resource in US maize breeding, IPRs here are of greater concern than restrictions due to bio-diversity legislation. The use of some popular parents such as Pioneer's com-mercial hybrids dropped significantly due to more stringent patent restrictions (Mikel and Dudley, 2006). The study thus shows that genetic concentration often reflects market concentration. Four companies (Dekalb Genetics, Hold-en's Foundation Seed, Pioneer Hi–Bred and Syngenta) were responsible for 75 per cent of the IPRs on the lines investigated. The authors' conclusion resonates with the quote by the breeder above: 'over the last 20 years [germ-plasm] has become increasingly constrained by the restrictions imposed through protection of corn inbreds' (Mikel and Dudley, 2006: 1194). In addi-tion, public breeders in most countries have experienced a decline in funding (Knight, 2003). Recent reports characterize the situation in the USA as 'a state of crisis' and say that the trends of decreasing funding and increasing constraints on access are interlinked (Shelton and Tracy, 2017).

The scientific report for the UK Foresight Project on the Future of Food also expressed concerns over the effect of the dominant IPR regimes on plant breeding and global food security:

The trend over recent decades is of a general decline in investment in technological innovation in food production (with some notable excep-tions, such as in China and Brazil) and a switch from public to private sources [ref.]. Fair returns on investment are essential for the proper func-tioning of the private sector, but the extension of the protection of intel-lectual property rights to biotechnology has led to a growing public perception in some countries that biotech research purely benefits com-mercial interests and offers no long-term public good. Just as seriously, it also led to a virtual monopoly of GM traits in some parts of the world,

by a restricted number of companies which limits innovation and investment in the technology. Finding ways to incentivise wide access and
sustainability, whilst encouraging a competitive and innovative private
sector to make best use of developing technology, is a major governance
challenge.

(Godfray *et al.*, 2010)

Similarly, in a rare glimpse behind the scenes of industry and the philan-
thropic politics underlying the Green Revolution for Africa, the Rockefeller
Foundation's Gary Toenniessen explicitly criticized the effect of the IPR
position of the industry: 'New crop genetic improvement technologies are no
longer flowing to the International Agricultural Research System due to IPR
constraints' (Schurman, 2017: 446).

The case of the global wheat–rust challenge

Unlike the crops themselves, crop diseases – especially if air-borne – know no
national boundaries. Phytosanitary regulations are preventive measures for
containing them, set up and required by most countries. However, in an epidemic situation, more measures are needed: access to knowledge, to pesticides and to resistant varieties or more generally germplasm.

Take the case of the *wheat-rust epidemics* that currently threaten wheat production worldwide (Singh *et al.*, 2016). Starting with the *Ug99* race of stem
rust in 1998, yellow rust has become increasingly severe, and smallholder
farmers are the most vulnerable (Singh *et al.*, 2015). Since spores need no
passports and can spread between continents in a matter of a year or two,
containing them requires concerted access and action from farmers, breeders
(including participatory plant breeding) and seed systems.

The Borlaug Global Rust Initiative (BGRI)[7] has been operating since
2008. Its director, Ronnie Coffman, has reviewed the impressive record of
the first decade of this effort (Coffman *et al.*, 2016). It has involved 60+ scientists from 30+ countries, conducting pathogen monitoring, germplasm
testing, or dissemination of varieties and breeding lines and of fungicides. By
2016, 400,000 wheat lines from 37 countries had been tested for resistance to
Kenyan or Ethiopian rusts. Year by year, more resistant lines are found, and
genes are passed on to breeders, with markers that aid their selection. Beside
race-specific genes, there is emphasis on the more durable slow-rusting genes,
like *Sr2Lr34, Lr46* and *Lr67*. That farmers get access to new varieties with
rust resistance can be seen as a very tangible form of benefit-sharing resulting
from breeding programs access to wheat genetic resources.

A case in point is the story of the Ethiopian landrace *Digalu* that was resistant to Ug99 and yellow rust and was recommended from 2010. By 2014, it
was being grown on 0.5 million ha, mostly among small farmers. The new
'Digalu race' of stem rust blew in from Yemen, but was intercepted in a
pathogen survey. The CIMMYT variety 'Kingbird', with resistance based on

the gene *Tmp* (*Triticum timopheevii*) from the Caucasus, was identified and rapidly supplied to farmers – and, fortunately there was no epidemic in 2015–2016. This race is still present in nine countries in the Middle East and the Caucasus. The Digalu case shows how *benefit-sharing* for Ethiopian small farmers is directly linked to smooth *access* unhampered by legal obstacles beyond the SMTA. One thing is certain: races of rusts may have their 'country of origin', but they need neither visas nor SMTA.[8]

Against this background, the experience of BGRI after two years of the Nagoya Protocol is worth noting: '[S]tringent regulations and country-specific control are stifling the germplasm exchanges. (…) It is not only the improved seeds that are subject to regulation, but isolates of country-specific disease organisms such as Ug99 (…) are subject to the Convention on Biological Diversity because they are "bio-resources"' (Coffman *et al.*, 2016: 419).

New map and new pathways to access

Prior to formal ownership, germplasm was generally accessed informally as the 'common heritage of mankind' (Halewood *et al.*, 2012). This was formally abolished through the CBD – but, in retrospect, we can note that the first legal deviations from this concept were the US patent law on clonally propagated crops in 1930 and the UPOV PVP system in 1961. The concepts of ABS – popularized as defeating 'biopiracy' – may be seen as countermeasures. To complain about the CBD and the Nagoya Protocol, while forgetting the problems created by IPRs, is not convincing – neither are complaints about 'biopiracy' without recognizing and protecting the key role of access for all types of plant breeding programs. The Plant Treaty may be considered a way to reconstitute the advantages of common heritage while respecting the 'country of origin' concept of the CBD. However, the polarized situation from the years of the Plant Treaty negotiations continues to haunt the debate; and the view that the multilateral system represents a 'fragmentation' of ABS is still present in the legal literature. For example, Tvedt has argued that in order to be effective ABS agreements must be 'negotiated and enforced as commercial contracts' (Tvedt, 2017: 21). But would not a contract law approach to ABS arrangements be an equivalent to the US contract law approach to IPR and lead to further enclosure of the genepool for breeders and farmers alike? In our view, rather than elaborating new ways to enforce the enclosures in the seed commons, more research is needed to test and evaluate pathways that can ensure breeders and farmers (and any constellation of collaboration between the two) access to as much relevant genetic resources as possible. Concerning these pathways, the Vavilov maps from the early twentieth century are less relevant than the maps generated by Khoury *et al.* (2016), which illustrate the global interdependency in crop diversity. In the latter maps, genebanks and breeding programs, both private and public, are the 'centres of diversity' of most immediate usefulness.

Pathways on private land

The private plant-breeding sector has taken some interesting initiatives to improve access to IPR protected material lately. Patent pools, like the one existing for vegetable crops,[9] and e-licensing, as practised by Syngenta (Kock, 2015) seek to promote access on fair, reasonable and non-discriminatory (FRAND) terms. These systems allow effective access *and* guarantee reasonable returns on investments. However, membership is voluntary,[10] as are the patents put in the pool. So far, this appears to be largely a European system. Five European countries also have a 'research exception' in their patent laws that allow breeders to use patented materials for breeding purposes without seeking the owners' permission. However, for marketing a variety, users of those materials need a license from the owner. In case the owner refuses to grant a license, Swiss law provides that a compulsory license can be issued, if the new variety has proven its economic (agronomic) merit. Prifti (2015) has thoroughly discussed the legal implications and applicability of this system.

As outlined by Bjørnstad (2016), the goal should be to ensure remuneration without hindering access: *toll roads – not road blocks* (Bjørnstad, 2016). Extensions of the European research exception may serve as a model here. But would it provide a sufficient incentive for an owner to license on FRAND terms? A major incentive would be the access to other people's IPR on the same terms. That would also counteract obviously unfair license terms. In case of denied license, the Swiss compulsory licensing procedure is available. In practical application, the IPRs used should be declared when registering new varieties. Kock (2015) has similarly phrased the Syngenta e-licensing approach as *Free access, but not access for free*. The experience from these cases has been that informed and mutual interest quickly arrives at FRAND terms.

The UPOV approach to IPR is in principle more flexible than the patent approach. In theory, access to protected varieties should be no problem for breeders, given the Breeders' Exemption in UPOV. And at least for some crops, such as wheat, commercial practice indicates that this is indeed the case (Gouache *et al.*, 2015). However, the UPOV system has no requirement about *marketing* a protected variety, which may be available to farmers only on contract terms. This undermines the benefit-sharing aspect of the system, a balance between rights (exclusive sales) and obligations. There is also widespread anecdotal experience of cases of obtaining seeds from private breeders for further breeding. Mandatory deposit of seeds[11] – like that demanded for US PVPs – would provide breeders with automatic access, and continuously update genebanks. Otherwise, genebanks become repositories for increasingly ancient germplasm. These simple measures would improve the multilateral system aspect of the UPOV system.

Two weighty arguments for IPRs and the need for returns are (1) The high development costs of new technologies and; (2) The short *lead times* (advance) in modern plant breeding. Since a competing breeder could access varieties only after their official release, it used to take a decade before new progeny varieties could be produced (Gouache *et al.*, 2015). Today, on

UPOV terms, a competitor may use markers, double haploids or speed breeding to produce a competing variety in half that time. In the case of tomato and broccoli patents, breeders have made significant progress by backcrossing and developing improved traits. Once these varieties are released, competitors may be 'free riders' on the work of the serious breeder. This weakness in the UPOV system could be dealt with by introducing, say, five years of extra lead time before a competing breeder could access a *deposited* variety under the Breeder's Exemption. Commenting on the related UPOV concept of Essentially Derived Varieties, Troyer and Rocheford have suggested that for maize inbreds, five years' protection would supply the necessary lead time (Troyer and Rocheford, 2002). After this, inbreds should be made available to others, rather than remaining patented for 20 years.

The private sector has also made some interesting proposals for navigating the ABS landscape lately. During the seventh governing body session of the Plant Treaty in 2017, the president of the International Seed Federation made a statement commending the ongoing process to enhance the multilateral system of the Plant Treaty (see next section) and presenting a 'Declaration of Commitment' to become 'subscribers' under a revised SMTA signed by 41 seed companies.[12] In the letter to the co-chairs of the working groups for the enhancement of the functioning of the multilateral system which first presented the declaration, the IFS president made it clear that the seed company offer was conditional on some 'business critical conditions' specified in the declaration.[13] The declaration states that the signatories share the ISF view that the SMTA should 'maintain multiple benefit-sharing options, one being the subscription system and the other being the single access system (payment based on use of accessed genetic resources)'. Further, the declaration contains a proposed subscription rate at '0.01% of a subscribers' seed revenues generated from crops included in Annex 1 minus 30 per cent as specified in the current revised draft'. In his statement to the governing body, the IFS president asserted that this 'would represent an important and predictable monetary flow from the seed industry to the Treaty's Benefit-sharing Fund'.

The proactive steps taken by the private sector discussed here must be seen in the context of the use of contract law and bag tags described in earlier sections. The contract law approach to IPR protection currently common in the USA is the main roadblock to access. Note should be taken of the widespread rejection of such practices on the part of the European seed industry (Gouache *et al.*, 2015, Louwaars *et al.*, 2009), as well as some governments and NGOs. As Kock (2015) has pointed out, it is not a wise future seed-business practice to be perceived as anti-social protectors of 'licensed technologies' relating to food or health security.

Pathways in the commons

The Open Source Seed Initiative (OSSI) was established with the objective of promoting and maintaining open access to plant varieties worldwide (Luby *et al.*, 2015). The initiative was launched in 2014; as of May 2019, there were 475 varieties of different crops on the list released by both public and private breeders under the open seed pledge:

> You have the freedom to use these OSSI-Pledged seeds in any way you choose. In return, you pledge not to restrict others' use of these seeds or their derivatives by patents or other means, and to include this Pledge with any transfer of these seeds or their derivatives.[14]

The OSSI is explicitly concerned with providing 'an alternative to pervasive IPR agreements' and '[to] the increasing difficulties in accessing genetic resources'. As the OSSI approach prohibits all types of IPRs on varieties in the system, there is no *toll road* option in this approach.

The countries that painstakingly negotiated the Plant Treaty managed to develop a new type of commons: the multilateral system (Halewood *et al.*, 2012). In practice, most of the genebanks that actually distribute seeds make no distinction whether crops are listed in Annex 1 or not: they routinely distribute all PGRFA under the SMTA. This practice, as well as the blurry border between the crop genepools included in the multilateral system and the more distantly related wild relatives, has become a bone of contention between proponents of the Nagoya Protocol and those who want the reach of the multilateral system to become as great as possible (Visser, 2013).

While the multilateral system is widely recognized as an excellent instrument for access (if countries abide by it), it has faced considerable criticism from various groups – ranging from Nagoya Protocol proponents, to NGOs that take a political standpoint against any capitalist involvement in agriculture, and to industry actors who reject the idea of paying royalties on material not under IPR. Some civil-society actors and critical scholars have articulated views that see the Plant Treaty more as part of the problem than part of the solution: La Via Campesina categorically scorns the Plant Treaty framework for the multilateral approach to collection, conservation and exchange of plant genetic resources as 'a contradictory and ambiguous treaty, which in the final analysis comes down on the side of theft' (Kloppenburg, 2014). The compromise for ABS embodied by the Plant Treaty has clearly not convinced all factions in the seed wars.

If monetary benefit-sharing in the form of 'mandatory payments' to the BSF is taken as the main proof of success for the Plant Treaty, some will say it has been a disappointment. According to the Plant Treaty's monitoring mechanism for the multilateral system, more than 4.6 million samples have been distributed under the SMTA,[15] but it was only in 2018 the first payment of royalties was made. That year the Dutch seed company Nunhems, a subsidiary company of Bayer, contributed 0.77 per cent of its sales of ten vegetable varieties developed from germplasm acquired from the Dutch and German genebanks.[16] The major source of income for the BSF so far has been voluntary payments from contracting parties. The intergovernmental process referred to above as the process to enhance the functioning of the multilateral system was launched to address some parties' discontent with this situation. The Ad hoc Open-ended Working Group appointed to deal with this task consists of 27 representatives from the 7 Plant Treaty regions. A draft revised SMTA was discussed in June 2019.[17] The two major issues for discussion were interlinked: the coverage of the multilateral system (whether the Annex 1 list of crops should be expanded, and if so, how) and the terms and conditions for ABS in the SMTA. The current draft of the revised SMTA has several alternatives still open, including various subscription models.[18] Unsurprisingly, the size of the subscription fee and how this is to be differentiated in cases when the products are available with or without 'restrictions to further research and breeding' is still contentious. It is difficult to see how this issue can be resolved unless ABS and IPRs are discussed at the same time.

Notes

1 Agreed at the Earth Summit in Rio de Janeiro in 1992, entered into force in 1993.
2 CBD website www.cbd.int/abs/nagoya-protocol/signatories/default.shtml accessed 12 February 2019.
3 FAO Plant Treaty website www.fao.org/plant-treaty/areas-of-work/benefit-sharing-fund/overview/en/ accessed 12 February 2019.
4 This figure refers to genebank accessions. In addition, the CGIAR breeding programs distribute three–five times more materials each year under the SMTA (Halewood, personal communication, May 2019).
5 Aegean Agricultural Research Institute, e-mail 4 December 2012.
6 Institute of Biodiversity Conservation, e-mail 22 October 2012.
7 www.globalrust.org/.
8 Now this 'initiative' must itself become durable. It is funded until 2020 by the Gates Foundation and the UK Department for International Development (DFID) – laudable, but paradoxical for something that is a quintessential public good.
9 ILP website www.ilp-vegetable.org accessed 12 February 2019.
10 Monsanto's vegetable breeding company Seminis has not joined, nor have any public research institutes. These models may be more attractive to symmetrical situations where parties are both providers and users.
11 Including F1-hybrid seed, not stored in many genebanks.
12 The statement is found in the report of the seventh session of the governing body of the international treaty on plant genetic resources for food and agriculture available from www.fao.org/3/MV606/mv606.pdf accessed 18 May 2019.
13 Letter to the Plant Treaty from ISF President Jean-Christophe Gouache, dated 28 August 2017 with Declaration of Commitment appended. The letter is available from https://haveyoursay.agriculture.gov.au/plant-genetics-in-food-ag-treaty/forum_topics/seed-companies-declaration-of-commitment accessed 18 May 2019.
14 OSSI webpage https://osseeds.org/what-is-an-ossi-pledged-variety/ accessed 12 February 2019.
15 Plant Treaty webpage: https://mls.planttreaty.org/itt/index.php?r=stats/pubStats accessed 12 February 2019.
16 FAO website www.fao.org/plant-treaty/news/news-detail/en/c/1143273/ accessed 12 February 2019.
17 The agendas, background documents and reports are posted at the Plant Treaty webpage www.fao.org/plant-treaty/areas-of-work/the-multilateral-system/policy-guidance/en/ accessed 18 May 2019.

18 The draft new SMTA is appended to the report of eight session of the governing body of the international treaty on plant genetic resources for food and agriculture available from www.fao.org/3/CA2216EN/ca2216en.pdf accessed 18 May 2019.

References

Atlin GN, Cairns JE, Das B, 2017. Rapid breeding and varietal replacement are critical to adaptation of cropping systems in the developing world to climate change. *Global Food Security* 12, 31–37.

Bjørnstad Å, 2016. 'Do not privatize the giant's shoulders': Rethinking patents in plant breeding. *Trends in Biotechnology* 34, 609–617.

Bjørnstad Å, Tekle S, Göransson M, 2013. 'Facilitated access' to plant genetic resources: Does it work? *Genetic Resources and Crop Evolution* 60, 1959–1965.

Bjørnstad Å, unpublished data from the study Bjørnstad Å, Tekle S, and Göransson M, 2013. 'Facilitated access' to plant genetic resources: Does it work? *Genetic Resources and Crop Evolution* 60 (7), 1959–1965.

Burke MB, Lobell DB, Guarino L, 2009. Shifts in African crop climates by 2050, and the implications for crop improvement and genetic resources conservation. *Global Environmental Change: Human and Policy Dimensions* 19, 317–325.

Challinor AJ, Koehler A-K, Ramirez-Villegas J, Whitfield S, Das B, 2016. Current warming will reduce yields unless maize breeding and seed systems adapt immediately. *Nature Climate Change* 6, 954–958.

Coffman R, Acevedo M, McCandless L, 2016. Rust, risk, and germplasm exchange: The Borlaug Global Rust Initiative. *Indian Journal of Plant Genetic Resources* 29 (3), 417–419.

Coffman W, Lesser WH, McCouch SR, 2003. Commercialization and the scientific research process: The example of plant breeding. Paper presented at the conference Science and the University, Cornell Higher Education Research Institute, Ithaca, NY, 20–21 May 2003.

FAO, 2010. *Second Report on the State of the World's Plant Genetic Resources*. Rome: FAO.

FAO, 2014. *Genebank Standards for Plant Genetic Resources for Food and Agriculture*, rev. edn. Rome: FAO.

Fowler C, Hodgkin T, 2004. Plant genetic resources for food and agriculture: Assessing global availability. *Annual Review of Environment and Resources* 29, 143–179.

Fowler C, Smale M, Gaiji S, 2001. Unequal exchange? Recent transfers of agricultural resources and their implications for developing countries. *Development Policy Review* 19, 181–204.

Galluzzi G, Halewood M, Noriega IL, Vernooy R, 2016. Twenty-five years of international exchanges of plant genetic resources facilitated by the CGIAR genebanks: A case study on global interdependence. *Biodiversity and Conservation* 25, 1421–1446.

Ghijsen H, 2009. Intellectual property rights and access rules for germplasm: Benefit or straitjacket? *Euphytica* 170 (1–2), 229.

Godfray HCJ, Beddington JR, Crute IR, Haddad L, Lawrence D, Muir JF, Pretty J, Robinson S, Thomas SM, Toulmin C, 2010. Food security: The challenge of feeding 9 billion people. *Science* 327, 812–818.

Gouache J-C, Desprez, F, Tebel, C, 2015. Amélioration des plantes: Il faut faire évoluer les outils de propriété industrielle. *Paysans* 354, 9–16.

Halewood M, Noriega IL, Louafi S, 2012. The global crop commons and access and benefit-sharing laws. *Crop Genetic Resources as a Global Commons*. London: Earthscan/Routledge.

Heisey P, Day-Rubenstein KD, 2015. Using crop genetic resources to help agriculture adapt to climate change: Economics and policy. United States Department of Agriculture, *Economic Research Service Economic Information Bulletin* 139.

Howard PH, 2015. Intellectual property and consolidation in the seed industry. *Crop Science* 55, 2489–2495.

Janis MD, Kesan JP, 2002. Intellectual property protection for plant innovation: Unresolved issues after JEM v. Pioneer. *Nature Biotechnology* 20, 1161–1164.

Khoury CK, Achicanoy HA, Bjorkman AD, Navarro-Racines C, Guarino L, Flores-Palacios X, Engels JMM, Wiersema JH, Dempewolf H, Sotelo S, Ramírez-Villegas J, Castañeda-Álvarez NP, Fowler C, Jarvis A, Rieseberg LH, Struik PC, 2016. Origins of food crops connect countries worldwide. *Proceedings of the Royal Society B*. https://doi.org/10.1098/rspb.2016.0792.

Kloppenburg J, 2014. Re-purposing the master's tools: The open source seed initiative and the struggle for seed sovereignty. *Journal of Peasant Studies* 41, 1225–1246.

Kloppenburg JR, 2005. *First the Seed: The Political Economy of Plant Biotechnology*. Madison: University of Wisconsin Press.

Knight J, 2003. Crop improvement: A dying breed. *Nature* 421, 568–570.

Kock MA, 2015. Plant breeding innovations: Free access, but not access for free. A new approach to facilitating FRAND licenses for plant-related patents. *Bioscience Law Review* 14, 123–129.

Louwaars N, Dons H, Van Overwalle G, Raven H, Arundel A, Eaton D, Nelis A, 2009. Breeding business: The future of plant breeding in the light of developments in patent rights and plant breeder's rights. *CGN Report* 14. Wageningen: Centre for Genetic Resources.

Luby CH, Kloppenburg J, Michaels TE, Goldman IL, 2015. Enhancing freedom to operate for plant breeders and farmers through open source plant breeding. *Crop Science* 55, 2481–2488.

Mikel MA, Dudley JW, 2006. Evolution of North American dent corn from public to proprietary germplasm. *Crop Science* 46, 1193–1205.

Prathapan KD, Pethiyagoda R, Bawa KS, Raven PH, Rajan PD, 2018. When the cure kills: CBD limits biodiversity research. *Science* 360, 1405–1406.

Prifti V, 2015. *The Breeder's Exception to Patent Rights. Analysis of Compliance with Article 30 of the TRIPS Agreement*. Cham: Springer International.

Schurman R, 2017. Building an alliance for biotechnology in Africa. *Journal of Agrarian Change* 17, 441–458.

Shelton AC, Tracy WF, 2017. Cultivar development in the US public sector. *Crop Science* 57, 1823–1835.

Singh RP, Hodson DP, Jin Y, Lagudah ES, Ayliffe MA, 2015. Emergence and spread of new races of wheat stem rust fungus: Continued threat to food security and prospects of genetic control. *Phytopathology* 105, 872–884.

Singh RP, Singh PK, Rutkoski J, Hodson DP, He X, *et al*. 2016. Disease impact on wheat yield potential and prospects of genetic control. *Annual Review of Phytopathology* 54, 303–322.

Troyer AF, Rocheford TR, 2002. Germplasm ownership: Related corn inbreds. *Crop Science* 42, 3–11.

Tvedt MW, 2017. Legal and ethnoecological components of bioprospecting. In R Paterson, N Lima (eds) *Bioprospecting: Success, Potential and Constraints*, 15–28. Cham: Springer.

Van Deynze A, Zamora P, Delaux PM, Heitmann C, Jayaraman D, 2018. Nitrogen fixation in a landrace of maize is supported by a mucilage-associated diazotrophic microbiota. *PLOS Biology* 16, e2006352.

Visser B, 2013 The moving scope of Annex I. In Halewood M, López-Noriega I, (eds) *Crop Genetic Resources as a Global Commons*, 265–282. Abingdon: Routledge.

Westengen OT, Skarbø K, Mulesa TH, Berg T, 2018. Access to genes: Linkages between genebanks and farmers' seed systems. *Food Security* 10, 9–25.

Index

Page numbers in **bold** denote tables, those in *italics* denote figures.